SCIENCE AND PHILOSOPHY IN THE TWENTIETH CENTURY

Basic Works of Logical Empiricism

SERIES EDITOR

SAHOTRA SARKAR

*Dibner Institute at MIT
and McGill University*

A GARLAND SERIES IN
READINGS IN PHILOSOPHY

ROBERT NOZICK, *Advisor*
Harvard University

SERIES CONTENTS

VOLUME

3

LOGIC, PROBABILITY, AND EPISTEMOLOGY

THE POWER OF SEMANTICS

Edited with introductions by

SAHOTRA SARKAR

Dibner Institute at MIT
and McGill University

Routledge
Taylor & Francis Group

NEW YORK AND LONDON

First published by Garland Publishing, Inc.
This edition published 2013 by Routledge
711 Third Avenue, New York, NY 10017
2 Park Square, Milton Park, Abingdon, Oxfordshire OX14 4RN

Routledge is an imprint of the Taylor & Francis Group, an informa business

Library of Congress Cataloging-in-Publication Data

Logic, probability, and epistemology : the power of semantics /
edited with introductions by Sahotra Sarkar.
   p.   cm. — (Science and philosophy in the twentieth
century ; v. 3)
   Includes bibliographical references.
   ISBN 0-8153-2264-X (alk. paper)
   1. Logical positivism.   2. Semantics (Philosophy)   I. Sarkar,
Sahotra.   II. Series.
B824.6.L622   1996
146'.42—dc20   95–51785
                                                            CIP

SET ISBN   9780815322610
POD ISBN   9780415628365
    Vol1   9780815322627
    Vol2   9780815322634
    Vol3   9780815322641
    Vol4   9780815322658
    Vol5   9780815322665
    Vol6   9780815322672

CONTENTS

SEMANTICS

PROBABILITY

CONFIRMATION AND INDUCTIVE LOGIC

THE STRUCTURE OF META-SCIENTIFIC CONCEPTS

COMMENTARIES

SERIES INTRODUCTION

The early years of the twentieth century saw remarkable developments in the sciences, particularly physics and biology. The century began with Planck's introduction of what came to be known as the "quantum hypothesis," followed by the work of Einstein, Bohr, and others, which paved the way for the development of quantum mechanics in the 1920s. It remains the most radical departure from the classical worldview that physics has seen. Not only were some physical quantities "quantized," that is, they could only have discrete values, but there were situations in which some of these values were indeterminate. Perhaps even worse, the basic dynamics of physical systems was indeterministic. The mechanical picture of the world, inherited from the seventeenth century, and already under attack during the nineteenth, finally collapsed beyond hope of recovery. Nevertheless, the new physics was unavoidable. Not only did atomic phenomena abide by its rules, but it provided a successful account of chemical bonding and valency. Meanwhile, in 1905, Einstein's special theory of relativity challenged classical notions of space and time. A decade later, general relativity replaced gravitation as a force by the curvature of space-time. Developments in astrophysics confirmed general relativity's unusual claims.

Also around 1900, biologists recovered the laws for the transmission of hereditary factors, or "genes." These laws, though published by Mendel in 1865, had remained largely unknown for a generation. By 1905, a new science called "genetics" had been created. For the first time, the phenomena of heredity were subsumed under exact (mathematical) laws. In the early 1920s, these laws were used by Fisher, Haldane, and Wright to formulate a quantitative, basically testable theory of evolution by natural selection. Around 1900 it also became clear that the transfer of chromosomes mediated the transmission of hereditary characters from parents to offspring. Between 1910 and 1920, genes were shown to be linearly positioned on chromosomes. The rudiments of a physical account of biological inheritance were in place by the mid-1920s. Eventually this work was integrated with other biological subdisciplines, especially biochemistry (itself largely a turn-of-the-century creation), to generate molecular biology, arguably the greatest triumph of science since 1950.

The philosophical response to the advances of early twentieth century science was schizophrenic. Some philosophers, especially in Germany, ignored scientific developments almost altogether and continued to elaborate extensive metaphysical systems having little contact with the physical world. Collectively, these projects came to be called phenomenology. In sharp contrast, another group of philosophers attempted to reform—or, perhaps, even replace—academic philosophy so as to bring it into consonance with modern science. At times, they claimed to have inherited the mantles of Aristotle and Descartes, Newton and Leibniz, Locke and Hume, Kant and Marx. More often, they claimed to be doing something altogether novel.

Most prominent among the latter group of philosophers were those who called themselves "logical positivists" or "logical empiricists." Many of them were associated, in their early years, with a group that met regularly in Vienna (starting in 1922) and called itself the Vienna Circle. The central figure was Moritz Schlick. (A complete list of members of the Vienna Circle will be found in their 1929 manifesto, which is reprinted in Volume 2.) The members of the Vienna Circle had an almost worshipful attitude towards the new physics though, in general, they seemed to have been completely ignorant of the equally fundamental changes taking place in biology. They were impressed by developments in logic, particularly Whitehead and Russell's attempt to carry out Frege's project of constructing mathematics from logic. Kurt Gödel, a member of the Vienna Circle, though hardly a logical empiricist in his philosophical leanings, probed the foundations of logic and showed that any relatively complex system of mathematics must allow statements to be formulated that can neither be proved nor disproved using formalized rules of proof—this is Gödel's famous incompleteness theorem.

Meanwhile, in Berlin, a smaller group around Hans Reichenbach came to a similar philosophical orientation and concentrated on probing the foundations of physics. In Poland, an eminent group of logicians, with Alfred Tarski as the central figure, began equally important investigations of logical notions. There was considerable intellectual exchange between these different groups. These exchanges led to convergence on many points—the philosophical theses that were most commonly advanced will be described below (and in the introductions to Volumes 1–4).

To return to the historical story, most of the logical empiricists had relatively progressive politics. A few, notably Otto Neurath, were avowed Marxists. Others, including Rudolf Carnap and Hans Hahn, were socialists. With the rise of nazism and fascism in Europe in the 1930s, many of the logical empiricists emigrated to Britain and, especially, to the United States. There they eventually came to establish a temporary hegemony over academic philosophy. Reichenbach moved to the University of California at Los Angeles; Herbert Feigl to the University of Minnesota; and Carnap to the University of Chicago. Meanwhile, during his youthful days, W.V.O. Quine was already preaching the logical empiricist gospel at Harvard. Of the major figures, only Neurath remained in Europe. (Hans Hahn had died in 1934 and Schlick had been murdered in 1936—see the introduction to Volume 2.)

Because of its migration to the U.S., logical empiricism became part of the Anglo-American tradition in philosophy, in spite of its European origins. It is at least arguable that as a movement it matured in the U.S. However, in spite of being relatively organized compared to other philosophical movements, the logical empiricists did not present a unified system of universally held theses—a point that seems to elude their modern critics—though they generally exhibited a coherent attitude to the analysis of philosophical problems. This attitude can be traced back to the 1920s. They *generally* accepted an *a priori* faith in logic, though they were sometimes known to disagree on what logic could be. Other than in logic (and in mathematics, which, for most logical empiricists, could be derived from logic), the logical empiricists endorsed a thoroughgoing empiricism—hence their name. All factual (that is, nonlogical) knowledge was ultimately empirical. A sharp distinction between empirical, *a posteriori*, synthetic claims on one hand and *a priori*, analytic claims on the other was a cherished doctrine for most (but not all) logical empiricists. Its rejection by Quine and others in the 1950s was a significant event in the decline of logical empiricism (see Volume 5).

Any claim that was neither logic nor able to be adjudicated by empirical means was rejected by the logical empiricists as "meaningless" or "cognitively insignificant," whatever its noncognitive (for instance, emotional) appeal. Logic escaped this fate by being true by virtue of meaning (of the logical connectives such as "not" and "and" and operators such as "all," "any," and "some") or of conventions. Mathematics was true because it could be reduced to, or constructed from, logic. Besides logic, the logical empiricists generally did not accept any other normative discipline as consisting of meaningful claims. (Ethical claims, according to some of them, were only devices to evoke appropriate emotive responses from others.)

Given these positions, there did not remain much metaphysics to be done (at least insofar as "metaphysics" was interpreted by the academic philosophers). Some logical empiricists, notably Carnap, claimed to have successfully eliminated metaphysics. In practice, metaphysics was replaced by attempts—rarely profound—at the analysis and interpretation of scientific concepts. Those logical empiricists who were particularly enamored of the technical apparatus of mathematical logic, again, most notably Carnap, interpreted this endeavor as describing the syntax and elaborating a semantics for the language of science. (In the case of logic itself, the logical empiricists achieved some important successes in their interpretive efforts in the 1930s—see Volume 3.) Metaphysics cast aside, the logical empiricists turned to epistemology; in particular, to the possibility of quantifying the extent to which different scientific claims were grounded in experience. The project turned out to be far more complex—and convoluted—than initially envisioned. By the time logical empiricism disappeared as an explicit movement within philosophy, little progress had been made towards this end.

An enumeration of positions advanced—or of successes and failures—only barely captures the spirit of logical empiricism. Within their self-proclaimed framework of accepting only logic and empirical knowledge, they venerated a critical attitude. This included continual self-criticism. Much has been written about the untenability of the doctrines espoused by the logical empiricists—what unfortunately goes unrecognized is that the most severe (and the most relevant) criticisms almost always came from within the movement or, at least, from individuals schooled in the movement (notably Quine). There were significant disagreements among the logical empiricists (for instance, between Carnap and Reichenbach on epistemology). There were also significant disagreements within the Vienna Circle: Kurt Gödel probably rejected most of the tenets in the Vienna Circle manifesto; Karl Menger refused to reject metaphysics on logical grounds (see his paper in Volume 2). These cases, however, may only show that not all members of that circle should be regarded as logical empiricists. Nonetheless, and most importantly, the logical empiricists believed philosophy to be a collective enterprise, like the natural sciences, and one in which progress could be made.

The logical empiricists' domination of Anglo-American philosophy was never complete and whatever hegemony they established was brief. Even within their chosen subdisciplines, such as the philosophy of science or logic or mathematics, their positions came under attack in the 1950s. Cherished doctrines such as the analytic-synthetic distinction were abandoned by a new generation of philosophers. The value of their type of conceptual analysis was sometimes derided by the later Wittgenstein's followers and by the so-called "ordinary language" philosophers. Metaphysics returned with a vengeance and, arguably, the influence of the logical empiricists was largely confined to the philosophy of science after the 1950s. But the 1960s saw logical empiricism under attack even among philosophers of science. It is probably reasonable to say that by around 1970, a new generation of philosophers of science had decided that the analyses offered by the logical empiricists were largely superficial and were to be replaced by more sophisticated work. The most popular position of those days was "scientific realism," a return to exactly the kind of metaphysics that the logical empiricists had found devoid of cognitive content.

Significant interest in logical empiricism resurfaced again in the early 1980s. This did not indicate any general return to the positions the logical empiricists advocated. Rather, the source of the interest was largely historical, part of a desire to understand the history of twentieth-century philosophy. It was aided by a new interest among philosophers in the history of the philosophy of science. Carnap and Reichenbach were probably the only prominent logical empiricists who had continued to be read during the 1960s and 1970s; now the works of Schlick and Neurath, among others, were once again read (and, sometimes, translated into English for the first time). Archives began to be mined to expose the intricate details of the relationships between the logical empiricists, and between them and other social and cultural movements of the 1920s and 1930s. This new work took place not only in the U.S., but also in Austria, Germany, and to a lesser extent, elsewhere in Europe. Slowly, as this historical work has progressed, a more positive philosophical assessment of the movement than was usually found in the 1960s and 1970s has also emerged (Sarkar 1992). These developments are far too recent for any assessment to be made of their lasting value. While the historical interest is neither hard to explain nor appreciate, it is less clear why, but perhaps even more interesting that, this positive reassessment is taking place.

There seem to be at least three reasons for the relatively positive reassessment that deserve mention: (1) since more than a generation had passed between the heyday of the movement and the mid- and late-1980s, the new commentators found it easier to have a more balanced view of both the contributions and the failures of logical empiricism than those—especially in the 1960s—who felt that they had to react to its dominance; (2) historical exploration—and exegesis—has revealed that the logical empiricists held a variety of views that are both more complex and more interesting than what their critics attributed to them (see, for example, Suppe 1974); and (3) arguably, the various alternatives to logical empiricism as a philosophy of science that were formulated in the 1960s and 1970s have not delivered on their promises. Going further, and much more controversially, these alternatives (including scientific realism) have proved less fertile and less robust than logical empiricism.

In this new intellectual context, it seems appropriate to make available, to as wide an audience as possible, some of the basic works of logical empiricism, as well as some of the new commentaries that have followed the renewal of interest in the movement. Many of the original pieces are not easily available and there is, at present, neither a detailed history of logical empiricism nor an annotated guide to its most important writings. An important old collection is Ayer (1959), which has a fairly comprehensive bibliography of work up to that point. Many valuable collections devoted to individual figures have been published. Schilpp (1963) collects many important critical pieces on Carnap, with Carnap's responses. The basic works of Feigl (1981), Hahn (1980), Kraft (1981), Menger (1979), Neurath (1973, 1987), Reichenbach (1978), Schlick (1979, 1987), and Waismann (1977) have been published as part of the Vienna Circle Collection. Collections of articles on logical positivism from the 1960s and 1970s include Achinstein and Barker (1969) and Hintikka (1975). Recent works of interest include Coffa (1991), Haller (1982), Menger (1994), and Uebel (1991, 1992). However, a detailed history of logical empiricism remains to be written.

What makes this series different from these works is an attempt to present a global picture of logical empiricism, including the influences that led to its initiation and the criticisms that were responsible for its decline. The emphasis here is on issues rather than on individual figures even though some of the most influential figures—especially Carnap and Reichenbach—feature prominently. However, for most of the topics treated, all the historically and conceptually important exchanges on that topic are collected together. Finally, modern commentaries are also included to bring the series up to date. In general, complete papers (in English whenever translations are available) are included over book sections in an effort to present complete arguments as far as possible. Volume 1 deals with the initial influences on logical empiricism and with the Vienna Circle period. Volume 2 concerns primarily the 1930s, when logical empiricism was at its most confident phase, when its adherents truly believed that they were reforming philosophy for all future times. Volume 3 includes pieces that reflect logical empiricism in its mature phase, after self-criticism and technical developments induced more sophisticated doctrines than those produced in the 1930s. Volume 4 shows how logical empiricism analyzed the special sciences. Volume 5 consists of the most important criticisms of logical empiricism and its responses. It marks the decline of logical empiricism. All of these volumes, except Volume 4, include a concluding section with modern commentaries. Volume 6 consists entirely of these commentaries. Each volume is introduced with an editorial note that puts the contents in perspective. Thanks are due to Richard Creath and Alan Richardson for advice on selecting the pieces for this series, and to Gregg Jaeger for help in assembling them and for commenting on the introductions. Work on these volumes was done while the editor was a Fellow at the Dibner Institute for the History of Science at MIT. Thanks are due to it for its support.

Further Reading

Achinstein, P. and Barker, S., eds. 1969. *The Legacy of Logical Positivism*. Baltimore: Johns Hopkins University Press.

Ayer, A.J., ed. 1959. *Logical Positivism*. New York: Free Press.

Coffa, A. 1991. *The Semantic Tradition from Kant to Carnap: To the Vienna Station*. Cambridge, UK: Cambridge University Press.

Feigl, H. 1981. *Inquiries and Provocations: Selected Writings, 1929 – 1974*. Dordrecht: Kluwer.

Hahn, H. 1980. *Empiricism, Logic, and Mathematics: Philosophical Papers*. Dordrecht: Kluwer.

Haller, R., ed. 1982. *Schlick und Neurath—Ein Symposion. Grazer philosophische Studien* 16 –17.

Hintikka, J., ed. 1975. *Rudolf Carnap, Logical Empiricist*. Dordrecht: Reidel.

Kraft, V. 1981. *Foundations for a Scientific Analysis of Value*. Dordrecht: Kluwer.

Menger, K. 1979. *Selected Papers in Logic, Foundations, Didactics, Economics*. Dordrecht: Kluwer.

Menger, K. 1994. *Reminiscences of the Vienna Circle and the Mathematical Colloquium*. Dordrecht: Kluwer.

Neurath, O. 1973. *Empiricism and Sociology*. Dordrecht: Kluwer.

Neurath, O. 1987. *Unified Science*. Dordrecht: Kluwer.

Reichenbach, H. 1978. *Selected Writings, 1909–1953*. Vols. 1, 2. Dordrecht: Kluwer.

Sarkar, S., ed. 1991. *Rudolf Carnap—A Centenary Reappraisal*. *Synthese* 93.

Schilpp, P. A., ed. 1963. *The Philosophy of Rudolf Carnap*. La Salle, IL: Open Court.

Schlick, M. 1979. *Philosophical Papers*. Vols. 1, 2. Dordrecht: Kluwer.

Schlick, M. 1987. *The Problems of Philosophy in Their Interconnection*. Dordrecht: Kluwer.

Suppe, F., ed. 1974. *The Structure of Scientific Theories*. Urbana: University of Illinois Press.

Uebel, T. E., ed. 1991. *Rediscovering the Forgotten Vienna Circle*. Dordrecht: Kluwer.

Uebel, T. E. 1992. *Overcoming Logical Positivism from Within: The Emergence of Neurath's Naturalism in the Vienna Circle's Protocol Sentence Debate*. Amsterdam: Rodopi.

Waismann, F. 1977. *Philosophical Papers*. Dordrecht: Kluwer.

INTRODUCTION

Tarski's (1935) demonstration that the concept of truth for a language can be formalized in a suitably constructed metalanguage was a critical development for the logical empiricists. (Coffa [1991] provides a history of these developments—see also his paper in Volume 6.) The logical empiricists had, by and large, already made the linguistic move in the sense that they had begun to talk of logic and science as formalized languages. Now their conception of the philosophical enterprise took its final form. As Carnap (1942) put it, philosophy was to become, or be replaced by, the analysis of the syntax and semantics of the language of science. The choice of a scientific language was to be governed by pragmatic considerations. By 1940, even Reichenbach, whose work had remained far more grounded in actual scientific practice than the work of Carnap or any other logical empiricist, also accepted semantics and, less explicitly, the linguistic orientation.

Within logic, Carnap (1942) clarified what semantics could do. It allowed talk of truth and reference. The new view of logic, as consisting of syntax and semantics, enabled the codification of concepts such as derivation (or deduction) and consequence, theorem and logical truth, consistency and satisfiability. This codification is used even today, and it might well be the logical empiricists' most significant philosophical contribution. More controversially, as it turned out, Carnap used semantics to explicate analyticity and formalize the analytic-synthetic distinction. (The problems that arose are dealt with in Volume 5.) Not only did the semantic move allow Carnap to accept truth as a systematic rather than pragmatic concept, but, by about 1940, it convinced him that confirmation could also be codified. Thus began his series of studies on probability and inductive logic (Carnap 1950, 1952, 1971, 1980). Reichenbach, of course, had been pursuing a theory of induction (broadly conceived) at least since the early 1930s (see, for example, Reichenbach 1938). The final stage of logical empiricism had begun. Reichenbach also continued to work on the philosophical problems of physics. His work ended with his death in 1953. Carnap continued to work on inductive logic with many collaborators until his death in 1970.

This volume collects together the basic papers of logical empiricism from this period. It includes papers on semantics, the interpretations of probability, and the developments of inductive logic (including the nontrivial problems it immediately encountered). It also reprints a few pieces on the nature of scientific theories and explanation. It excludes the specific analyses of the empirical sciences, which will be the subject of Volume 4.

The first section consists of two papers. The first, by Tarski, is a succinct philosophical exposition of the principles of semantics that lie at the basis of much of the work recorded in this volume. The second, by Carnap, marks the resurrection of modal logic, which had almost entirely been ignored in the 1920s and 1930s. It shows how, with clearly formulated semantics, various types of modal logic can be classified. Carnap (1947) explored the philosophical foundations for modal logic in greater detail.

The second section reprints four papers on the interpretations of probability. Reichenbach's first piece, his response to Nagel's review of Reichenbach (1935), shows how early he had committed himself to a frequency interpretation of probability, even in the context of the confirmation of scientific theories. It also shows his acceptance of semantics, as formulated by Tarski and Carnap. His second piece gives a more systematic account of his views. Carnap's paper makes a distinction between a logical concept of probability to be used in confirmation contexts and a frequency concept, which is relevant in many other contexts. Carnap also attempts to respond to Reichenbach's position that only the frequency concept was consistent with empiricism. Finally, Popper's piece is included because it summarizes a third important interpretation of probability, the propensity interpretation, which has been very influential in the philosophy of science. This is not to suggest that Popper was a logical empiricist—in deference to his explicit claim of distance from that movement, his work has been excluded from this series except for this piece, and another critical one in Volume 5.

The third section consists of some of the more important papers on confirmation and inductive logic. It begins with what is called Reichenbach's vindication of induction (see, also Salmon's and Putnam's papers in Volume 6 in this context). Carnap's first paper shows how his view of semantics provides a framework for inductive logic. His last—and influential—paper outlines what inductive logic must achieve if it is to be regarded as successful. Kemeny's paper attempts to explicate simplicity and the role it plays in induction. The rest of the section consists of exchanges among Hempel, Goodman, and Carnap on problems that arose with inductive logic. The papers are presented in chronological order. Hempel's first paper (in two parts) presents his account of induction and also contains an influential discussion of the paradoxes of confirmation; his second analyzes the raven paradox in some detail. Goodman's first paper presents the paradox of confirmation for nonprojectible properties (the "grue" paradox). His second paper records his dissatisfaction with Carnap's attempt to resolve the paradox. Carnap's final response is also included.

The fourth section reprints three papers. That by Hempel and Oppenheim is a classic of logical empiricism. It lays out the structure of deductive-nomological explanation (also known as the "covering law" model), which the logical empiricists and their predecessors had implicitly assumed (but, apparently, had not explicitly analyzed). It attempts to analyze the power of theories. Nagel's paper presents what became the standard model of theory reduction, especially after this analysis was extended and reprinted in his influential *The Structure of Science* (1961). (Woodger [1952] independently formulated the same model.) Basically, this model claims that theory reduction is a type of deductive-nomological explanation where what is to be explained is also a law, rather than a singular fact. Braithwaite's paper is not strictly written from within the logical empiricists' camp, though it is consistent with most of their doctrines. It is included here because it contains a rare discussion of the role of models in science.

The two commentaries that are included in the last section are important for different reasons. Hilpinen's paper provides a clear resumé of Carnap's (1971, 1980) "new system" of inductive logic, otherwise not represented in these volumes. (However, the paper by Skyrms and the second paper by Jeffrey in Volume 6 contain more on this topic.) The second paper, by Creath, questions how important the move to semantics really was, at least in Carnap's case. Many concerns, both formal and pragmatic, remained the same. Creath emphasizes continuity in these intellectual developments (see also Sarkar's paper in Volume 2).

FURTHER READING

Carnap, R. 1942. *Introduction to Semantics*. Cambridge, MA: Harvard University Press.

Carnap, R. 1947. *Meaning and Necessity: A Study in the Semantics of Modal Logic*. Chicago: University of Chicago Press.

Carnap, R. 1950. *Logical Foundations of Probability*. Chicago: University of Chicago Press. Carnap, R. 1952. *The Continuum of Inductive Methods*. Chicago: University of Chicago Press.

Carnap, R. 1971. "The Basic System of Inductive Logic, Part I." In Carnap, R. and Jeffery, R., eds., *Studies in Inductive Logic and Probability*, vol. 1. Berkeley: University of California Press, pp. 33 -165.

Carnap, R. 1980. "The Basic System of Inductive Logic, Part II." In Jeffery, R., ed., *Studies in Inductive Logic and Probability*, vol. 2. Berkeley: University of California Press, pp. 7–155.

Coffa, A. 1991. *The Semantic Tradition from Kant to Carnap: To the Vienna Station*. Cambridge, UK: Cambridge University Press.

Nagel, E. 1961. *The Structure of Science*. New York: Harcourt, Brace.

Reichenbach, H. 1935. *Wahrscheinlichkeitslehre: eine Untersuchung uber die logischen und mathematischen Grundlagen der Wahrscheinlichkeitsrechnung*. Leiden: A.W. Sijthoff's.

Reichenbach, H. 1938. *Experience and Prediction: An Analysis of the Foundations and the Structure of Knowledge*. Chicago: University of Chicago Press.

Tarski, A. 1935. "Der Wahrheitsbegriff in den formalisierten Sprachen." *Studia Philosophica* 1: 261–405.

Woodger, J.H. 1952. *Biology and Language*. Cambridge, UK: Cambridge University Press.

THE SEMANTIC CONCEPTION OF TRUTH

AND THE FOUNDATIONS OF SEMANTICS

This paper consists of two parts; the first has an expository character, and the second is rather polemical.

In the first part I want to summarize in an informal way the main results of my investigations concerning the definition of truth and the more general problem of the foundations of semantics. These results have been embodied in a work which appeared in print several years ago.[1] Although my investigations concern concepts dealt with in classical philosophy, they happen to be comparatively little known in philosophical circles, perhaps because of their strictly technical character. For this reason I hope I shall be excused for taking up the matter once again.[2]

Since my work was published, various objections, of unequal value, have been raised to my investigations; some of these appeared in print, and others were made in public and private discussions in which I took part.[3] In the second part of the paper I should like to express my views regarding these objections. I hope that the remarks which will be made in this context will not be considered as purely polemical in character, but will be found to contain some constructive contributions to the subject.

In the second part of the paper I have made extensive use of material graciously put at my disposal by Dr. Marja Kokoszyńska (University of Lwów). I am especially indebted and grateful to Professors Ernest Nagel (Columbia University) and David Rynin (University of California, Berkeley) for their help in preparing the final text and for various critical remarks.

I. EXPOSITION

1. THE MAIN PROBLEM—A SATISFACTORY DEFINITION OF TRUTH. Our discussion will be centered around the notion[4] of *truth*. The main problem is that of giving a *satisfactory definition* of this notion, i.e., a definition which is *materially adequate* and *formally correct*. But such a formulation of the problem, because of its generality, cannot be considered unequivocal, and requires some further comments.

In order to avoid any ambiguity, we must first specify the conditions under which the definition of truth will be considered adequate from the material point of view. The desired definition does not aim to specify the meaning of a familiar word used to denote a novel notion; on the contrary, it aims to catch hold of the actual meaning of an old notion. We must then characterize this notion precisely enough to enable anyone to determine whether the definition actually fulfills its task.

1

Secondly, we must determine on what the formal correctness of the definition depends. Thus, we must specify the words or concepts which we wish to use in defining the notion of truth; and we must also give the formal rules to which the definition should conform. Speaking more generally, we must describe the formal structure of the language in which the definition will be given.

The discussion of these points will occupy a considerable portion of the first part of the paper.

2. THE EXTENSION OF THE TERM "TRUE." We begin with some remarks regarding the extension of the concept of truth which we have in mind here.

The predicate "*true*" is sometimes used to refer to psychological phenomena such as judgments or beliefs, sometimes to certain physical objects, namely, linguistic expressions and specifically sentences, and sometimes to certain ideal entities called "propositions." By "sentence" we understand here what is usually meant in grammar by "declarative sentence"; as regards the term "proposition," its meaning is notoriously a subject of lengthy disputations by various philosophers and logicians, and it seems never to have been made quite clear and unambiguous. For several reasons it appears most convenient to *apply the term "true" to sentences*, and we shall follow this course.[5]

Consequently, we must always relate the notion of truth, like that of a sentence, to a specific language; for it is obvious that the same expression which is a true sentence in one language can be false or meaningless in another.

Of course, the fact that we are interested here primarily in the notion of truth for sentences does not exclude the possibility of a subsequent extension of this notion to other kinds of objects.

3. THE MEANING OF THE TERM "TRUE." Much more serious difficulties are connected with the problem of the meaning (or the intension) of the concept of truth.

The word "*true*," like other words from our everyday language, is certainly not unambiguous. And it does not seem to me that the philosophers who have discussed this concept have helped to diminish its ambiguity. In works and discussions of philosophers we meet many different conceptions of truth and falsity, and we must indicate which conception will be the basis of our discussion.

We should like our definition to do justice to the intuitions which adhere to the *classical Aristotelian conception of truth*—intuitions which find their expression in the well-known words of Aristotle's *Metaphysics*:

2

*To say of what is that it is not, or of what is not that it is, is false, while
to say of what is that it is, or of what is not that it is not, is true.*

If we wished to adapt ourselves to modern philosophical terminology,
we could perhaps express this conception by means of the familiar formula:

The truth of a sentence consists in its agreement with (or correspondence to) reality.

(For a theory of truth which is to be based upon the latter formulation the
term "correspondence theory" has been suggested.)

If, on the other hand, we should decide to extend the popular usage of
the term *"designate"* by applying it not only to names, but also to sentences, and if we agreed to speak of the designata of sentences as "states
of affairs," we could possibly use for the same purpose the following
phrase:

A sentence is true if it designates an existing state of affairs.[6]

However, all these formulations can lead to various misunderstandings,
for none of them is sufficiently precise and clear (though this applies
much less to the original Aristotelian formulation than to either of the
others); at any rate, none of them can be considered a satisfactory definition of truth. It is up to us to look for a more precise expression of our
intuitions.

4. A CRITERION FOR THE MATERIAL ADEQUACY OF THE DEFINITION.[7]
Let us start with a concrete example. Consider the sentence *"snow is
white."* We ask the question under what conditions this sentence is true
or false. It seems clear that if we base ourselves on the classical conception of truth, we shall say that the sentence is true if snow is white, and
that it is false if snow is not white. Thus, if the definition of truth is to
conform to our conception, it must imply the following equivalence:

The sentence "snow is white" is true if, and only if, snow is white.

Let me point out that the phrase *"snow is white"* occurs on the left
side of this equivalence in quotation marks, and on the right without
quotation marks. On the right side we have the sentence itself, and on
the left the name of the sentence. Employing the medieval logical terminology we could also say that on the right side the words *"snow is white"*
occur in *suppositio formalis*, and on the left in *suppositio materialis*. It is
hardly necessary to explain why we must have the name of the sentence,
and not the sentence itself, on the left side of the equivalence. For, in the
first place, from the point of view of the grammar of our language, an
expression of the form *"X is true"* will not become a meaningful sentence
if we replace in it 'X' by a sentence or by anything other than a name—

since the subject of a sentence may be only a noun or an expression functioning like a noun. And, in the second place, the fundamental conventions regarding the use of any language require that in any utterance we make about an object it is the name of the object which must be employed, and not the object itself. In consequence, if we wish to say something about a sentence, for example, that it is true, we must use the name of this sentence, and not the sentence itself.[8]

It may be added that enclosing a sentence in quotation marks is by no means the only way of forming its name. For instance, by assuming the usual order of letters in our alphabet, we can use the following expression as the name (the description) of the sentence *"snow is white"*:

the sentence constituted by three words, the first of which consists of the 19th, 14th, 15th, and 23rd letters, the second of the 9th and 19th letters, and the third of the 23rd, 8th, 9th, 20th, and 5th letters of the English alphabet.

We shall now generalize the procedure which we have applied above. Let us consider an arbitrary sentence; we shall replace it by the letter '*p*.' We form the name of this sentence and we replace it by another letter, say '*X*.' We ask now what is the logical relation between the two sentences "*X is true*" and '*p*.' It is clear that from the point of view of our basic conception of truth these sentences are equivalent. In other words, the following equivalence holds:

(T)   *X is true if, and only if, p.*

We shall call any such equivalence (with '*p*' replaced by any sentence of the language to which the word "*true*" refers, and '*X*' replaced by a name of this sentence) an "*equivalence of the form* (T)."

Now at last we are able to put into a precise form the conditions under which we will consider the usage and the definition of the term "*true*" as adequate from the material point of view: we wish to use the term "*true*" in such a way that all equivalences of the form (T) can be asserted, and *we shall call a definition of truth "adequate" if all these equivalences follow from it.*

It should be emphasized that neither the expression (T) itself (which is not a sentence, but only a schema of a sentence) nor any particular instance of the form (T) can be regarded as a definition of truth. We can only say that every equivalence of the form (T) obtained by replacing '*p*' by a particular sentence, and '*X*' by a name of this sentence, may be considered a partial definition of truth, which explains wherein the truth of this one individual sentence consists. The general definition has to be, in a certain sense, a logical conjunction of all these partial definitions.

(The last remark calls for some comments. A language may admit the construction of infinitely many sentences; and thus the number of partial definitions of truth referring to sentences of such a language will also be infinite. Hence to give our remark a precise sense we should have to explain what is meant by a "logical conjunction of infinitely many sentences"; but this would lead us too far into technical problems of modern logic.)

5. TRUTH AS A SEMANTIC CONCEPT. I should like to propose the name *"the semantic conception of truth"* for the conception of truth which has just been discussed.

Semantics is a discipline which, speaking loosely, *deals with certain relations between expressions of a language and the objects* (or "states of affairs") *"referred to" by those expressions.* As typical examples of semantic concepts we may mention the concepts of *designation, satisfaction,* and *definition* as these occur in the following examples:

the expression *"the father of his country" designates (denotes) George Washington;*

snow satisfies the sentential function (the condition) "x is white";

the equation "$2 \cdot x = 1$*" defines (uniquely determines) the number 1/2.*

While the words *"designates," "satisfies,"* and *"defines"* express relations (between certain expressions and the objects "referred to" by these expressions), the word *"true"* is of a different logical nature: it expresses a property (or denotes a class) of certain expressions, viz., of sentences. However, it is easily seen that all the formulations which were given earlier and which aimed to explain the meaning of this word (cf. Sections 3 and 4) referred not only to sentences themselves, but also to objects "talked about" by these sentences, or possibly to "states of affairs" described by them. And, moreover, it turns out that the simplest and the most natural way of obtaining an exact definition of truth is one which involves the use of other semantic notions, e.g., the notion of satisfaction. It is for these reasons that we count the concept of truth which is discussed here among the concepts of semantics, and the problem of defining truth proves to be closely related to the more general problem of setting up the foundations of theoretical semantics.

It is perhaps worth while saying that semantics as it is conceived in this paper (and in former papers of the author) is a sober and modest discipline which has no pretensions of being a universal patent-medicine for all the ills and diseases of mankind, whether imaginary or real. You will not find in semantics any remedy for decayed teeth or illusions of grandeur or class conflicts. Nor is semantics a device for establishing that everyone except the speaker and his friends is speaking nonsense.

From antiquity to the present day the concepts of semantics have played an important role in the discussions of philosophers, logicians, and philologists. Nevertheless, these concepts have been treated for a long time with a certain amount of suspicion. From a historical standpoint, this suspicion is to be regarded as completely justified. For although the meaning of semantic concepts as they are used in everyday language seems to be rather clear and understandable, still all attempts to characterize this meaning in a general and exact way miscarried. And what is worse, various arguments in which these concepts were involved, and which seemed otherwise quite correct and based upon apparently obvious premises, led frequently to paradoxes and antinomies. It is sufficient to mention here the *antinomy of the liar*, Richard's *antinomy of definability* (by means of a finite number of words), and Grelling-Nelson's *antinomy of heterological terms*.[9]

I believe that the method which is outlined in this paper helps to overcome these difficulties and assures the possibility of a consistent use of semantic concepts.

6. LANGUAGES WITH A SPECIFIED STRUCTURE. Because of the possible occurrence of antinomies, the problem of specifying the formal structure and the vocabulary of a language in which definitions of semantic concepts are to be given becomes especially acute; and we turn now to this problem.

There are certain general conditions under which the structure of a language is regarded as *exactly specified*. Thus, to specify the structure of a language, we must characterize unambiguously the class of those words and expressions which are to be considered *meaningful*. In particular, we must indicate all words which we decide to use without defining them, and which are called *"undefined* (or *primitive) terms"*; and we must give the so-called *rules of definition* for introducing new or *defined terms*. Furthermore, we must set up criteria for distinguishing within the class of expressions those which we call *"sentences."* Finally, we must formulate the conditions under which a sentence of the language can be *asserted*. In particular, we must indicate all *axioms* (or *primitive sentences*), i.e., those sentences which we decide to assert without proof; and we must give the so-called *rules of inference* (or *rules of proof*) by means of which we can deduce new asserted sentences from other sentences which have been previously asserted. Axioms, as well as sentences deduced from them by means of rules of inference, are referred to as *"theorems"* or *"provable sentences."*

If in specifying the structure of a language we refer exclusively to the form of the expressions involved, the language is said to be *formalized*. In such a language theorems are the only sentences which can be asserted.

At the present time the only languages with a specified structure are the formalized languages of various systems of deductive logic, possibly enriched by the introduction of certain non-logical terms. However, the field of application of these languages is rather comprehensive; we are able, theoretically, to develop in them various branches of science, for instance, mathematics and theoretical physics.

(On the other hand, we can imagine the construction of languages which have an exactly specified structure without being formalized. In such a language the assertability of sentences, for instance, may depend not always on their form, but sometimes on other, non-linguistic factors. It would be interesting and important actually to construct a language of this type, and specifically one which would prove to be sufficient for the development of a comprehensive branch of empirial science; for this would justify the hope that languages with specified structure could finally replace everyday language in scientific discourse.)

The problem of the definition of truth obtains a precise meaning and can be solved in a rigorous way only for those languages whose structure has been exactly specified. For other languages—thus, for all natural, "spoken" languages—the meaning of the problem is more or less vague, and its solution can have only an approximate character. Roughly speaking, the approximation consists in replacing a natural language (or a portion of it in which we are interested) by one whose structure is exactly specified, and which diverges from the given language "as little as possible."

7. THE ANTINOMY OF THE LIAR. In order to discover some of the more specific conditions which must be satisfied by languages in which (or for which) the definition of truth is to be given, it will be advisable to begin with a discussion of that antinomy which directly involves the notion of truth, namely, the antinomy of the liar.

To obtain this antinomy in a perspicuous form,[10] consider the following sentence:

The sentence printed in this paper on p. 347, l. 31, is not true.

For brevity we shall replace the sentence just stated by the letter '*s*.'

According to our convention concerning the adequate usage of the term "*true*," we assert the following equivalence of the form (T):

(1) '*s*' *is true if, and only if, the sentence printed in this paper on p. 347, l. 31, is not true.*

On the other hand, keeping in mind the meaning of the symbol '*s*,' we establish empirically the following fact:

(2) '*s*' *is identical with the sentence printed in this paper on p. 347, l. 31.*

Now, by a familiar law from the theory of identity (Leibniz's law), it follows from (2) that we may replace in (1) the expression *"the sentence printed in this paper on p. 347, l. 31"* by the symbol " 's.' " We thus obtain what follows:

(3)   *'s' is true if, and only if, 's' is not true.*

In this way we have arrived at an obvious contradiction.

In my judgment, it would be quite wrong and dangerous from the standpoint of scientific progress to depreciate the importance of this and other antinomies, and to treat them as jokes or sophistries. It is a fact that we are here in the presence of an absurdity, that we have been compelled to assert a false sentence (since (3), as an equivalence between two contradictory sentences, is necessarily false). If we take our work seriously, we cannot be reconciled with this fact. We must discover its cause, that is to say, we must analyze premises upon which the antinomy is based; we must then reject at least one of these premises, and we must investigate the consequences which this has for the whole domain of our research.

It should be emphasized that antinomies have played a preeminent role in establishing the foundations of modern deductive sciences. And just as class-theoretical antinomies, and in particular Russell's antinomy (of the class of all classes that are not members of themselves), were the starting point for the successful attempts at a consistent formalization of logic and mathematics, so the antinomy of the liar and other semantic antinomies give rise to the construction of theoretical semantics.

8. THE INCONSISTENCY OF SEMANTICALLY CLOSED LANGUAGES.[7] If we now analyze the assumptions which lead to the antinomy of the liar, we notice the following:

(I) We have implicitly assumed that the language in which the antinomy is constructed contains, in addition to its expressions, also the names of these expressions, as well as semantic terms such as the term *"true"* referring to sentences of this language; we have also assumed that all sentences which determine the adequate usage of this term can be asserted in the language. A language with these properties will be called *"semantically closed."*

(II) We have assumed that in this language the ordinary laws of logic hold.

(III) We have assumed that we can formulate and assert in our language an empirical premise such as the statement (2) which has occurred in our argument.

It turns out that the assumption (III) is not essential, for it is possible to reconstruct the antinomy of the liar without its help [11]   But the assumptions (I) and (II) prove essential. Since every language which satisfies both of these assumptions is inconsistent, we must reject at least one of them.

It would be superfluous to stress here the consequences of rejecting the assumption (II), that is, of changing our logic (supposing this were possible) even in its more elementary and fundamental parts. We thus consider only the possibility of rejecting the assumption (I). Accordingly, we decide *not to use any language which is semantically closed* in the sense given.

This restriction would of course be unacceptable for those who, for reasons which are not clear to me, believe that there is only one "genuine" language (or, at least, that all "genuine" languages are mutually translatable). However, this restriction does not affect the needs or interests of science in any essential way. The languages (either the formalized languages or—what is more frequently the case—the portions of everyday language) which are used in scientific discourse do not have to be semantically closed. This is obvious in case linguistic phenomena and, in particular, semantic notions do not enter in any way into the subject-matter of a science; for in such a case the language of this science does not have to be provided with any semantic terms at all. However, we shall see in the next section how semantically closed languages can be dispensed with even in those scientific discussions in which semantic notions are essentially involved.

The problem arises as to the position of everyday language with regard to this point. At first blush it would seem that this language satisfies both assumptions (I) and (II), and that therefore it must be inconsistent. But actually the case is not so simple. Our everyday language is certainly not one with an exactly specified structure. We do not know precisely which expressions are sentences, and we know even to a smaller degree which sentences are to be taken as assertible. Thus the problem of consistency has no exact meaning with respect to this language. We may at best only risk the guess that a language whose structure has been exactly specified and which resembles our everyday language as closely as possible would be inconsistent.

9. OBJECT-LANGUAGE AND META-LANGUAGE. Since we have agreed not to employ semantically closed languages, we have to use two different languages in discussing the problem of the definition of truth and, more generally, any problems in the field of semantics. The first of these languages is the language which is "talked about" and which is the subject-matter of the whole discussion; the definition of truth which we are seeking

applies to the sentences of this language. The second is the language in which we "talk about" the first language, and in terms of which we wish, in particular, to construct the definition of truth for the first language. We shall refer to the first language as *"the object-language,"* and to the second as *"the meta-language."*

It should be noticed that these terms "object-language" and "meta-language" have only a relative sense. If, for instance, we become interested in the notion of truth applying to sentences, not of our original object-language, but of its meta-language, the latter becomes automatically the object-language of our discussion; and in order to define truth for this language, we have to go to a new meta-language—so to speak, to a meta-language of a higher level. In this way we arrive at a whole hierarchy of languages.

The vocabulary of the meta-language is to a large extent determined by previously stated conditions under which a definition of truth will be considered materially adequate. This definition, as we recall, has to imply all equivalences of the form (T):

(T)   *X is true if, and only if, p.*

The definition itself and all the equivalences implied by it are to be formulated in the meta-language. On the other hand, the symbol '*p*' in (T) stands for an arbitrary sentence of our object-language. Hence it follows that every sentence which occurs in the object-language must also occur in the meta-language; in other words, the meta-language must contain the object-language as a part. This is at any rate necessary for the proof of the adequacy of the definition—even though the definition itself can sometimes be formulated in a less comprehensive meta-language which does not satisfy this requirement.

(The requirement in question can be somewhat modified, for it suffices to assume that the object-language can be translated into the meta-language; this necessitates a certain change in the interpretation of the symbol '*p*' in (T). In all that follows we shall ignore the possibility of this modification.)

Furthermore, the symbol '*X*' in (T) represents the name of the sentence which '*p*' stands for. We see therefore that the meta-language must be rich enough to provide possibilities of constructing a name for every sentence of the object-language.

In addition, the meta-language must obviously contain terms of a general logical character, such as the expression "if, and only if."[12]

It is desirable for the meta-language not to contain any undefined terms except such as are involved explicitly or implicitly in the remarks above, i.e.: terms of the object-language; terms referring to the form of the expressions of the object-language, and used in building names for these expressions; and terms of logic. In particular, we desire *semantic terms* (referring to the object-language) *to be introduced into the meta-language only by definition.* For, if this postulate is satisfied, the definition of truth, or of any other semantic concept, will fulfill what we intuitively expect from every definition; that is, it will explain the meaning of the term being defined in terms whose meaning appears to be completely clear and unequivocal. And, moreover, we have then a kind of guarantee that the use of semantic concepts will not involve us in any contradictions.

We have no further requirements as to the formal structure of the object-language and the meta-language; we assume that it is similar to that of other formalized languages known at the present time. In particular, we assume that the usual formal rules of definition are observed in the meta-language.

10. CONDITIONS FOR A POSITIVE SOLUTION OF THE MAIN PROBLEM. Now, we have already a clear idea both of the conditions of material adequacy to which the definition of truth is subjected, and of the formal structure of the language in which this definition is to be constructed. Under these circumstances the problem of the definition of truth acquires the character of a definite problem of a purely deductive nature.

The solution of the problem, however, is by no means obvious, and I would not attempt to give it in detail without using the whole machinery of contemporary logic. Here I shall confine myself to a rough outline of the solution and to the discussion of certain points of a more general interest which are involved in it.

The solution turns out to be sometimes positive, sometimes negative. This depends upon some formal relations between the object-language and its meta-language; or, more specifically, upon the fact whether the meta-language in its logical part is *"essentially richer"* than the object-language or not. It is not easy to give a general and precise definition of this notion of "essential richness." If we restrict ourselves to languages based on the logical theory of types, the condition for the meta-language to be "essentially richer" than the object-language is that it contain variables of a higher logical type than those of the object-language.

If the condition of "essential richness" is not satisfied, it can usually be shown that an interpretation of the meta-language in the object-language is possible; that is to say, with any given term of the meta-language a well-determined term of the object-language can be correlated in such a way that the assertible sentences of the one language turn out to be correlated with assertible sentences of the other. As a result of this interpretation, the hypothesis that a satisfactory definition of truth has been formulated in the meta-language turns out to imply the possibility of reconstructing in that language the antinomy of the liar; and this in turn forces us to reject the hypothesis in question.

(The fact that the meta-language, in its non-logical part, is ordinarily more comprehensive than the object-language does not affect the possibility of interpreting the former in the latter. For example, the names of expressions of the object-language occur in the meta-language, though for the most part they do not occur in the object-language itself; but, nevertheless, it may be possible to interpret these names in terms of the object-language.)

Thus we see that the condition of "essential richness" is necessary for the possibility of a satisfactory definition of truth in the meta-language. If we want to develop the theory of truth in a meta-language which does not satisfy this condition, we must give up the idea of defining truth with the exclusive help of those terms which were indicated above (in Section 8). We have then to include the term "*true*," or some other semantic term, in the list of undefined terms of the meta-language, and to express fundamental properties of the notion of truth in a series of axioms. There is nothing essentially wrong in such an axiomatic procedure, and it may prove useful for various purposes.[13]

It turns out, however, that this procedure can be avoided. For *the condition of the "essential richness" of the meta-language proves to be, not only necessary, but also sufficient for the construction of a satisfactory definition of truth;* i.e., if the meta-language satisfies this condition, the notion of truth can be defined in it. We shall now indicate in general terms how this construction can be carried through.

11. THE CONSTRUCTION (IN OUTLINE) OF THE DEFINITION.[14] A definition of truth can be obtained in a very simple way from that of another semantic notion, namely, of the notion of *satisfaction*.

Satisfaction is a relation between arbitrary objects and certain expressions called "*sentential functions*." These are expressions like "*x is white*," "*x is greater than y*," etc. Their formal structure is analogous to that of sentences; however, they may contain the so-called free variables (like '*x*' and '*y*' in "*x is greater than y*"), which cannot occur in sentences.

In defining the notion of a sentential function in formalized languages, we usually apply what is called a "recursive procedure"; i.e., we first describe sentential functions of the simplest structure (which ordinarily presents no difficulty), and then we indicate the operations by means of which compound functions can be constructed from simpler ones. Such an operation may consist, for instance, in forming the logical disjunction or conjunction of two given functions, i.e., by combining them by the word "*or*" or "*and.*" A sentence can now be defined simply as a sentential function which contains no free variables.

As regards the notion of satisfaction, we might try to define it by saying that given objects satisfy a given function if the latter becomes a true sentence when we replace in it free variables by names of given objects. In this sense, for example, snow satisfies the sentential function "*x is white*" since the sentence "*snow is white*" is true. However, apart from other difficulties, this method is not available to us, for we want to use the notion of satisfaction in defining truth.

To obtain a definition of satisfaction we have rather to apply again a recursive procedure. We indicate which objects satisfy the simplest sentential functions; and then we state the conditions under which given objects satisfy a compound function—assuming that we know which objects satisfy the simpler functions from which the compound one has been constructed. Thus, for instance, we say that given numbers satisfy the logical disjunction "*x is greater than y or x is equal to y*" if they satisfy at least one of the functions "*x is greater than y*" or "*x is equal to y.*"

Once the general definition of satisfaction is obtained, we notice that it applies automatically also to those special sentential functions which contain no free variables, i.e., to sentences. It turns out that for a sentence only two cases are possible: a sentence is either satisfied by all objects, or by no objects. Hence we arrive at a definition of truth and falsehood simply by saying that *a sentence is true if it is satisfied by all objects, and false otherwise.*[15]

(It may seem strange that we have chosen a roundabout way of defining the truth of a sentence, instead of trying to apply, for instance, a direct recursive procedure. The reason is that compound sentences are constructed from simpler sentential functions, but not always from simpler sentences; hence no general recursive method is known which applies specifically to sentences.)

From this rough outline it is not clear where and how the assumption of the "essential richness" of the meta-language is involved in the discussion; this becomes clear only when the construction is carried through in a detailed and formal way.[16]

12. CONSEQUENCES OF THE DEFINITION. The definition of truth which was outlined above has many interesting consequences.

In the first place, the definition proves to be not only formally correct, but also materially adequate (in the sense established in Section 4); in other words, it implies all equivalences of the form (T). In this connection it is important to notice that the conditions for the material adequacy of the definition determine uniquely the extension of the term "*true.*"

Therefore, every definition of truth which is materially adequate would necessarily be equivalent to that actually constructed. The semantic conception of truth gives us, so to speak, no possibility of choice between various non-equivalent definitions of this notion.

Moreover, we can deduce from our definition various laws of a general nature. In particular, we can prove with its help the *laws of contradiction and of excluded middle*, which are so characteristic of the Aristotelian conception of truth; i.e., we can show that one and only one of any two contradictory sentences is true. These semantic laws should not be identified with the related logical laws of contradiction and excluded middle; the latter belong to the sentential calculus, i.e., to the most elementary part of logic, and do not involve the term *"true"* at all.

Further important results can be obtained by applying the theory of truth to formalized languages of a certain very comprehensive class of mathematical disciplines; only disciplines of an elementary character and a very elementary logical structure are excluded from this class. It turns out that for a discipline of this class *the notion of truth never coincides with that of provability;* for all provable sentences are true, but there are true sentences which are not provable.[17] Hence it follows further that every such discipline is consistent, but incomplete; that is to say, of any two contradictory sentences at most one is provable, and—what is more—there exists a pair of contradictory sentences neither of which is provable.[18]

13. EXTENSION OF THE RESULTS TO OTHER SEMANTIC NOTIONS. Most of the results at which we arrived in the preceding sections in discussing the notion of truth can be extended with appropriate changes to other semantic notions, for instance, to the notion of satisfaction (involved in our previous discussion), and to those of *designation* and *definition*.

Each of these notions can be analyzed along the lines followed in the analysis of truth. Thus, criteria for an adequate usage of these notions can be established; it can be shown that each of these notions, when used in a semantically closed language according to those criteria, leads necessarily to a contradiction;[19] a distinction between the object-language and the meta-language becomes again indispensable; and the "essential richness" of the meta-language proves in each case to be a necessary and sufficient condition for a satisfactory definition of the notion involved. Hence the results obtained in discussing one particular semantic notion apply to the general problem of the foundations of theoretical semantics.

Within theoretical semantics we can define and study some further notions, whose intuitive content is more involved and whose semantic origin is less obvious; we have in mind, for instance, the important notions of *consequence*, *synonymity*, and *meaning*.[20]

We have concerned ourselves here with the theory of semantic notions related to an individual object-language (although no specific properties of this language have been involved in our arguments). However, we could also consider the problem of developing *general semantics* which applies to a comprehensive class of object-languages. A considerable part of our previous remarks can be extended to this general problem; however, certain new difficulties arise in this connection, which will not be discussed here. I shall merely observe that the axiomatic method (mentioned in Section 10) may prove the most appropriate for the treatment of the problem.[21]

II. POLEMICAL REMARKS

14. IS THE SEMANTIC CONCEPTION OF TRUTH THE "RIGHT" ONE? I should like to begin the polemical part of the paper with some general remarks.

I hope nothing which is said here will be interpreted as a claim that the semantic conception of truth is the "right" or indeed the "only possible" one. I do not have the slightest intention to contribute in any way to those endless, often violent discussions on the subject: "What is the right conception of truth?"[22] I must confess I do not understand what is at stake in such disputes; for the problem itself is so vague that no definite solution is possible. In fact, it seems to me that the sense in which the phrase "the right conception" is used has never been made clear. In most cases one gets the impression that the phrase is used in an almost mystical sense based upon the belief that every word has only one "real" meaning (a kind of Platonic or Aristotelian idea), and that all the competing conceptions really attempt to catch hold of this one meaning; since, however, they contradict each other, only one attempt can be successful, and hence only one conception is the "right" one.

Disputes of this type are by no means restricted to the notion of truth. They occur in all domains where—instead of an exact, scientific terminology—common language with its vagueness and ambiguity is used; and they are always meaningless, and therefore in vain.

It seems to me obvious that the only rational approach to such problems would be the following: We should reconcile ourselves with the fact that we are confronted, not with one concept, but with several different concepts which are denoted by one word; we should try to make these concepts as clear as possible (by means of definition, or of an axiomatic procedure, or in some other way); to avoid further confusions, we should agree to use different terms for different concepts; and then we may proceed to a quiet and systematic study of all concepts involved, which will exhibit their main properties and mutual relations.

Referring specifically to the notion of truth, it is undoubtedly the case that in philosophical discussions—and perhaps also in everyday usage—

some incipient conceptions of this notion can be found that differ essentially from the classical one (of which the semantic conception is but a modernized form). In fact, various conceptions of this sort have been discussed in the literature, for instance, the pragmatic conception, the coherence theory, etc.[6]

It seems to me that none of these conceptions have been put so far in an intelligible and unequivocal form. This may change, however; a time may come when we find ourselves confronted with several incompatible, but equally clear and precise, conceptions of truth. It will then become necessary to abandon the ambiguous usage of the word *"true,"* and to introduce several terms instead, each to denote a different notion. Personally, I should not feel hurt if a future world congress of the "theoreticians of truth" should decide—by a majority of votes—to reserve the word *"true"* for one of the non-classical conceptions, and should suggest another word, say, *"frue,"* for the conception considered here. But I cannot imagine that anybody could present cogent arguments to the effect that the semantic conception is "wrong" and should be entirely abandoned.

15. FORMAL CORRECTNESS OF THE SUGGESTED DEFINITION OF TRUTH. The specific objections which have been raised to my investigations can be divided into several groups; each of these will be discussed separately.

I think that practically all these objections apply, not to the special definition I have given, but to the semantic conception of truth in general. Even those which were leveled against the definition actually constructed could be related to any other definition which conforms to this conception.

This holds, in particular, for those objections which concern the formal correctness of the definition. I have heard a few objections of this kind; however, I doubt very much whether anyone of them can be treated seriously.

As a typical example let me quote in substance such an objection.[23] In formulating the definition we use necessarily sentential connectives, i.e., expressions like *"if . . . , then," "or,"* etc. They occur in the definiens; and one of them, namely, the phrase *"if, and only if"* is usually employed to combine the definiendum with the definiens. However, it is well known that the meaning of sentential connectives is explained in logic with the help of the words *"true"* and *"false"*; for instance, we say that an equivalence, i.e., a sentence of the form *"p if, and only if, q,"* is true if either both of its members, i.e., the sentences represented by *'p'* and *'q,'* are true or both are false. Hence the definition of truth involves a vicious circle.

If this objection were valid, no formally correct definition of truth would be possible; for we are unable to formulate any compound sentence without using sentential connectives, or other logical terms defined with their help. Fortunately, the situation is not so bad.

It is undoubtedly the case that a strictly deductive development of logic is often preceded by certain statements explaining the conditions under which sentences of the form "*if p, then q*," etc., are considered true or false. (Such explanations are often given schematically, by means of the so-called truth-tables.) However, these statements are outside of the system of logic, and should not be regarded as definitions of the terms involved. They are not formulated in the language of the system, but constitute rather special consequences of the definition of truth given in the meta-language. Moreover, these statements do not influence the deductive development of logic in any way. For in such a development we do not discuss the question whether a given sentence is true, we are only interested in the problem whether it is provable.[24]

On the other hand, the moment we find ourselves within the deductive system of logic—or of any discipline based upon logic, e.g., of semantics— we either treat sentential connectives as undefined terms, or else we define them by means of other sentential connectives, but never by means of semantic terms like "*true*" or "*false*." For instance, if we agree to regard the expressions "*not*" and "*if . . . , then*" (and possibly also "*if, and only if*") as undefined terms, we can define the term "*or*" by stating that a sentence of the form "*p or q*" is equivalent to the corresponding sentence of the form "*if not p, then q*." The definition can be formulated, e.g., in the following way:

$$(p \text{ or } q) \textit{ if, and only if, } (\textit{if not } p, \textit{ then } q).$$

This definition obviously contains no semantic terms.

However, a vicious circle in definition arises only when the definiens contains either the term to be defined itself, or other terms defined with its help. Thus we clearly see that the use of sentential connectives in defining the semantic term "*true*" does not involve any circle.

I should like to mention a further objection which I have found in the literature and which seems also to concern the formal correctness, if not of the definition of truth itself, then at least of the arguments which lead to this definition.[25]

The author of this objection mistakenly regards scheme (T) (from Section 4) as a definition of truth. He charges this alleged definition with "inadmissible brevity, i.e., incompleteness," which "does not give us the means of deciding whether by 'equivalence' is meant a logical-formal, or a non-logical and also structurally non-describable relation." To remove this "defect" he suggests supplementing (T) in one of the two following ways:

(T′)   *X is true if, and only if, p is true,*

or

(T″) *X is true if, and only if, p is the case (i.e., if what p states is the case).*

17

Then he discusses these two new "definitions," which are supposedly free from the old, formal "defect," but which turn out to be unsatisfactory for other, non-formal reasons.

This new objection seems to arise from a misunderstanding concerning the nature of sentential connectives (and thus to be somehow related to that previously discussed). The author of the objection does not seem to realize that the phrase "*if, and only if*" (in opposition to such phrases as "*are equivalent*" or "*is equivalent to*") expresses no relation between sentences at all since it does not combine names of sentences.

In general, the whole argument is based upon an obvious confusion between sentences and their names. It suffices to point out that —in contradistinction to (T)—schemata (T') and (T″) do not give any meaningful expressions if we replace in them '*p*' by a sentence; for the phrases "*p is true*" and "*p is the case*" (i.e., "*what p states is the case*") become meaningless if '*p*' is replaced by a sentence, and not by the name of a sentence (cf. Section 4).[26]

While the author of the objection considers schema (T) "inadmissible brief," I am inclined, on my part, to regard schemata (T') and (T″) as "inadmissibly long." And I think even that I can rigorously prove this statement on the basis of the following definition: An expression is said to be "inadmissibly long" if (i) it is meaningless, and (ii) it has been obtained from a meaningful expression by inserting superfluous words.

16. REDUNDANCY OF SEMANTIC TERMS—THEIR POSSIBLE ELIMINATION. The objection I am going to discuss now no longer concerns the formal correctness of the definition, but is still concerned with certain formal features of the semantic conception of truth.

We have seen that this conception essentially consists in regarding the sentence "*X is true*" as equivalent to the sentence denoted by '*X*' (where '*X*' stands for a name of a sentence of the object-language). Consequently, the term "*true*" when occurring in a simple sentence of the form "*X is true*" can easily be eliminated, and the sentence itself, which belongs to the meta-language, can be replaced by an equivalent sentence of the object-language; and the same applies to compound sentences provided the term "*true*" occurs in them exclusively as a part of the expressions of the form "*X is true*."

Some people have therefore urged that the term "*true*" in the semantic sense can always be eliminated, and that for this reason the semantic conception of truth is altogether sterile and useless. And since the same considerations apply to other semantic notions, the conclusion has been drawn that semantics as a whole is a purely verbal game and at best only a harmless hobby.

18

But the matter is not quite so simple.[27]   The sort of elimination here discussed cannot always be made.   It cannot be done in the case of universal statements which express the fact that all sentences of a certain type are true, or that all true sentences have a certain property.   For instance, we can prove in the theory of truth the following statement:

All consequences of true sentences are true.

However, we cannot get rid here of the word *"true"* in the simple manner contemplated.

Again, even in the case of particular sentences having the form *"X is true"* such a simple elimination cannot always be made.   In fact, the elimination is possible only in those cases in which the name of the sentence which is said to be true occurs in a form that enables us to reconstruct the sentence itself.   For example, our present historical knowledge does not give us any possibility of eliminating the word *"true"* from the following sentence:

The first sentence written by Plato is true.

Of course, since we have a definition for truth and since every definition enables us to replacé the definiendum by its definiens, an elimination of the term *"true"* in its semantic sense is always theoretically possible.   But this would not be the kind of simple elimination discused above, and it would not result in the replacement of a sentence in the meta-language by a sentence in the object-language.

If, however, anyone continues to urge that—because of the theoretical possibility of eliminating the word *"true"* on the basis of its definition— the concept of truth is sterile, he must accept the further conclusion that all defined notions are sterile.   But this outcome is so absurd and so unsound historically that any comment on it is unnecessary.   In fact, I am rather inclined to agree with those who maintain that the moments of greatest creative advancement in science frequently coincide with the introduction of new notions by means of definition.

17. CONFORMITY OF THE SEMANTIC CONCEPTION OF TRUTH WITH PHILOSOPHICAL AND COMMON-SENSE USAGE.   The question has been raised whether the semantic conception of truth can indeed be regarded as a precise form of the old, classical conception of this notion.

Various formulations of the classical conception were quoted in the early part of this paper (Section 3).   I must repeat that in my judgment none of them is quite precise and clear.   Accordingly, the only sure way of settling the question would be to confront the authors of those statements with our new formulation, and to ask them whether it agrees with their intentions. Unfortunately, this method is impractical since they died quite some time ago.

As far as my own opinion is concerned, I do not have any doubts that our formulation does conform to the intuitive content of that of Aristotle. I am less certain regarding the later formulations of the classical conception, for they are very vague indeed.[28]

Furthermore, some doubts have been expressed whether the semantic conception does reflect the notion of truth in its common-sense and everyday usage. I clearly realize (as I already indicated) that the common meaning of the word *"true"*—as that of any other word of everyday language—is to some extent vague, and that its usage more or less fluctuates. Hence the problem of assigning to this word a fixed and exact meaning is relatively unspecified, and every solution of this problem implies necessarily a certain deviation from the practice of everyday language.

In spite of all this, I happen to believe that the semantic conception does conform to a very considerable extent with the common-sense usage— although I readily admit I may be mistaken. What is more to the point, however, I believe that the issue raised can be settled scientifically, though of course not by a deductive procedure, but with the help of the statistical questionnaire method. As a matter of fact, such research has been carried on, and some of the results have been reported at congresses and in part published.[29]

I should like to emphasize that in my opinion such investigations must be conducted with the utmost care. Thus, if we ask a highschool boy, or even an adult intelligent man having no special philosophical training, whether he regards a sentence to be true if it agrees with reality, or if it designates an existing state of affairs, it may simply turn out that he does not understand the question; in consequence his response, whatever it may be, will be of no value for us. But his answer to the question whether he would admit that the sentence *"it is snowing"* could be true although it is not snowing, or could be false although it is snowing, would naturally be very significant for our problem.

Therefore, I was by no means surprised to learn (in a discussion devoted to these problems) that in a group of people who were questioned only 15% agreed that *"true"* means for them *"agreeing with reality,"* while 90% agreed that a sentence such as *"it is snowing"* is true if, and only if, it is snowing. Thus, a great majority of these people seemed to reject the classical conception of truth in its "philosophical" formulation, while accepting the same conception when formulated in plain words (waiving the question whether the use of the phrase "the same conception" is here justified).

18. THE DEFINITION IN ITS RELATION TO "THE PHILOSOPHICAL PROBLEM OF TRUTH" AND TO VARIOUS EPISTEMOLOGICAL TRENDS. I have heard it remarked that the formal definition of truth has nothing to do with "the philosophical problem of truth."[30]. However, nobody has ever pointed out to me in an intelligible way just what this problem is. I have been informed in this connection that my definition, though it states necessary and sufficient conditions for a sentence to be true, does not really grasp the "essence" of this concept. Since I have never been able to understand what the "essence" of a concept is, I must be excused from discussing this point any longer.

In general, I do not believe that there is such a thing as "the philosophical problem of truth." I do believe that there are various intelligible and interesting (but not necessarily philosophical) problems concerning the notion of truth, but I also believe that they can be exactly formulated and possibly solved only on the basis of a precise conception of this notion.

While on the one hand the definition of truth has been blamed for not being philosophical enough, on the other a series of objections have been raised charging this definition with serious philosophical implications, always of a very undesirable nature. I shall discuss now one special objection of this type; another group of such objections will be dealt with in the next section.

It has been claimed that—due to the fact that a sentence like "snow is white" is taken to be semantically true if snow is *in fact* white (italics by the critic)—logic finds itself involved in a most uncritical realism.[31]

If there were an opportunity to discuss the objection with its author, I should raise two points. First, I should ask him to drop the words "*in fact*," which do not occur in the original formulation and which are misleading, even if they do not affect the content. For these words convey the impression that the semantic conception of truth is intended to establish the conditions under which we are warranted in asserting any given sentence, and in particular any empirical sentence. However, a moment's reflection shows that this impression is merely an illusion; and I think that the author of the objection falls victim to the illusion which he himself created.

In fact, the semantic definition of truth implies nothing regarding the conditions under which a sentence like (1):

(1) *snow is white*

can be asserted. It implies only that, whenever we assert or reject this sentence, we must be ready to assert or reject the correlated sentence (2):

(2) *the sentence "snow is white" is true.*

Thus, we may accept the semantic conception of truth without giving up any epistemological attitude we may have had; we may remain naive realists, critical realists or idealists, empiricists or metaphysicians—whatever we were before. The semantic conception is completely neutral toward all these issues.

In the second place, I should try to get some information regarding the conception of truth which (in the opinion of the author of the objection) does not involve logic in a most naive realism. I would gather that this conception must be incompatible with the semantic one. Thus, there must be sentences which are true in one of these conceptions without being true in the other. Assume, e.g., the sentence (1) to be of this kind. The truth of this sentence in the semantic conception is determined by an equivalence of the form (T):

> The sentence "snow is white" is true if, and only if, snow is white.

Hence in the new conception we must reject this equivalence, and consequently we must assume its denial:

> The sentence "snow is white" is true if, and only if, snow is not white (or perhaps: snow, in fact, is not white).

This sounds somewhat paradoxical. I do not regard such a consequence of the new conception as absurd; but I am a little fearful that someone in the future may charge this conception with involving logic in a "most sophisticated kind of irrealism." At any rate, it seems to me important to realize that every conception of truth which is incompatible with the semantic one carries with it consequences of this type.

I have dwelt a little on this whole question, not because the objection discussed seems to me very significant, but because certain points which have arisen in the discussion should be taken into account by all those who for various epistemological reasons are inclined to reject the semantic conception of truth.

19. ALLEGED METAPHYSICAL ELEMENTS IN SEMANTICS. The semantic conception of truth has been charged several times with involving certain metaphysical elements. Objections of this sort have been made to apply not only to the theory of truth, but to the whole domain of theoretical semantics.[32]

I do not intend to discuss the general problem whether the introduction of a metaphysical element into a science is at all objectionable. The only point which will interest me here is whether and in what sense metaphysics is involved in the subject of our present discussion.

The whole question obviously depends upon what one understands by "metaphysics." Unfortunately, this notion is extremely vague and equivocal. When listening to discussions in this subject, sometimes one gets the impression that the term "metaphysical" has lost any objective meaning, and is merely used as a kind of professional philosophical invective.

For some people metaphysics is a general theory of objects (ontology)—a discipline which is to be developed in a purely empirical way, and which differs from other empirical sciences only by its generality. I do not know whether such a discipline actually exists (some cynics claim that it is customary in philosophy to baptize unborn children); but I think that in any case metaphysics in this conception is not objectionable to anybody, and has hardly any connections with semantics.

For the most part, however, the term "metaphysical" is used as directly opposed—in one sense or another—to the term "empirical"; at any rate, it is used in this way by those people who are distressed by the thought that any metaphysical elements might have managed to creep into science. This general conception of metaphysics assumes several more specific forms.

Thus, some people take it to be symptomatic of a metaphysical element in a science when methods of inquiry are employed which are neither deductive nor empirical. However, no trace of this symptom can be found in the development of semantics (unless some metaphysical elements are involved in the object-language to which the semantic notions refer). In particular, the semantics of formalized languages is constructed in a purely deductive way.

Others maintain that the metaphysical character of a science depends mainly on its vocabulary and, more specifically, on its primitive terms. Thus, a term is said to be metaphysical if it is neither logical nor mathematical, and if it is not associated with an empirical procedure which enables us to decide whether a thing is denoted by this term or not. With respect to such a view of metaphysics it is sufficient to recall that a meta-language includes only three kinds of undefined terms: (i) terms taken from logic, (ii) terms of the corresponding object-language, and (iii) names of expressions in the object-language. It is thus obvious that no metaphysical undefined terms occur in the meta-language (again, unless such terms appear in the object-language itself).

There are, however, some who believe that, even if no metaphysical terms occur among the primitive terms of a language, they may be introduced by definitions; namely, by those definitions which fail to provide us with general criteria for deciding whether an object falls under the defined concept. It is argued that the term *"true"* is of this kind, since no universal criterion of truth follows immediately from the definition of this term, and since it is generally believed (and in a certain sense can even be proved) that such a criterion will never be found. This comment on the actual character of the notion of truth seems to be perfectly just. However, it should be noticed that the notion of truth does not differ in this respect from many notions in logic, mathematics, and theoretical parts of various empirical sciences, e.g., in theoretical physics.

In general, it must be said that if the term "metaphysical" is employed in so wide a sense as to embrace certain notions (or methods) of logic, mathematics, or empirical sciences, it will apply *a fortiori* to those of semantics. In fact, as we know from Part I of the paper, in developing the semantics of a language we use all the notions of this language, and we apply even a stronger logical apparatus than that which is used in the language itself. On the other hand, however, I can summarize the arguments given above by stating that in no interpretation of the term "metaphysical" which is familiar and more or less intelligible to me does semantics involve any metaphysical elements peculiar to itself.

I should like to make one final remark in connection with this group of objections. The history of science shows many instances of concepts which were judged metaphysical (in a loose, but in any case derogatory sense of this term) before their meaning was made precise; however, once they received a rigorous, formal definition, the distrust in them evaporated. As typical examples we may mention the concepts of negative and imaginary numbers in mathematics. I hope a similar fate awaits the concept of truth and other semantic concepts; and it seems to me, therefore, that those who have distrusted them because of their alleged metaphysical implications should welcome the fact that precise definitions of these concepts are now available. If in consequence semantic concepts lose philosophical interest, they will only share the fate of many other concepts of science, and this need give rise to no regret.

20. APPLICABILITY OF SEMANTICS TO SPECIAL EMPIRICAL SCIENCES. We come to the last and perhaps the most important group of objections. Some strong doubts have been expressed whether semantic notions find or can find applications in various domains of intellectual activity. For the most part such doubts have concerned the applicability of semantics to the field of empirical science—either to special sciences or to the general methodology of this field; although similar skepticism has been expressed regarding possible applications of semantics to mathematical sciences and their methodology.

I believe that it is possible to allay these doubts to a certain extent, and that some optimism with respect to the potential value of semantics for various domains of thought is not without ground.

To justify this optimism, it suffices I think to stress two rather obvious points. First, the development of a theory which formulates a precise definition of a notion and establishes its general properties provides *eo ipso* a firmer basis for all discussions in which this notion is involved; and, therefore, it cannot be irrelevant for anyone who uses this notion, and desires to do so in a conscious and consistent way. Secondly, semantic notions are actually involved in various branches of science, and in particular of empirical science.

The fact that in empirical research we are concerned only with natural languages and that theoretical semantics applies to these languages only with certain approximation, does not affect the problem essentially. However, it has undoubtedly this effect that progress in semantics will have but a delayed and somewhat limited influence in this field. The situation with which we are confronted here does not differ essentially from that which arises when we apply laws of logic to arguments in everyday life—or, generally, when we attempt to apply a theoretical science to empirical problems.

Semantic notions are undoubtedly involved, to a larger or smaller degree, in psychology, sociology, and in practically all the humanities. Thus, a psychologist defines the so-called intelligence quotient in terms of the numbers of *true* (right) and *false* (wrong) answers given by a person to certain questions; for a historian of culture the range of objects for which a human race in successive stages of its development possesses adequate *designations* may be a topic of great significance; a student of literature may be strongly interested in the problem whether a given author always uses two given words with the same *meaning*. Examples of this kind can be multiplied indefinitely.

The most natural and promising domain for the applications of theoretical semantics is clearly linguistics—the empirical study of natural languages. Certain parts of this science are even referred to as "semantics," sometimes with an additional qualification. Thus, this name is occasionally given to that portion of grammar which attempts to classify all words of a language into parts of speech, according to what the words mean or designate. The study of the evolution of meanings in the historical development of a language is sometimes called "historical semantics." In general, the totality of investigations on semantic relations which occur in a natural language is referred to as "descriptive semantics." The relation between theoretical and descriptive semantics is analogous to that between pure and applied mathematics, or perhaps to that between theoretical and empirical physics; the role of formalized languages in semantics can be roughly compared to that of isolated systems in physics.

It is perhaps unnecessary to say that semantics cannot find any direct applications in natural sciences such as physics, biology, etc.; for in none of these sciences are we concerned with linguistic phenomena, and even less with semantic relations between linguistic expressions and objects to which these expressions refer. We shall see, however, in the next section that semantics may have a kind of indirect influence even on those sciences in which semantic notions are not directly involved.

21. APPLICABILITY OF SEMANTICS TO THE METHODOLOGY OF EMPIRICAL SCIENCE. Besides linguistics, another important domain for possible applications of semantics is the methodology of science; this term is used here in a broad sense so as to embrace the theory of science in general. Independent of whether a science is conceived merely as a system of statements or as a totality of certain statements and human activities, the study of scientific language constitutes an essential part of the methodological discussion of a science. And it seems to me clear that any tendency to eliminate semantic notions (like those of truth and designation) from this discussion would make it fragmentary and inadequate.[33] Moreover, there is no reason for such a tendency today, once the main difficulties in using semantic terms have been overcome. The semantics of scientific language should be simply included as a part in the methodology of science.

I am by no means inclined to charge methodology and, in particular, semantics—whether theoretical or descriptive—with the task of clarifying the meanings of all scientific terms. This task is left to those sciences in which the terms are used, and is actually fulfilled by them (in the same way in which, e.g., the task of clarifying the meaning of the term *"true"* is left to, and fulfilled by, semantics). There may be, however, certain special problems of this sort in which a methodological approach is desirable or indeed necessary (perhaps, the problem of the notion of causality is a good example here); and in a methodological discussion of such problems semantic notions may play an essential role. Thus, semantics may have some bearing on any science whatsoever.

The question arises whether semantics can be helpful in solving general and, so to speak, classical problems of methodology. I should like to discuss here with some detail a special, though very important, aspect of this question.

One of the main problems of the methodology of empirical science consists in establishing conditions under which an empirical theory or hypothesis should be regarded as acceptable. This notion of acceptability must be relativized to a given stage of the development of a science (or to a given amount of presupposed knowledge). In other words, we may consider it as provided with a time coefficient; for a theory which is acceptable today may become untenable tomorrow as a result of new scientific discoveries.

It seems *a priori* very plausible that the acceptability of a theory somehow depends on the truth of its sentences, and that consequently a methodologist in his (so far rather unsuccessful) attempts at making the notion of acceptability precise, can expect some help from the semantic theory of truth. Hence we ask the question: Are there any postulates which can be reasonably imposed on acceptable theories and which involve the notion of truth? And, in particular, we ask whether the following postulate is a reasonable one:

An acceptable theory cannot contain (or imply) any false sentences.

The answer to the last question is clearly negative. For, first of all, we are practically sure, on the basis of our historical experience, that every empirical theory which is accepted today will sooner or later be rejected and replaced by another theory. It is also very probable that the new theory will be incompatible with the old one; i.e., will imply a sentence which is contradictory to one of the sentences contained in the old theory. Hence, at least one of the two theories must include false sentences, in spite of the fact that each of them is accepted at a certain time. Secondly, the postulate in question could hardly ever be satisfied in practice; for we do not know, and are very unlikely to find, any criteria of truth which enable us to show that no sentence of an empirical theory is false.

The postulate in question could be at most regarded as the expression of an ideal limit for successively more adequate theories in a given field of research; but this hardly can be given any precise meaning.

Nevertheless, it seems to me that there is an important postulate which can be reasonably imposed on acceptable empirical theories and which involves the notion of truth. It is closely related to the one just discussed, but is essentially weaker. Remembering that the notion of acceptability is provided with a time coefficient, we can give this postulate the following form:

As soon as we succeed in showing that an empirical theory contains (or implies) false sentences, it cannot be any longer considered acceptable.

In support of this postulate, I should like to make the following remarks.

I believe everybody agrees that one of the reasons which may compel us to reject an empirical theory is the proof of its inconsistency: a theory becomes untenable if we succeed in deriving from it two contradictory sentences. Now we can ask what are the usual motives for rejecting a theory on such grounds. Persons who are acquainted with modern logic are inclined to answer this question in the following way: A well-known logical law shows that a theory which enables us to derive two contradictory sentences enables us also to derive every sentence; therefore, such a theory is trivial and deprived of any scientific interest.

I have some doubts whether this answer contains an adequate analysis of the situation. I think that people who do not know modern logic are as little inclined to accept an inconsistent theory as those who are thoroughly familiar with it; and probably this applies even to those who regard (as some still do) the logical law on which the argument is based as a highly controversial issue, and almost as a paradox. I do not think that our attitude toward an inconsistent theory would change even if we decided for some reasons to weaken our system of logic so as to deprive ourselves of the possibility of deriving every sentence from any two contradictory sentences.

It seems to me that the real reason of our attitude is a different one: We know (if only intuitively) that an inconsistent theory must contain false sentences; and we are not inclined to regard as acceptable any theory which has been shown to contain such sentences.

There are various methods of showing that a given theory includes false sentences. Some of them are based upon purely logical properties of the theory involved; the method just discussed (i.e., the proof of inconsistency) is not the sole method of this type, but is the simplest one, and the one which is most frequently applied in practice. With the help of certain assumptions regarding the truth of empirical sentences, we can obtain methods to the same effect which are no longer of a purely logical nature. If we decide to accept the general postulate suggested above, then a successful application of any such method will make the theory untenable.

22. APPLICATIONS OF SEMANTICS TO DEDUCTIVE SCIENCE. As regards the applicability of semantics to mathematical sciences and their methodology, i.e., to meta-mathematics, we are in a much more favorable position than in the case of empirical sciences. For, instead of advancing reasons which justify some hopes for the future (and thus making a kind of pro-semantics propaganda), we are able to point out concrete results already achieved.

Doubts continue to be expressed whether the notion of a true sentence—as distinct from that of a provable sentence—can have any significance for mathematical disciplines and play any part in a methodological discussion of mathematics. It seems to me, however, that just this notion of a true sentence constitutes a most valuable contribution to meta-mathematics by semantics. We already possess a series of interesting meta-mathematical results gained with the help of the theory of truth. These results concern the mutual relations between the notion of truth and that of provability; establish new properties of the latter notion (which, as well known, is one of the basic notions of meta-mathematics); and throw some light on the fundamental problems of consistency and completeness. The most significant among these results have been briefly discussed in Section 12.[34]

Furthermore, by applying the method of semantics we can adequately define several important meta-mathematical notions which have been used so far only in an intuitive way—such as, e.g., the notion of definability or that of a model of an axiom system; and thus we can undertake a systematic study of these notions. In particular, the investigations on definability have already brought some interesting results, and promise even more in the future.[35]

We have discussed the applications of semantics only to meta-mathematics, and not to mathematics proper. However, this distinction between mathematics and meta-mathematics is rather unimportant. For meta-mathematics is itself a deductive discipline and hence, from a certain point of view, a part of mathematics; and it is well known that—due to the formal character of deductive method—the results obtained in one deductive discipline can be automatically extended to any other discipline in which the given one finds an interpretation. Thus, for example, all meta-mathematical results can be interpreted as results of number theory. Also from a practical point of view there is no clear-cut line between meta-mathematics and mathematics proper; for instance, the investigations on definability could be included in either of these domains.

23. FINAL REMARKS. I should like to conclude this discussion with some general and rather loose remarks concerning the whole question of the evaluation of scientific achievements in terms of their applicability. I must confess I have various doubts in this connection.

Being a mathematician (as well as a logician, and perhaps a philosopher of a sort), I have had the opportunity to attend many discussions between specialists in mathematics, where the problem of applications is especially acute, and I have noticed on several occasions the following phenomenon: If a mathematician wishes to disparage the work of one of his colleagues, say, A, the most effective method he finds for doing this is to ask where the results can be applied. The hard pressed man, with his back against the wall, finally unearths the researches of another mathematician B as the locus of the application of his own results. If next B is plagued with a similar question, he will refer to another mathematician C. After a few steps of this kind we find ourselves referred back to the researches of A, and in this way the chain closes.

Speaking more seriously, I do not wish to deny that the value of a man's work may be increased by its implications for the research of others and for practice. But I believe, nevertheless, that it is inimical to the progress of science to measure the importance of any research exclusively or chiefly in terms of its usefulness and applicability. We know from the history of science that many important results and discoveries have had to wait centuries before they were applied in any field. And, in my opinion, there are also other important factors which cannot be disregarded in determining the value of a scientific work.   It seems to me that there is a special domain of very profound and strong human needs related to scientific research, which are similar in many ways to aesthetic and perhaps religious needs. And it also seems to me that the satisfaction of these needs should be considered an important task of research.   Hence, I believe, the question of the value of any research cannot be adequately answered without taking into account the intellectual satisfaction which the results of that research bring to those who understand it and care for it.   It may be unpopular and out-of-date to say—but I do not think that a scientific result which gives us a better understanding of the world and makes it more harmonious in our eyes should be held in lower esteem than, say, an invention which reduces the cost of paving roads, or improves household plumbing.

It is clear that the remarks just made become pointless if the word "application" is used in a very wide and liberal sense.   It is perhaps not less obvious that nothing follows from these general remarks concerning the specific topics which have been discussed in this paper; and I really do not know whether research in semantics stands to gain or lose by introducing the standard of value I have suggested.

NOTES

[1] Compare Tarski [2] (see bibliography at the end of the paper).   This work may be consulted for a more detailed and formal presentation of the subject of the paper, especially of the material included in Sections 6 and 9–13.   It contains also references to my earlier publications on the problems of semantics (a communication in Polish, 1930; the article Tarski [1] in French, 1931; a communication in German, 1932; and a book in Polish, 1933).   The expository part of the present paper is related in its character to Tarski [3].   My investigations on the notion of truth and on theoretical semantics have been reviewed or discussed in Hofstadter [1], Juhos [1], Kokoszyńska [1] and [2], Kotarbiński [2], Scholz [1], Weinberg [1], et al.

[2] It may be hoped that the interest in theoretical semantics will now increase, as a result of the recent publication of the important work Carnap [2].

[3] This applies, in particular, to public discussions during the I. International Congress for the Unity of Science (Paris, 1935) and the Conference of International Congresses for the Unity of Science (Paris, 1937); cf., e.g., Neurath [1] and Gonseth [1].

[4] The words "notion" and "concept" are used in this paper with all of the vagueness and ambiguity with which they occur in philosophical literature.   Thus, sometimes they refer simply to a term, sometimes to what is meant by a term, and in other cases to what is denoted by a term.   Sometimes it is irrelevant which of these interpretations is meant; and in certain cases perhaps none of them applies adequately. While on principle I share the tendency to avoid these words in any exact discussion, I did not consider it necessary to do so in this informal presentation.

[5] For our present purposes it is somewhat more convenient to understand by "expressions," "sentences," etc., not individual inscriptions, but classes of inscriptions of similar form (thus, not individual physical things, but classes of such things).

[6] For the Aristotelian formulation see Aristotle [1]; Γ, 7, 27. The other two formulations are very common in the literature, but I do not know with whom they originate. A critical discussion of various conceptions of truth can be found, e.g., in Kotarbiński [1] (so far available only in Polish), pp. 123 ff., and Russell [1], pp. 362 ff.

[7] For most of the remarks contained in Sections 4 and 8, I am indebted to the late S. Leśniewski who developed them in his unpublished lectures in the University of Warsaw (in 1919 and later). However, Leśniewski did not anticipate the possibility of a rigorous development of the theory of truth, and still less of a definition of this notion; hence, while indicating equivalences of the form (T) as premisses in the antinomy of the liar, he did not conceive them as any sufficient conditions for an adequate usage (or definition) of the notion of truth. Also the remarks in Section 8 regarding the occurrence of an empirical premiss in the antinomy of the liar, and the possibility of eliminating this premiss, do not originate with him.

[8] In connection with various logical and methodological problems involved in this paper the reader may consult Tarski [6].

[9] The antinomy of the liar (ascribed to Eubulides or Epimenides) is discussed here in Sections 7 and 8. For the antinomy of definability (due to J. Richard) see, e.g., Hilbert-Bernays [1], vol. 2, pp. 263 ff.; for the antinomy of heterological terms see Grelling-Nelson [1], p. 307.

[10] Due to Professor J. Łukasiewicz (University of Warsaw).

[11] This can roughly be done in the following way. Let S be any sentence beginning with the words *"Every sentence."* We correlate with S a new sentence S^* by subjecting S to the following two modifications: we replace in S the first word, *"Every,"* by *"The"*; and we insert after the second word, *"sentence,"* the whole sentence S enclosed in quotation marks. Let us agree to call the sentence S *"(self-)applicable"* or *"non-(self-)applicable"* dependent on whether the correlated sentence S^* is true or false. Now consider the following sentence:

Every sentence is non-applicable.

It can easily be shown that the sentence just stated must be both applicable and non-applicable; hence a contradiction. It may not be quite clear in what sense this formulation of the antinomy does not involve an empirical premiss; however, I shall not elaborate on this point.

[12] The terms "logic" and "logical" are used in this paper in a broad sense, which has become almost traditional in the last decades; logic is assumed here to comprehend the whole theory of classes and relations (i.e., the mathematical theory of sets). For many different reasons I am personally inclined to use the term "logic" in a much narrower sense, so as to apply it only to what is sometimes called "elementary logic," i.e., to the sentential calculus and the (restricted) predicate calculus.

[13] Cf. here, however, Tarski [3], pp. 5 f.

[14] The method of construction we are going to outline can be applied—with appropriate changes—to all formalized languages that are known at the present time; although it does not follow that a language could not be constructed to which this method would not apply.

[15] In carrying through this idea a certain technical difficulty arises. A sentential function may contain an arbitrary number of free variables; and the logical nature of the notion of satisfaction varies with this number. Thus, the notion in question when applied to functions with one variable is a binary relation between these functions and single objects; when applied to functions with two variables it becomes a ternary relation between functions and couples of objects; and so on. Hence, strictly

speaking, we are confronted, not with one notion of satisfaction, but with infinitely many notions; and it turns out that these notions cannot be defined independently of each other, but must all be introduced simultaneously.

To overcome this difficulty, we employ the mathematical notion of an infinite sequence (or, possibly, of a finite sequence with an arbitrary number of terms). We agree to regard satisfaction, not as a many-termed relation between sentential functions and an indefinite number of objects, but as a binary relation between functions and sequences of objects. Under this assumption the formulation of a general and precise definition of satisfaction no longer presents any difficulty; and a true sentence can now be defined as one which is satisfied by every sequence.

[16] To define recursively the notion of satisfaction, we have to apply a certain form of recursive definition which is not admitted in the object-language. Hence the "essential richness" of the meta-language may simply consist in admitting this type of definition. On the other hand, a general method is known which makes it possible to eliminate all recursive definitions and to replace them by normal, explicit ones. If we try to apply this method to the definition of satisfaction, we see that we have either to introduce into the meta-language variables of a higher logical type than those which occur in the object-language; or else to assume axiomatically in the meta-language the existence of classes that are more comprehensive than all those whose existence can be established in the object-language. See here Tarski [2], pp. 393 ff., and Tarski [5], p. 110.

[17] Due to the development of modern logic, the notion of mathematical proof has undergone a far-reaching simplification. A sentence of a given formalized discipline is provable if it can be obtained from the axioms of this discipline by applying certain simple and purely formal rules of inference, such as those of detachment and substitution. Hence to show that all provable sentences are true, it suffices to prove that all the sentences accepted as axioms are true, and that the rules of inference when applied to true sentences yield new true sentences; and this usually presents no difficulty.

On the other hand, in view of the elementary nature of the notion of provability, a precise definition of this notion requires only rather simple logical devices. In most cases, those logical devices which are available in the formalized discipline itself (to which the notion of provability is related) are more than sufficient for this purpose. We know, however, that as regards the definition of truth just the opposite holds. Hence, as a rule, the notions of truth and provability cannot coincide; and since every provable sentence is true, there must be true sentences which are not provable.

[18] Thus the theory of truth provides us with a general method for consistency proofs for formalized mathematical disciplines. It can be easily realized, however, that a consistency proof obtained by this method may possess some intuitive value—i.e., may convince us, or strengthen our belief, that the discipline under consideration is actually consistent—only in case we succeed in defining truth in terms of a meta-language which does not contain the object-language as a part (cf. here a remark in Section 9). For only in this case the deductive assumptions of the meta-language may be intuitively simpler and more obvious than those of the object-language—even though the condition of "essential richness" will be formally satisfied. Cf. here also Tarski [3], p. 7.

The incompleteness of a comprehensive class of formalized disciplines constitutes the essential content of a fundamental theorem of K. Gödel; cf. Gödel [1], pp. 187 ff. The explanation of the fact that the theory of truth leads so directly to Gödel's theorem is rather simple. In deriving Gödel's result from the theory of truth we make an essential use of the fact that the definition of truth cannot be given in a meta-language which is only as "rich" as the object-language (cf. note 17); however, in establishing this fact, a method of reasoning has been applied which is very closely related to that used (for the first time) by Gödel. It may be added that Gödel was clearly guided in his proof by certain intuitive considerations regarding the notion of truth, although this notion does not occur in the proof explicitly; cf. Gödel [1], pp. 174 f.

[19] The notions of designation and definition lead respectively to the antinomies of Grelling-Nelson and Richard (cf. note 9). To obtain an antinomy for the notion of satisfaction, we construct the following expression:

The sentential function X does not satisfy X.

A contradiction arises when we consider the question whether this expression, which is clearly a sentential function, satisfies itself or not.

[20] All notions mentioned·in this section can be defined in terms of satisfaction. We can say, e.g., that a given term designates a given object if this object satisfies the sentential function "x is identical with . T" where 'T' stands for the given term. Similarly, a sentential function is said to define a given object if the latter is the only object which satisfies this function. For a definition of consequence see Tarski [4], and for that of synonymity—Carnap [2].

[21] General semantics is the subject of Carnap [2]. Cf. here also remarks in Tarski [2], pp. 388 f.

[22] Cf. various quotations in Ness [1], pp. 13 f.

[23] The names of persons who have raised objections will not be quoted here, unless their objections have appeared in print.

[24] It should be emphasized, however, that as regards the question of an alleged vicious circle the situation would not change even if we took a different point of view, represented, e.g., in Carnap [2]; i.e., if we regarded the specification of conditions under which sentences of a language are true as an essential part of the description of this language. On the other hand, it may be noticed that the point of view represented in the text does not exclude the possibility of using truth-tables in a deductive development of logic. However, these tables are to be regarded then merely as a formal instrument for checking the provability of certain sentences; and the symbols 'T' and 'F' which occur in them and which are usually considered abbreviations of "*true*" and "*false*" should not be interpreted in any intuitive way.

[25] Cf. Juhos [1]. I must admit that I do not clearly understand von Juhos' objections and do not know how to classify them; therefore, I confine myself here to certain points of a formal character. Von Juhos does not seem to know my definition of truth; he refers only to an informal presentation in Tarski [3] where the definition has not been given at all. If he knew the actual definition, he would have to change his argument. However, I have no doubt that he would discover in this definition some "defects" as well. For he believes he has proved that "on ground of principle it is impossible to give such a definition at all."

[26] The phrases "p is *true*" and "p is *the case*" (or better "*it is true that p*" and "*it is the case that p*") are sometimes used in informal discussions, mainly for stylistic reasons; but they are considered then as synonymous with the sentence represented by 'p'. On the other hand, as far as I understand the situation, the phrases in question cannot be used by von Juhos synonymously with 'p'; for otherwise the replacement of (T) by (T') or (T'') would not constitute any "improvement."

[27] Cf. the discussion of this problem in Kokoszyńska [1], pp. 161 ff.

[28] Most authors who have discussed my work on the notion of truth are of the opinion that my definition does conform with the classical conception of this notion; see, e.g., Kotarbiński [2] and Scholz [1].

[29] Cf. Ness [1]. Unfortunately, the results of that part of Ness' research which is especially relevant for our problem are not discussed in his book; compare p. 148, footnote 1.

[30] Though I have heard this opinion several times, I have seen it in print only once and, curiously enough, in a work which does not have a philosophical character—in fact, in Hilbert-Bernays [1], vol. II, p. 269 (where, by the way, it is not expressed as any kind of objection). On the other hand, I have not found any remark to this effect in discussions of my work by professional philosophers (cf. note 1).

[31] Cf. Gonseth [1], pp. 187 f.

[32] See Nagel [1], and Nagel [2], pp. 471 f. A remark which goes, perhaps, in the same direction is also to be found in Weinberg [1], p. 77; cf., however, his earlier remarks, pp. 75 f.

[33] Such a tendency was evident in earlier works of Carnap (see, e.g., Carnap [1], especially Part V) and in writings of other members of Vienna Circle. Cf. here Kokoszyńska [1] and Weinberg [1].

[34] For other results obtained with the help of the theory of truth see Gödel [2]; Tarski [2], pp. 401 ff.; and Tarski [5], pp. 111 f.

[35] An object—e.g., a number or a set of numbers—is said to be definable (in a given formalism) if there is a sentential function which defines it; cf. note 20. Thus, the term "definable," though of a meta-mathematical (semantic) origin, is purely mathematical as to its extension, for it expresses a property (denotes a class) of mathematical objects. In consequence, the notion of definability can be re-defined in purely mathematical terms, though not within the formalized discipline to which this notion refers; however, the fundamental idea of the definition remains unchanged. Cf. here—also for further bibliographic references—Tarski [1]; various other results concerning definability can also be found in the literature, e.g., in Hilbert-Bernays [1], vol. I, pp. 354 ff., 369 ff., 456 ff., etc., and in Lindenbaum-Tarski [1]. It may be noticed that the term "definable" is sometimes used in another, meta-mathematical (but not semantic), sense; this occurs, for instance, when we say that that a term is definable in other terms (on the basis of a given axiom system). For a definition of a model of an axiom system see Tarski [4].

BIBLIOGRAPHY

Only the books and articles actually referred to in the paper will be listed here.

Aristotle [1]. *Metaphysica.* (*Works*, vol. VIII.) English translation by W. D. Ross. Oxford, 1908.

Carnap, R. [1]. *Logical Syntax of Language.* London and New York, 1937.

Carnap, R. [2]. *Introduction to Semantics.* Cambridge, 1942.

Gödel, K. [1]. "Über formal unentscheidbare Sätze der *Principia Mathematica* und verwandter Systeme, I." *Monatshefte für Mathematik und Physik*, vol. XXXVIII, 1931, pp. 173–198.

Gödel, K. [2]. "Über die Länge von Beweisen." *Ergebnisse eines mathematischen Kolloquiums*, vol. VII, 1936, pp. 23–24.

Gonseth, F. [1]. "Le Congrès Descartes. Questions de Philosophie scientifique." *Revue thomiste*, vol. XLIV, 1938, pp. 183–193.

Grelling, K., and Nelson, L. [1]. "Bemerkungen zu den Paradoxien von Russell und Burali-Forti." *Abhandlungen der Fries'schen Schule*, vol. II (new series), 1908, pp. 301–334.

Hofstadter, A. [1]. "On Semantic Problems." *The Journal of Philosophy*, vol. XXXV, 1938, pp. 225–232.

Hilbert, D., and Bernays, P. [1]. *Grundlagen der Mathematik.* 2 vols. Berlin, 1934–1939.

Juhos, B. von. [1]. "The Truth of Empirical Statements." *Analysis*, vol. IV, 1937, pp. 65–70.

Kokoszyńska, M. [1]. "Über den absoluten Wahrheitsbegriff und einige andere semantische Begriffe." *Erkenntnis*, vol. VI, 1936, pp. 143–165.

Kokoszyńska, M. [2]. "Syntax, Semantik und Wissenschaftslogik." *Actes du Congrès International de Philosophie Scientifique*, vol. III, Paris, 1936, pp. 9–14.

Kotarbiński, T. [1]. *Elementy teorji poznania, logiki formalnej i metodologji nauk.* (*Elements of Epistemology, Formal Logic, and the Methodology of Sciences*, in Polish.) Lwów, 1929.

Kotarbiński, T. [2]. "W sprawie pojęcia prawdy." (*"Concerning the Concept of Truth,"* in Polish.) *Przegląd filozoficzny*, vol. XXXVII, pp. 85–91.

Lindenbaum, A., and Tarski, A. [1]. "Über die Beschränktheit der Ausdrucksmittel deduktiver Theorien." *Ergebnisse eines mathematischen Kolloquiums*, vol. VII, 1936, pp. 15–23.

Nagel, E. [1]. Review of Hofstadter [1]. *The Journal of Symbolic Logic*, vol. III, 1938, p. 90.

Nagel, E. [2]. Review of Carnap [2]. *The Journal of Philosophy*, vol. XXXIX, 1942, pp. 468–473.

Ness, A. [1]. " 'Truth' As Conceived by Those Who Are Not Professional Philosophers." *Skrifter utgitt av Det Norske Videnskaps-Akademi i Oslo, II. Hist.-Filos. Klasse*, vol. IV, Oslo, 1938.

Neurath, O. [1]. "Erster Internationaler Kongress für Einheit der Wissenschaft in Paris 1935." *Erkenntnis*, vol. V, 1935, pp. 377–406.

Russell, B. [1]. *An Inquiry Into Meaning and Truth.* New York, 1940.

Scholz, H. [1]. Review of *Studia philosophica*, vol. I. *Deutsche Literaturzeitung*, vol. LVIII, 1937, pp. 1914–1917.

Tarski, A. [1]. "Sur les ensembles définissables de nombres réels. I." *Fundamenta mathematicae*, vol. XVII, 1931, pp. 210–239.

Tarski, A. [2]. "Der Wahrheitsbegriff in den formalisierten Sprachen." (German translation of a book in Polish, 1933.) *Studia philosophica*, vol. I, 1935, pp. 261–405.

Tarski, A. [3]. "Grundlegung der wissenschaftlichen Semantik." *Actes du Congrès International de Philosophie Scientifique*, vol. III, Paris, 1936, pp. 1–8.

Tarski, A. [4]. "Über den Begriff der logischen Folgerung." *Actes du Congrès International de Philosophie Scientifique*, vol. VII, Paris, 1937, pp. 1–11.

Tarski, A. [5]. "On Undecidable Statements in Enlarged Systems of Logic and the Concept of Truth." *The Journal of Symbolic Logic*, vol. IV, 1939, pp. 105–112.

Tarski, A. [6]. *Introduction to Logic.* New York, 1941.

Weinberg, J. [1]. Review of *Studia philosophica*, vol. I. *The Philosophical Review*, vol. XLVII, pp. 70–77.

ALFRED TARSKI.

University of California, Berkeley.

THE JOURNAL OF SYMBOLIC LOGIC
Volume 11, Number 2, June 1946

MODALITIES AND QUANTIFICATION

RUDOLF CARNAP

1. The problems of modal logic. The purpose of this article is to give a survey of some results I have found in investigations concerning logical modalities. The results refer: (1) to semantical systems, i.e., symbolic language systems for which semantical rules of interpretation are laid down; (2) to corresponding calculi, i.e., syntactical systems with primitive sentences and a rule of inference; (3) to relations between a semantical system and the corresponding calculus.

The semantical systems to be dealt with are the following: propositional logic (PL), functional logic (FL), and the corresponding modal systems, viz. modal propositional logic (MPL) and modal functional logic (MFL). MPL is built out of PL by the addition of the symbol 'N' for logical necessity; likewise MFL out of FL. In terms of Lewis's symbol '$\Diamond$' for logical possibility, 'Np' means the same as '$\sim\Diamond\sim p$'. All other logical modalities can, of course, be defined on the basis of 'N'; e.g., impossibility by 'N$\sim p$', possibility by '$\sim$N$\sim p$', contingency by '$\sim$N$p.\sim$N$\sim p$', etc.

The calculi corresponding to these semantical systems are the following: the propositional calculus (PC), the functional calculus (FC), and the modal calculi (MPC and MFC) again constructed by the addition of 'N'.

Lewis's systems of strict implication[1] are forms of MPC. So far, no forms of MFC have been constructed, and the construction of such a system is our chief aim. The corresponding semantical systems MPL and MFL are constructed chiefly for the purpose of enabling us to show that the modal calculi MPC and MFC are adequate, i.e., that every sentence provable in them is L-true (analytic). With the help of a normal form, it can further be shown that for MPC the inverse holds also; MPC is complete in the sense that every sentence which is L-true in MPL is provable in MPC. The reduction to the normal form constitutes a decision method for MPC and MPL. For MFC likewise a method of reduction to a normal form will be given. This reduction removes all occurrences of 'N' of higher order, i.e., such that the scope of one 'N' contains another 'N'. A decision method for MFC is of course not possible; however, the reduction makes it possible to apply to MFC the known decision methods for special cases in FC.

The semantical systems FL and MFL contain an infinite number of individual constants. Therefore the representation of these systems requires a very strong metalanguage, dealing with classes of classes of sentences. Consequently, the semantical concepts defined, e.g., L-truth, are indefinite (non-effective) to a high degree. The chief reasons for constructing corresponding calculi are here, as usually in the case of logical calculi, the following two: (1) avoidance of any reference to the meanings of the signs and sentences, (2) use of basic concepts which are effective. The second purpose is here, as generally in the case of calculi without transfinite rules, fulfilled in the following sense. Al-

---

Received November 26, 1945.

[1] C. I. Lewis and C. H. Langford, *Symbolic logic*, 1932; the systems are developed from those in Lewis's earlier book (1918).

though C-truth (provability) is not itself an effective concept, it is defined on the basis of two effective concepts, viz. the concept of primitive sentence, given by a finite list of primitive sentence schemata, and the concept of direct derivability, defined by the rule of inference.

For lack of space, this article will state only a few of the relevant theorems, most of them without proofs. For the same reason, this article will be restricted to the technical aspects of the systems dealt with and will not contain any discussion of the more general problems connected with logical modalities.

The guiding idea in our constructions of systems of modal logic is this: a proposition p is logically necessary if and only if a sentence expressing p is logically true. That is to say, the modal concept of the logical necessity of a proposition and the semantical concept of the logical truth or analyticity of a sentence correspond to each other. Both concepts have been used in logic and philosophy, mostly, however, without exact rules. If we succeed in explicating one of these two concepts, that is, in finding an exact concept, which we call the explicatum, to take the place of the given inexact concept, the explicandum, then this leads, on the basis of the parallelism stated between the two concepts, to an explication for the other concept. Now it is easy to give, with the help of the semantical concepts of state-description and range, an exact definition for 'L-true' as an explicatum for logical truth with respect to the systems PL and FL, as we shall see. Therefore it seems natural to interpret 'N' in such a way that the following convention is always fulfilled:

C1-1. If ' $\cdots$ ' is any sentence in a system S containing 'N', then the corresponding sentence 'N($\cdots$)' is to be taken as true if and only if ' $\cdots$ ' is L-true in S.

This convention determines our interpretation of 'N', but it is not a definition for 'N'. The sentence 'N($\cdots$)' cannot be transformed by definition into the sentence " ' $\cdots$ ' is L-true in S," because the first sentence belongs to the object-language S, the second to the metalanguage M; but the first sentence holds, according to the convention, if and only if the second holds. We shall not define 'N' (it cannot be defined on the basis of the ordinary truth-functional connectives and quantifiers for individuals) but shall take it as a primitive sign in MPL and MFL. However, we shall frame the semantical rules of these systems in such a manner that the convention is fulfilled.[2, 3]

C1-1 gives a sufficient and necessary condition for the truth of 'N($\cdots$)'. Now the following two questions remain: (1) if 'N($\cdots$)' is true, is it L-true? If so, 'NN($\cdots$)' is likewise true; in other words, 'Np $\supset$ NNp' is always true.

---

[2] I shall hereafter refer to the following publications of mine by the signs in square brackets:

[Syntax] *The logical syntax of language,* (1934) 1937.

[I] *Introduction to semantics,* 1942.

[II] *Formalization of logic,* 1943.

[3] I have indicated the parallelism between the modal concept of the necessity of a proposition and the meta-concept of the analyticity of a sentence first in [Syntax] §69 (where, however, 'analytic' was still regarded as a syntactical term), and, more clearly, in [I] pp. 91 ff.

(2) If 'N($\cdots$)' is false, is it L-false? ('L-false' is taken as the explicatum for 'logically false,' 'self-contradictory.') If so, '$\sim$N(...)' is L-true and hence 'N$\sim$N($\cdots$)' is true; in other words, '$\sim Np \supset N \sim Np$' is always true.

At the present moment, these questions are not meant with respect to any given system, but as pre-systematic questions, concerning the inexact, pre-systematic explicandum rather than the exact explicatum. The purpose of the following considerations is merely to make the vague meaning of logical necessity or logical truth clearer to ourselves, so as to lead to a convention more specific than C1-1 concerning the use of 'N'. This convention will then later guide us in constructing our systems. Once the systems are constructed, the two questions can be answered in an exact way. At the present stage, however, our considerations, as always in tasks of self-clarification, are necessarily inexact and, in a certain sense, even circular.

In order to make clearer what is meant by the explicandum of logical, necessary truth, we will distinguish two kinds of data concerning any sentence 'C' as follows:

I. The meaning of 'C' is given, that is to say, the interpretation assigned to 'C' by the semantical rules. (In technical terms, the rules may either be formulated so as to determine the proposition expressed by 'C' or so as to determine the range of 'C'; the rules of our system will have the latter form.)

II. Information concerning the facts relevant for 'C' is given, that is to say, concerning the properties and relations of the individuals involved.

If the answer to a given question is merely dependent upon data of the kind I but independent of those of kind II, we call it a logical question; if in addition, data of the kind II are required, we call it a factual question. In particular, if a sentence 'C' is true in such a way that its truth is based on I alone, we regard it as logically true; if its truth is dependent upon II also, we regard it as factually, contingently true. This conception of the distinction between logical and factual truth as explicandum will guide our choice of the definition of 'L-true' as explicatum. It seems to me that this conception is in agreement with customary conceptions.

Let us take as an example the sentence 'Pa$.\sim$Qb', which we abbreviate by 'A'. We learn from the semantical rules what individuals are named by 'a' and 'b' and what properties are designated by 'P' and 'Q'; we learn further that 'A' says that a is P and b is not Q. This is all we can obtain from data of the kind I. In order to establish the truth-value of 'A' we need data of the kind II, viz., information whether or not a is P and whether or not b is Q. Thus, 'A' is neither L-true nor L-false; we say that it is L-indeterminate or factual.

(i) Now consider the sentence 'A $\lor \sim$A'. We can find that it is true by using merely the semantical rules for '$\lor$' and '$\sim$' (in our system, the rules of ranges D7-5c and b, which correspond to the customary truth-tables for the two connectives); we need no factual information concerning the individuals a and b occurring in the sentence. Therefore, 'A $\lor \sim$A' is L-true. Hence, according to our convention C1-1, 'N(A $\lor \sim$A)' is true. The question is whether it is L-true. Now we can easily see that it must be, because the truth of this N-sentence follows from those semantical rules by which we established the truth and hence the L-truth of 'A ∨ ∼A' together with the semantical rule for 'N' which is to be laid down in accordance with C1-1. Thus no factual knowledge is required for establishing that 'N(A ∨ ∼A)' is true; hence it is L-true.

(ii) Similarly, the falsity of 'A.∼A' can be established by the semantical rules alone. Therefore, this sentence is L-false and not L-true. Hence, according to C1-1, 'N(A.∼A)' is false and '∼N(A.∼A)' is true.

(iii) Finally, let us go back to the sentence 'A' itself, i.e., 'Pa.∼Qb'. We found that 'A' is neither L-true nor L-false by merely using semantical rules, not using any factual knowledge concerning the individuals occurring in 'A'. Therefore we see that, according to C1-1, 'N(A)' is false and '∼N(A)' is true. These results are based merely on the semantical rules for the signs occurring in 'A' and for 'N'. Therefore, 'N(A)' is L-false and '∼N(A)' is L-true.

The results found for these simple examples can be generalized. Let 'C' be an abbreviation for a given sentence of any form with or without 'N'.

(i) Suppose that 'N(C)' is true. Then, according to C1-1, 'C' must be L-true. Hence the truth of 'C' is determined by certain semantical rules. Then these same rules together with the rule for 'N' determine the truth of 'N(C)'. Therefore, 'N(C)' is L-true, and hence 'NN(C)' is true. Thus our earlier question (1) is answered in the affirmative.

(ii) Let us now suppose that 'C' is L-false and hence 'N(C)' is false. Then those semantical rules which determine the falsity of 'C' together with the rule for 'N' determine the falsity of 'N(C)'. Therefore, 'N(C)' is L-false, and '∼N(C)' is L-true.

(iii) Finally, let us suppose that 'N(C)' is false but 'C' is not L-false. Then 'C' is neither L-true nor L-false. The decisive question here is this: is the result that 'C' is not L-true determined by data I alone or are data II, i.e., factual knowledge, required? Data II are certainly relevant for the truth-value of 'C', but they cannot be relevant for the character of 'C' being L-indeterminate, factual, contingent. It would be absurd to assume such a relevance, to say, for example: " 'C' is contingent because the individual c happens to have the property Q; if this were not so then 'C' would not be contingent but L-true." Thus contingent facts, by being relevant for the contingency of 'C' would also be relevant for L-truth or L-falsity, in contradiction to our explanation of these concepts. Since now data I alone determine that 'C' is not L-true, they determine that 'N(C)' is false and '∼N(C)' is true. Therefore, 'N(C)' is L-false, and '∼N(C)' is L-true.

From (ii) and (iii) together we see that, if 'N(C)' is false, it is L-false. Thus our earlier question (2) is answered in the affirmative. Together with the result under (i), this leads to the following convention, which is more specific than C1-1.

C1-2. If '···' is L-true, 'N(···)' is L-true; otherwise 'N(···)' is L-false.

We shall later construct the rule for 'N' (D9-5i) in such a manner that this convention is fulfilled (T9-1).

In the preceding analysis, I have repeatedly referred to a certain result as "following from" or "determined by" certain data. This is not meant in the sense that the result can be derived from the data with the help of deductive means which are systematized in a given metalanguage; still less is it implied that there is an effective method for this derivation. What I mean is rather that, if the data hold, the result cannot possibly fail to hold. This is the wide, non-systematized concept of logical implication which logicians have in mind before they construct their systems and of which only a part can be grasped in any one fixed system. This concept is necessarily inexact; but it is clear enough for practical purposes of pre-systematic discussions. Logicians refer to it as their explicandum before they offer their explicatum in the form of a system with exact rules; so does, for example, Lewis when he explains that his explicatum 'strict implication' is intended to systematize the common concept of logic, implication, deducibility, entailment.

The opinions of logicians on the two questions (1) and (2) mentioned earlier seem to differ, with no clear arguments on either side. Our affirmative answers to these questions do not mean that a negative answer to either question is wrong but only that it must be based on an interpretation of 'N' different from ours. It seems to me that the usual discussions of the validity of various systems of modal logic are inconclusive because no clear interpretation of the modal signs are offered. The interpretation here suggested, based on the parallelism between the necessity of propositions and the L-truth of sentences and on the distinction between data of kinds I and II, leads to clear solutions of the controversial questions. This interpretation seems to be in agreement with customary conceptions; it is, indeed, nothing else than a clarification of these conceptions. I am not aware of any clear and simple interpretation which leads to one of the alternative systems, e.g., to a negative answer to either of the questions (1) or (2).

Since we intend to combine modalities with quantification, we have also to decide how to interpret a sentence of the form '$(x)[N(\cdots x \cdots)]$'. This is done by the following convention.

C1-3. 'N' is to be interpreted in such a way that any sentence of the form '$(x)[N(\cdots x \cdots)]$' is regarded as L-equivalent to (i.e., meaning the same as) the corresponding sentence '$N[(x)(\cdots x \cdots)]$'.

The reasons for this convention (that is, "the motives for its choice," not "the proof of its validity") are as follows. We adopt the principle, which seems generally accepted, that any sentence with a universal quantifier, no matter whether it contains modal signs or not, is to be interpreted as a joint assertion for all values of the variable. Thus, if 'x' has only three values, say a, b, c, then '$(x)[N(\cdots x \cdots)]$' means the same as '$N(\cdots a \cdots).N(\cdots b \cdots).N(\cdots c \cdots)$'. The latter sentence is L-equivalent to '$N[(\cdots a \cdots).(\cdots b \cdots).(\cdots c \cdots)]$' (see below, T4-1e), because a conjunction is L-true if and only if all of its components are L-true. Finally, the sentence last mentioned is L-equivalent to '$N[(x)(\cdots x \cdots)]$', because, in virtue of the above principle, '$(x)(\cdots x \cdots)$' means the same as '$(\cdots a \cdots).(\cdots b \cdots).(\cdots c \cdots)$'. It seems natural to transfer this result also to variables with a denumerably infinite range of values, as in MFL. I think that the same convention could be applied even to variables with a non-denumerable range, e.g., real number variables, although in this case not all values are expressible in the language; this problem, however, need not concern us in the present context. [It may be remarked that the foregoing discussion oversimplifies the situation. The actual situation is complicated by the fact that the values of individual variables in a modal system are not individuals but

individual concepts; compare the remarks at the end of §12.   However, for the
present purposes we may leave aside this distinction.]

2. **Propositional logic (PL) and propositional calculus (PC).**   We shall not
construct PL as a separate system.   Instead, we shall say of certain systems
that they "contain PL" (roughly speaking, if they contain the ordinary connectives and semantical rules for them corresponding to the ordinary truth-tables)
and of some of their sentences that they are "L-true by PL" (roughly speaking,
if their truth can be shown by the truth tables alone).[4]   For the sake of brevity,
we omit here the exact definitions.   The definition for "S contains PL" requires
that S contain rules like D7-1a, b, c, D7-2c, d, e, D7-3, D7-5b, c, d, h.   The
definition for "L-true by PL" corresponds to D7-6a but is framed in such a way
that it applies only if the universality of a range is based on the mentioned
rules of ranges (D7-5b, c, d) which correspond to the ordinary truth-tables.
Thus we shall see that FL, and likewise MFL, contain PL.

As primitive signs in all our systems we shall use the connectives '$\sim$', '$\vee$',
and '$.$'; and also the tautologous sentence 't'.   ('$.$' could of course be defined;
likewise 't', e.g., by '$Pa \vee \sim Pa$'.   We take them, nevertheless, as primitive
in order to be able to write the normal forms in primitive notation; see D3-1.)
In the following discussions, we make use also of '$\supset$' and '$\equiv$'; they do not belong
to the systems themselves but serve as shorthand in the customary manner.
We adopt the customary conventions for the omission of parentheses; in particular, we write 'Np' instead of '$N(p)$' (but not if a compound sentence takes the
place of 'p').

We use *German letters* as signs of the metalanguage: '$\mathfrak{S}$' for sentences (all
sentences in our system are closed, i.e., without free variables); '$\mathfrak{M}$' for matrices
(by which we mean here always sentential matrices (Quine), sometimes called
"sentential functions"; they include the sentences); '$\mathfrak{in}$' for individual constants;
'$\mathfrak{i}$' for individual variables; and occasionally others.   Any expression containing
a German letter belongs to the metalanguage and denotes the corresponding
expression of the object-language in the customary way; e.g., '$\mathfrak{S}_i \vee \mathfrak{S}_j$' denotes
the disjunction with the components $\mathfrak{S}_i$ and $\mathfrak{S}_j$.

We say that $\mathfrak{S}_i$ is L-false by PL, that $\mathfrak{S}_i$ L-implies $\mathfrak{S}_j$ by PL, or that $\mathfrak{S}_i$
is L-equivalent to $\mathfrak{S}_j$ by PL if and only if $\sim\mathfrak{S}_i$, $\mathfrak{S}_i \supset \mathfrak{S}_j$, or $\mathfrak{S}_i \equiv \mathfrak{S}_j$, respectively, is L-true by PL.   (These simple definitions can be used here because all
sentences are closed.)

In analogy to PL, we do not take PC as a separate calculus, since our systems will not contain propositional variables.   We define instead "the calculus
K contains PC."   The definition requires that K contain the signs, matrices
and sentences required for PL; further the rule of implication (as D4-2e), and
certain primitive sentences (e.g., 't' and all sentences formed by substitution
from Bernays's[5] four axioms of the propositional calculus).   We say that $\mathfrak{S}_i$
is C-true by PC in K if $\mathfrak{S}_i$ is C-true (provable) in a sub-calculus K' of K contain-

---

[4] For this concept, the term 'tautologous' is sometimes used; see W. V. Quine, *Mathematical logic,* 1940, p. 50, and [I], p. 240.   In [II] D11-30, I have used the term 'L-true by
NTT.'

[5] D. Hilbert and W. Ackermann, *Grundzüge der theoretischen Logik,* (1928) 1938.   P.
Bernays, *Mathematische Zeitschrift,* vol. 25 (1926).

ing all sentences of K but only the rule of implication and those primitive sentences required for PC.

Definitions for other C-concepts (which we shall not state here) are constructed in such a way that they lead to the following results: $\mathfrak{S}_i$ is C-false (refutable) by PC, $\mathfrak{S}_i$ C-implies $\mathfrak{S}_j$ ($\mathfrak{S}_j$ is derivable from $\mathfrak{S}_i$) by PC, or $\mathfrak{S}_i$ is C-equivalent to $\mathfrak{S}_j$ (mutually derivable) by PC if and only if $\sim\mathfrak{S}_i$, $\mathfrak{S}_i \supset \mathfrak{S}_j$, or $\mathfrak{S}_i \equiv \mathfrak{S}_j$, respectively, is C-true by PC.

The following theorem T2-1a is well known; it is easily proved by examining the primitive sentences and the rule of inference of PC.

T2-1. Let the semantical system S and the calculus K contain the same signs and the same sentences. Let S contain PL and K contain PC.

a. Any sentence which is C-true by PC is L-true by PL.

b. Whenever C-falsity by PC holds, then L-falsity by PL holds. Analogously with C-implication and L-implication, and with C-equivalence and L-equivalence. Hence, in a certain sense, PL is an L-true interpretation of PC ([I] D34-1, [II] §14). (We add "in a certain sense," because PL and PC have not been defined here as independent systems.)

3. P-reduction. We shall now lay down rules of reduction leading to a normal form (definition D3-1). This is, essentially, the customary transformation into a conjunctive normal form. The present procedure is, however, considerably simplified, because by the use of 't' we get rid of many parts of a sentence which otherwise make the transformation very cumbersome. This method of reduction is applicable both to PL and to PC; therefore we use the neutral term 'P-reduction.' The rules are formulated so as to apply not only to sentences but to all matrices; this will be needed later in FC and MFC (see D8-1a). We presuppose that the customary notation of multiple disjunctions and conjunctions is used; this simplifies the procedure. (Thus we presuppose, for example, that instead of '(A ∨ B) ∨ C', the simple form 'A ∨ B ∨ C' is written; and further, that 'A ∨ (B ∨ C)' is transformed, according to the associative law, into '(A ∨ B) ∨ C' and hence likewise simplified to 'A ∨ B ∨ C'.)

D3-1. The *P-reduction* of a matrix $\mathfrak{M}_i$ is its transformation according to the following rules. At any step in the transformation, the first of these rules that can be applied must be applied. The replacement applies to any part having the specified form, provided this part is either the whole matrix or one of those components out of which the whole is built up with connectives. The final result to which no rule is applicable any more is called the *P-reductum* of $\mathfrak{M}_i$.

a. Any disjunction containing among its components a matrix and its negation is replaced by 't'.

b. Any conjunction containing among its components a matrix and its negation is replaced by '$\sim$t'.

c. $\sim\sim\mathfrak{M}_k$ is replaced by $\mathfrak{M}_k$.

d. If a disjunction or a conjunction contains among its components several occurrences of the same matrix, then all occurrences except the first one are omitted. (This rule is inessential but simplifies the form.)

e. If 't' is a component of a disjunction, then all other components of this disjunction are omitted.

f. If '$\sim$t' is a component of a disjunction, it is omitted.

g. If 't' is a component of a conjunction, it is omitted.

h. If '∼t' is a component of a conjunction, then all other components of this conjunction are omitted.

i. The negation of a disjunction is replaced by the conjunction of the negations of the components.

j. The negation of a conjunction is replaced by the disjunction of the negations of the components.

k. If a conjunction occurs as a component of a disjunction, it is distributed (e.g., $\mathfrak{M}_h \vee (\mathfrak{M}_{k1}.\mathfrak{M}_{k2}) \vee \mathfrak{M}_l$ is replaced by $(\mathfrak{M}_h \vee \mathfrak{M}_{k1} \vee \mathfrak{M}_l).(\mathfrak{M}_h \vee \mathfrak{M}_{k2} \vee \mathfrak{M}_l)$).

T3-1. Let S and K be as in T2-1. Let $\mathfrak{S}_i$ be any sentence in S and K, and let $\mathfrak{S}_j$ be its P-reductum; then the following holds.

a. $\mathfrak{S}_i$ and $\mathfrak{S}_j$ are C-equivalent by PC. (This follows by induction from the fact that C-equivalence holds for each application of one of the rules of reduction in D3-1.)

b. $\mathfrak{S}_i$ and $\mathfrak{S}_j$ are L-equivalent by PL. (From (a) and T2-1b.)

c. If $\mathfrak{S}_j$ is 't', $\mathfrak{S}_i$ is L-true by PL. (From (b).)

d. $\mathfrak{S}_i$ is L-true by PL if and only if $\mathfrak{S}_j$ is 't'.

Proof. Let $\mathfrak{S}_i$ be L-true by PL (tautologous). It is well known[6] that the application of the reduction rules D1c, i, j, k to $\mathfrak{S}_i$ leads to a conjunctive normal form, in which every disjunction contains a sentence and its negation as components. Therefore, the rules D1a and g lead to 't'.—The converse is (c).

e. If $\mathfrak{S}_j$ is 't', $\mathfrak{S}_i$ is C-true by PC. (From (a).)

f. *Completeness of PC.*[7] If $\mathfrak{S}_i$ is L-true by PL, it is C-true by PC. (From (d), (e).)

g. Whenever L-falsity by PL holds, then C-falsity by PC holds. Analogously with L-implication and C-implication, and with L-equivalence and C-equivalence. Hence, in a certain sense, PC is an L-exhaustive calculus for PL ([I] D36-3). (From (f).)

4. Modal propositional logic (MPL) and modal propositional calculus (MPC). MPL consists of PL with the addition of the modal symbol 'N' for logical necessity. Here again, we do not define MPL as an independent system but only: "The system S contains MPL." The definition (which will not be given here) requires the same rules as those for PL but with the following addition: the modal symbol 'N' is added (D9-1h, D9-2g); and the rule of ranges for it (as D9-5i) says that the range of $N\mathfrak{S}_i$ is the universal range if the range of $\mathfrak{S}_i$ is the universal range; otherwise it is the null range. This is in accord with our previous convention C1-2, since L-truth is defined by the universality of the range (D7-6a). Our system MFL (§9) will contain MPL.

In analogy to 'L-true by PL' we now define 'L-true by MPL':

D4-1. $\mathfrak{S}_i$ is *L-true by MPL* in $S =_{\text{Df}} S$ contains MPL; $\mathfrak{S}_i$ belongs to S; every sentence formed out of $\mathfrak{S}_i$ in the following way is L-true in S: the ultimate MPL-components of $\mathfrak{S}_i$ (i.e., those sentences out of which $\mathfrak{S}_i$ is built up with

---

[6] See, e.g., Hilbert and Ackermann, op.cit., Kap. I, §§3 and 4, or Hilbert and Bernays, *Grundlagen der Mathematik*, vol. I, 1934, pp. 53f.

[7] The completeness of PC was first proved by E. L. Post, *American journal of mathematics*, vol. 43 (1921). See W. V. Quine, this JOURNAL, vol. 3 (1938), pp. 37ff.

the help of connectives and 'N', which themselves, however, are not thus built up out of other sentences) are replaced by any sentences of S (occurrences of the same component to be replaced by occurrences of the same sentence).

L-falsity by MPL, L-implication by MPL, and L-equivalence by MPL are defined in analogy to the corresponding concepts for PL in §2.

Now we shall define: "The calculus K contains MPC." This will later be applied to MFC. We here make use of 'p', 'q', etc. as *auxiliary variables*; that is to say, these letters do not belong to the signs of our calculi but are merely used (following Quine) for the description of certain forms of sentences. We say, e.g., that a sentence of K has the form '$p \supset q$' if it is formed out of the auxiliary formula '$p \supset q$' (which does not belong to K) by substituting for 'p' and 'q' any sentences of K (which do not contain 'p', 'q', etc.); for instance, 'Pa $\supset$ N(Pb)'. We write '$p \strictimp q$' and '$p \equiv q$' as abbreviations for 'N($p \supset q$)' and 'N($p \equiv q$)', respectively. Thus '$\strictimp$' is a symbol of strict implication (corresponding to Lewis's '$\dashv$'); and '$\equiv$' is a symbol of strict equivalence (or identity of propositions, corresponding to Lewis's '$=$'). The essential features of the form of MPC here stated are due to M. Wajsberg.[8]

D4-2. K contains MPC $=_{Df}$ K is a calculus fulfilling the following conditions.

a. The following *signs* are among the signs of K:

    a1. Connectives: '$\sim$', '$\lor$', '$.$'.

    a2. Parentheses: '(', ')'.

    a3. 't'.

    a4. 'N'.

b. If $\mathfrak{M}_i$ and $\mathfrak{M}_j$ are matrices in K, then all expressions of the following forms are *matrices* in K:

    b1. 't'.

    b2. $\sim(\mathfrak{M}_i)$.

    b3. $(\mathfrak{M}_i) \lor (\mathfrak{M}_j)$.

    b4. $(\mathfrak{M}_i).(\mathfrak{M}_j)$.

    b5. N($\mathfrak{M}_i$).

c. All the closed matrices in K, and only these, are *sentences* in K.

---

[8] An earlier system MPC, which I constructed in 1940, was slightly different from the one here given; I constructed a proof for its completeness with the help of the reduction procedure explained in the next section. I found later that my system was equivalent to, but simpler than, Lewis's system S5. While writing this article, I found that M. Wajsberg had given a still simpler form (*Ein erweiterter Klassenkalkül, Monatshefte für Mathematik und Physik*, vol. 40 (1933), pp. 113–126); therefore I now adopt (in D4-2d) his axioms, with the following two inessential changes. (1) I take (d1), where Wajsberg takes the four axioms of the propositional calculus of Hilbert and Ackermann with the symbol of necessity added to each. (2) In (d3) I have '$\supset$', while Wajsberg has a symbol corresponding to '$\strictimp$'; this change is due to the fact that I do not use, like Wajsberg, a rule of strict implication but a rule of material implication in order to have MPC and MFC contain PC. Wajsberg's calculus is primarily intended as a class calculus with a symbol for class universality added to it, but he remarks that it can also be interpreted as an extended propositional calculus corresponding to Lewis's systems of strict implication. In this interpretation, Wajsberg's '$|X|$' corresponds to 'Np', and therefore his '$X < Y$' to my '$p \strictimp q$' and to Lewis's '$p \dashv q$'. In the same paper, Wajsberg gave a proof for the completeness of his calculus (see below, T6-2f).

    d. *K* contains among its primitive sentences all sentences of the following forms; we call them *primitive sentences of MPC* in *K*:

        d1. $N\mathfrak{S}_i$, where $\mathfrak{S}_i$ is any sentence whose P-reductum is 't'.

        d2. '$(p \ni q) \ni (Np \ni Nq)$'.

        d3. '$Np \supset p$'.

        d4. '$\sim Np \ni N\sim Np$'.

    e. *K* contains among its rules of inference the *rule of implication*: $\mathfrak{S}_j$ is a direct C-implicate of (directly derivable from) $\mathfrak{S}_i$ and $\mathfrak{S}_i \supset \mathfrak{S}_j$.

    f. *Rule of refutation*: the class of all sentences is directly C-false in *K*.

    (f) is the rule of refutation of the simplest form (see [II] §20). If we are willing to dispense with C-falsity, in order to have a form more similar to customary calculi, we may omit this rule.

    C-truth (provability) by MPC is defined in the customary way with reference to the primitive sentences (d) and the rule of inference (e). C-falsity, C-implication, and C-equivalence are defined in such a way that the conditions mentioned in §2 are fulfilled.

    We shall now state some theorems of MPC without proofs. We use again the auxiliary variables 'p' etc. for convenience and easier comparison with modal calculi of other authors. ('d1' etc. refer to D4-2d1 etc.)

    T4-1. Let K be a calculus containing MPC. Every sentence in *K* of any of the following forms is *C-true by MPC* in *K*.

    a. '$(p \ni q) \supset (Np \ni Nq)$. (From d2, d3.)

    b. '$\sim Np \equiv N\sim Np$'. (From d4, d3.)

    c. '$(p \ni q) \supset (Np \supset Nq)$'. (From (a), d3.)

    d. '$(p \ni q) \supset [(r \ni p) \supset (r \ni q)]$'. (From d1, (c).)

    e. '$Np.Nq \equiv N(p.q)$'. (From d1, (c).)

    f. '$Np \supset N(q \supset p)$'. (From d1, (c).)

    g. '$Np \equiv NNp$'. (From (a), (c), d4; d3.)

    h. '$(Np \ni q) \supset (Np \supset Nq)$'. (From (c), (g).)

    i. '$\sim p \supset \sim Np$'. (From d3.)

    j. '$q \supset \sim N\sim q$'. (From (i).)

    k. '$N\sim p \supset \sim Np$'. (From d3, (i).)

    l. '$N(\sim p_1 \vee \sim p_2 \vee \cdots \vee \sim p_n \vee q) \supset \sim Np_1 \vee \sim Np_2 \vee \cdots \vee \sim Np_n \vee Nq$'. (From (c).)

    m. '$N(\sim p_1 \vee \cdots \vee \sim p_n) \supset \sim Np_1 \vee \cdots \vee \sim Np_n$'. (From (l), (k).)

    n. '$N(Np \vee q) \supset Np \vee Nq$'. (From d1, (c), (b).)

    p. $NN\mathfrak{S}_i$ where $\mathfrak{S}_i$ is any sentence whose P-reductum is 't'. (From d1, (g).)

    q. $N\mathfrak{S}_i \supset N\mathfrak{S}_j$ where $\mathfrak{S}_i$ and $\mathfrak{S}_j$ are such that the P-reductum of $\mathfrak{S}_i \supset \mathfrak{S}_j$ is 't'. (From d1, (c).)

    r. '$Np \supset N(p \vee q)$'. (From (q).)

    s. '$Np \vee Nq \supset N(p \vee q)$'. (From (r), (q).)

    t. '$(Np \supset Nq) \equiv (Np \ni q)$'. (From (r), (b), (q); (h).)

    u. '$Np \ni p$'. (From (t).)

    v. '$Np \vee Nq \equiv N(Np \vee q)$'. (From (r), (b), (q); (n).)

    x. '$\sim N\sim Np \supset Np$'. (From (b).)

    y. '$Np \vee Nq \equiv N(Np \vee Nq)$'. (From (g), (v).)

For proofs of many of these theorems, see Wajsberg (op. cit.).

T4-2. Let K contain MPC. Let $\mathfrak{S}_i$ and $\sim\mathfrak{S}_j$ be C-true by MPC in K. Then the sentences of the following forms are likewise C-true by MPC in K.

a. $N\mathfrak{S}_i$.

Proof. If we prefix 'N' to every sentence in the proof of $\mathfrak{S}_i$, then it can be shown, with the help of T4-1g, u, and a, that every sentence thus resulting is C-true by MPC.

b. $\mathfrak{S}_i \equiv N\mathfrak{S}_i$. (From (a).)

c. $N\sim\mathfrak{S}_j$. (From (a).)

d. $\sim N\mathfrak{S}_j$. (From (c), T4-1i.)

e. $\sim\mathfrak{S}_j \equiv N\sim\mathfrak{S}_j$. (From (c).)

f. $\sim\mathfrak{S}_j \equiv \sim N\mathfrak{S}_j$. (From (d).)

g. $\mathfrak{S}_j \equiv N\mathfrak{S}_j$. (From (f).)

h. $\mathfrak{S}_j \equiv \sim N\sim\mathfrak{S}_j$. (From (e).)

6. MP-reduction. We shall now explain a reduction method which leads to a normal form for both MPL and MPC; therefore we call it MP-reduction. This method is an extension of P-reduction. It is similar to the method used by Wajsberg (op. cit.).

It is easy to construct a decision method for MPL if we require only that it be *theoretically* effective, i.e., that it lead to a decision in a finite number of steps, no matter how large. [If we have a modal sentence $\mathfrak{S}_i$ with n ultimate components, then they determine $p = 2^n$ smallest ranges. Thus there are $2^p - 1$ possibilities for none or some (but not all) of these ultimate ranges to be null (see the explanation of basic modal functions in §6). If we find, on the basis of the rules of ranges for MPL, that for each of these possibilities the range of $\mathfrak{S}_i$ is universal, then $\mathfrak{S}_i$ is L-true by MPL.] MP-reduction will yield a decision method which is, moreover, practicable, i.e., sufficiently short for modal sentences of ordinary length. This reduction method further leads to the result that MPC is complete in this sense: every sentence which is L-true by MPL is C-true by MPC.

D5-1. The *MP-reduction* of a matrix $\mathfrak{M}_i$ is its transformation according to the following rules. At any step in the transformation, the first of these rules that can be applied must be applied. The replacement applies to any part having the specified form, provided this part is either the whole matrix or one of those components out of which the whole is built up with connectives and 'N'. Here we have two kinds of rules: (a) to (n), and (o) to (p). If none of the rules (a) to (n) is any more applicable, the result is called the *first MP-reductum* of $\mathfrak{M}_i$. If none of all the rules (a) to (p) is any more applicable, the result is called the second MP-reductum or, briefly, the *MP-reductum* of $\mathfrak{M}_i$.

a to k, as in D3-1.

l. Omission of 'N'. $N\mathfrak{M}_k$ is replaced by $\mathfrak{M}_k$ if $\mathfrak{M}_k$ has one of the following four forms: (1) $N\mathfrak{M}_h$ (with any $\mathfrak{M}_h$), (2) $\sim N\mathfrak{M}_h$, (3) 't', (4) '$\sim$t'.

m. $N(\mathfrak{M}_{k1}\cdot\cdots\cdot\mathfrak{M}_{kn})$ is replaced by $N(\mathfrak{M}_{k1})\cdot\cdots\cdot N(\mathfrak{M}_{kn})$.

n. Suppose that $N\mathfrak{M}_h$ occurs where $\mathfrak{M}_h$ is a disjunction with n components ($n \geq 2$) such that there is at least one component of the form 'N($\cdots$)' or '$\sim$N($\cdots$)'. Let $\mathfrak{M}_k$ be the first of the components having either of these forms. Let $\mathfrak{M}_l$ be either (if $n > 2$) the disjunction of the remaining components (in the order in which they occur in $\mathfrak{M}_h$) or (if $n = 2$) the one remaining component. Then $\mathrm{N}\mathfrak{M}_h$ is replaced by $\mathfrak{M}_k \vee \mathrm{N}\mathfrak{M}_l$.

These are the rules of the first kind; the following two are those of the second kind.

o. Suppose that $\mathfrak{M}_k$ is either the whole matrix or a conjunctive component of the whole, and has the form $\sim\mathrm{N}\mathfrak{M}_{k1} \vee \cdots \vee \sim\mathrm{N}\mathfrak{M}_{km}$ ($m \geqq 1$; for $m = 1$, $\mathfrak{M}_k$ is $\sim\mathrm{N}\mathfrak{M}_{k1}$). Then $\mathfrak{M}_k$ is replaced by $\mathrm{N}(\sim\mathfrak{M}_{k1} \vee \cdots \vee \sim\mathfrak{M}_{km})$.

p. Suppose that $\mathfrak{M}_h$ is either the whole matrix or a conjunctive component of the whole and is a disjunction (with $m + n + p$ components) which has the following form or can be brought into this form by merely changing the order of the components: $\sim\mathrm{N}\mathfrak{M}_{j1} \vee \cdots \vee \sim\mathrm{N}\mathfrak{M}_{jm} \vee \mathrm{N}\mathfrak{M}_{k1} \vee \cdots \vee \mathrm{N}\mathfrak{M}_{kn} \vee \mathfrak{M}_{l1} \vee \cdots \vee \mathfrak{M}_{lp}$, where the p $\mathfrak{M}_l$-components have neither the form '$\sim\mathrm{N}(\cdots)$' nor '$\mathrm{N}(\cdots)$', and where $m \geqq 1, n \geqq 0, p \geqq 0$, but $n + p \geqq 1$, hence $m + n + p \geqq 2$. Let $\mathfrak{M}_j$ be $\sim\mathfrak{M}_{j1} \vee \cdots \vee \sim\mathfrak{M}_{jm}$, and $\mathfrak{M}_l$ be $\mathfrak{M}_{l1} \vee \cdots \vee \mathfrak{M}_{lp}$. Then $\mathfrak{M}_h$ he replaced by $\mathrm{N}(\mathfrak{M}_j \vee \mathfrak{M}_{k1}) \vee \mathrm{N}(\mathfrak{M}_j \vee \mathfrak{M}_{k2}) \vee \cdots \vee \mathrm{N}(\mathfrak{M}_j \vee \mathfrak{M}_{kn}) \vee \mathrm{N}(\mathfrak{M}_j \vee \mathfrak{M}_l)$. (If $p = 0$, the last of these $n + 1$ components disappears; if $n = 0$, all components except the last disappear.)

(Concerning the difference between the rules of the first and the second kind, see T6-1e, f, g.)

With respect to this reduction method, the following theorems can be proved. The proofs cannot be given here.

T5-1. Let $\mathfrak{M}_k$ be a matrix in a semantical system containing MPL or in a calculus containing MPC, such that none of the rules of MP-reduction (D1a to o) can be applied to $\mathfrak{M}_k$. Then $\mathfrak{M}_k$ has exactly one of the following forms (a) to (h).

a. $\mathfrak{M}_k$ is 't'.

b. $\mathfrak{M}_k$ is '$\sim$t'.

c. $\mathfrak{M}_k$ has a form different from the five forms listed in D4-2b.

d. $\mathfrak{M}_k$ is $\sim\mathfrak{M}_h$, where $\mathfrak{M}_h$ has form (c).

e. $\mathfrak{M}_k$ is a disjunction of two or more components of the forms (c), (d); no component is the negation of another component; no two components are alike.

f. $\mathfrak{M}_k$ is $\mathrm{N}(\mathfrak{M}_h)$, where $\mathfrak{M}_h$ has one of the forms (c), (d), (e).

g. $\mathfrak{M}_k$ is a disjunction of two or more components of the forms (c), (d), (f), at least one being of the form (f); no component is the negation of another component; no two components are alike.

h. $\mathfrak{M}_k$ is a conjunction of two or more components of the forms (c), (d), (e), (f), (g); no component is the negation of another component; no two components are alike.

6. Relations between MPC and MPL. If a calculus is constructed for the purpose of a formalization of a given logical or empirical theory, then two questions may be raised. (1) Is the calculus in accord with the given theory? In technical terms, is the theory a true (or L-true) interpretation for the calculus? (2) Is the calculus strong enough to yield all the statements of the theory? In technical terms, is the calculus an exhaustive (or L-exhaustive) calculus for the theory? An affirmative answer to the first question is certainly required, because otherwise the calculus would not fulfill its purpose. An affirmative answer to the second question is, although desirable, in general not required. If we use only the customary means (excluding so-called transfinite rules), then for certain logical systems no L-exhaustive calculus can be constructed. (We leave aside the question of the requirement of full formalization in a sense still stronger than that of L-exhaustiveness; this question has been discussed for PC and FC in [II].)

For PC and PL, it is well known that both questions can be answered affirmatively (see above, T2-1 and T3-1). The same result will now be stated for MPC and MPL; it was first found by Wajsberg. The affirmative answer to the first question is easily found (T6-1c, d); it is based on a simple examination of the primitive sentences of MPC (T6-1a) and its rule of inference (T6-1b). The proof of the completeness of MPC, however, is more complicated. It makes use of the method of MP-reduction (D5-1). The result is reached in two steps: (i) every L-true sentence is reducible to 't' (T6-2d); (ii) if a sentence is reducible to 't,' it is C-true (T6-2e); hence, if a sentence is L-true, it is C-true (T6-2f).

T6-1. Let the semantical system S and the calculus K contain the same signs and the same sentences. Let S contain MPL, and K contain MPC.

a. Every primitive sentence of MPC (D4-2d) is L-true by MPL.

b. In every instance of direct C-implication by MPC (that is, of an application of the rule of implication D4-2e), L-implication by MPL holds.

c. Any sentence which is C-true by MPC is L-true by MPL. (From (a), (b).)

d. Whenever C-falsity by MPC holds, then L-falsity by MPL holds; analogously with C-implication and L-implication, and with C-equivalence and L-equivalence. (From (c).) Thus, in a certain sense, MPL is an L-true interpretation of MPC.

e. If one application of any of the rules of MP-reduction of the first kind (D5-1a to n) to a sentence $\mathfrak{S}_i$ leads to $\mathfrak{S}_j$, then $\mathfrak{S}_i$ and $\mathfrak{S}_j$ are C-equivalent by MPC and hence L-equivalent by MPL.

f. Let $\mathfrak{S}_j$ be formed from $\mathfrak{S}_i$ by an application of one of the rules of MP-reduction of the second kind (D5-1o or p). Then, if $\mathfrak{S}_i$ is L-true by MPL, $\mathfrak{S}_j$ is likewise. (The converse holds too but will not be needed. Note that, in contradistinction to (e), L-equivalence is here not asserted; it does not hold in general.)

g. Let $\mathfrak{S}_j$ and $\mathfrak{S}_i$ be as in (f). Then $\mathfrak{S}_j$ C-implies $\mathfrak{S}_i$ by MPC. (From T4-1m, D4-2d3.) [Between $\mathfrak{S}_i$ and $\mathfrak{S}_j$ C-implication by MPC does not hold generally, but only the weaker relation that, if $\mathfrak{S}_i$ is C-true by MPC, then so is $\mathfrak{S}_j$.]

T6-2. Let S and K be as in T6-1. Let $\mathfrak{S}_j$ be the (second) MP-reductum of $\mathfrak{S}_i$. Then the following holds.

a. If $\mathfrak{S}_i$ is L-true by MPL, then so is $\mathfrak{S}_j$. (From T6-1e, f.)

b. $\mathfrak{S}_j$ C-implies $\mathfrak{S}_i$ by MPC. (From T6-1e, g.)

c. $\mathfrak{S}_j$ L-implies $\mathfrak{S}_i$ by MPL. (From (b), T6-1d.)

d. If $\mathfrak{S}_i$ is L-true by MPL, then $\mathfrak{S}_j$ is 't'.

Proof. $\mathfrak{S}_j$ is L-true by MPL (a), and none of the rules of MP-reduction is applicable to $\mathfrak{S}_j$. An examination of the forms listed in T5-1 shows that all of them except the first, i.e., 't', are impossible in this case.

e. If $\mathfrak{S}_j$ is 't', then $\mathfrak{S}_i$ is C-true by MPC. (From (b), D4-2d1.)

f. *Completeness of MPC.* If $\mathfrak{S}_i$ is L-true by MPL, then it is C-true by MPC. (From (d), (e).) In this case, the rules of MP-reduction provide an effective method for the construction of a proof of $\mathfrak{S}_i$ by MPC (according to T6-1e, g).

g. L-truth by MPL holds if and only if C-truth by MPC holds. (From (f), T6-1c.) Analogously with L-falsity and C-falsity, with L-implication and C-implication, with L-equivalence and C-equivalence. Thus, in a certain sense, MPC is an L-exhaustive calculus for MPL.

h. A sufficient and necessary condition for $\mathfrak{S}_i$ to be L-true by MPL is that $\mathfrak{S}_j$ is 't', and likewise for $\mathfrak{S}_i$ to be C-true by MPC. (From (d), (e), (f), T6-1c.)

T6-2h shows that MP-reduction yields a *decision method* for both MPL and MPC. The rules of MP-reduction can be applied not only to a given sentence in S and K but also to any formula constructed out of auxiliary variables 'p', 'q', etc. with the help of connectives and 'N'. (In this case, the form T5-1c is a single variable.) The reduction of such a formula leads to 't' if and only if every sentence obtainable from the formula by substitutions is L-true by MPL and C-true by MPC. In the MP-reductum of any such formula, no 'N' occurs in the scope of another 'N'.

In the case of PL and PC, there are two well-known decision methods, viz. P-reduction (the conjunctive normal form) and the truth-table method. Analogously, in the case of MPL and MPC, there are, in addition to MP-reduction, decision methods using tables. These methods are likewise applicable both to sentences and to auxiliary formulas. If a sentence with n different ultimate MP-components (i.e., form T5-1c) or a formula with n different auxiliary variables is given, then first a truth-table of the customary kind with respect to these n components (hence containing 2^n lines) is constructed. Then a table of a new kind is to be formed which supplies the decision. It is easy to find a table method in which the second table consists of 2^{2^n} lines (thus, for $n = 3$, 256 lines); but this method is of course impracticable. I have found a method in which the second table has, in ordinary cases, only a few lines (for instance, for some theorems and postulates of Lewis's systems with $n = 3$ only three or four lines are needed). This method cannot be explained here.

In order to compare our system with those of other authors, let us consider a modification $\mathrm{MPC_v}$ of MPC: $\mathrm{MPC_v}$ is itself a calculus; it contains the variables 'p', 'q', etc.; these variables and 't' are the only ultimate components; it possesses the same rules as MPC, except that the rule of refutation is omitted and a rule of substitution for the variables is added. The method of MP-reduction is likewise applicable to $\mathrm{MPC_v}$. As mentioned earlier (footnote 8), this system $\mathrm{MPC_v}$ (except for containing 't', which corresponds to '$p \vee \sim p$') is equivalent to Lewis's system S5, which is the strongest of a series of systems investigated by Lewis.[9] Some of the reasons for which this system seems to me preferable to Lewis's other systems were indicated in §1. The completeness of this system is a further advantage. On the basis of the interpretation given by MPL, all sentences C-true (provable) in MPC and $\mathrm{MPC_v}$ are L-true, while those princi-

---

[9] See Lewis and Langford (op. cit.), especially Appendix II; for S5, see p. 501. Compare also W. T. Parry, *Modalities in the Survey system of strict implication*, this JOURNAL, vol. 4 (1939), pp. 137-154; concerning S5, see pp. 151 ff.

ples of other systems which go beyond Lewis's S5[10] are L-false. My chief reason for preferring the interpretation given by MPL is the simple parallelism between the modalities in this system and the semantical L-concepts, in particular between necessity and L-truth. It is true that these semantical concepts could be defined in a different way so as to correspond to a different conception of the modalities. However, the definition of L-truth here chosen, which is based on Wittgenstein's conception of the nature of logical truth, has the advantage of great simplicity, taking as criterion the universality of the range (D7-6a).

Let us consider all possible functions of n propositions p_1, p_2, $\cdots$, p_n. Any formula composed of n auxiliary variables with connectives and 'N' represents such a function. We will say that two such formulas $\mathfrak{M}_i$ and $\mathfrak{M}_j$ represent the *same* function if and only if they are L-equivalent, that is to say, if the P-reductum of $\mathfrak{M}_i \equiv \mathfrak{M}_j$ is 't' and therefore all sentences of this form are L-true by MPL and hence C-true by MPC. If a function can be represented by a formula without 'N', it is called a *truth-function* (or extensional function); otherwise we call it a *modal function*[11] (or intensional function). [N($p \vee \sim p$) represents a truth-function because it is L-equivalent to '$p \vee \sim p$' (T4-2b).] If a modal function can be represented by a formula in which all variables are under 'N' (i.e., within the scope of an 'N'), we call it a *purely modal function* (e.g., 'N$p \vee$ N$\sim p$'), otherwise a *mixed modal function* (e.g., '$p \vee$ N$\sim p$').

We shall now determine the number of functions of these kinds (without giving here exact proofs). The considerations are formulated for MPL (in terms of ranges); the results hold likewise for MPC. For n sentences 'A$_1$', $\cdots$, 'A$_n$', there are $p = 2^n$ conjunctions containing for every sentence in the given order either the sentence itself or its negation but not both; they correspond to the 2^n lines of the customary truth-table. Each of these conjunctions may be regarded as representing a basic truth-function of the n sentences. The truth-functions in general are represented by the disjunctions corresponding to the possible selections of some (or none or all) of these p conjunctions. (Here, by the disjunction of one sentence, the sentence itself is meant; by the null selection, any L-false sentence, e.g., 'A$_1$.$\sim$A$_1$.A$_2$.$\cdots$.A$_n$'). Thus there are $q = 2^p = 2^{2^n}$ truth-functions, as is well known.

Let the p conjunctions mentioned be abbreviated by 'C$_1$', $\cdots$, 'C$_p$'. They have the smallest ranges expressible by the n sentences. If some (or none, but not all) of these ranges are null, but the others are not, a specific logical relation holds among the sentences; in other words, a basic modal function holds among the propositions. Each of these functions can be represented by a conjunction of p components, namely either 'N$\sim$C$_i$' (if the range of 'C$_i$' is null), or '$\sim$N$\sim$ C$_i$' (if the range is not null) for $i = 1, 2, \cdots, p$; but the conjunction with all components of the form 'N$\sim$C$_i$' is excluded. Thus there are $2^p - 1$ or $q - 1$ basic modal functions. (The conjunction containing all components of the form 'N$\sim$C$_i$' says that all basic conjunctions are L-false, which is impossible; therefore this conjunction is L-equivalent to 'A.$\sim$A' and hence is not a modal

---

[10] See Parry, op. cit., pp. 152 ff.

[11] This use of the term 'modal function' is narrower than that of Parry, op. cit., p. 144, who takes it to include the truth-functions.

function.) The purely modal functions are the disjunctions of some (but not all) of these $q - 1$ basic modal functions. The disjunction of all of them must be excluded because it is L-true and hence not a modal function (it is L-equivalent to 'A $\vee \sim$A'); likewise the null disjunction (so to speak), which is L-false. Therefore the number of all purely modal functions is $2^{q-1} - 2$.

Now we shall determine the number of all functions of n propositions. We construct first the basic possibilities. If one of the p conjunctions is stated, say 'C_i', then this one cannot simultaneously be stated to be impossible (its range cannot be null); but any of the other $p - 1$ conjunctions may have a null range, possibly all of them or none of them. Thus there are in this case 2^{p-1} possibilities; each is represented by a conjunction containing 'C_i' and in addition, for every other conjunction 'C_j' either 'N$\sim$C$_j$' or '$\sim$N$\sim$C$_j$'. In the same way we may start with any other of the p conjunctions instead of 'C_i'. Let the conjunctions which we obtain in this way (e.g., 'C_1.N$\sim$C$_2$.$\cdots$') be 'M_1',$\cdots$,'M_r'. Their number r is $p \cdot 2^{p-1} = 2^n \cdot 2^{p-1} = 2^{p+n-1}$. They represent the basic functions of n propositions. Every function of n propositions is represented by a disjunction of some (possibly all or none) of these r conjunctions 'M_i'. (Here again, by the null disjunction an L-false sentence is meant.) Thus the number of all functions of n propositions is 2^r. By subtraction we find the number of modal functions to be $2^r - q$, and the number of mixed modal functions to be $2^r - q - 2^{q-1} + 2$.

The Number of Functions of n Propositions

$(p = 2^n; q = 2^p; r = 2^{p+n-1}.)$

		$n=1$ ($p=2; q=4;$ $r=4$)	$n=2$ ($p=4; q=16; r=32$)	
Purely modal functions:	$2^{q-1} - 2$	6	$2^{15} - 2 =$	32,766
Mixed modal functions:	$2^r - q - 2^{q-1} + 2$	6		4,294,934,514
Modal functions:	$2^r - q$	12		4,294,967,280
Truth-functions:	$q = 2^p$	4	$2^4 =$	16
All functions:	2^r	16	$2^{32} =$	4,294,967,296

Examples for $n = 1$. The four truth-functions are represented by '$p \vee \sim p$', 'p', '$\sim p$', '$p.\sim p$'. The six purely modal functions are: 'Np' (p is necessary), 'N$\sim p$' (impossible), '$\sim$N$p.\sim$N$\sim p$' (contingent), '$\sim$Np' (non-necessary), '$\sim$N$\sim p$' (possible), 'N$p \vee$ N$\sim p$' (non-contingent). The six mixed modal functions are: '$p.\sim$Np', '$\sim p.\sim$N$\sim p$', '$p \vee$ N$\sim p$', '$\sim p \vee$ Np', 'N$p \vee (\sim p.\sim$N$\sim p)$', 'N$\sim p \vee (p.\sim$N$p)$'.

In spite of the finite number of functions in MPL, there is no finite characteristic value-table ("matrix") for MPL or MPC, i.e., a value-table with a finite number of values (so-called truth-values) which attributes to a sentence (or formula with auxiliary variables) one of a set of specified "designated" values if and only if that sentence is L-true by MPL and hence C-true by MPC.[12]

---

[12] See J. Dugundji, this JOURNAL, vol. 5 (1940), pp. 150 f., with references to Gödel and McKinsey.

7. Functional logic (FL).   In preparation for the construction of the modal systems MFL and MFC, which will be the chief aim of this paper, we shall now briefly outline the ordinary, non-modal systems FL and FC.   The term 'functional calculus' ('FC') is here used for a certain form of the lower functional calculus; in Church's terminology, it is a simple applied functional calculus of first order with identity.   The individual variables are the only variables.   The term 'functional logic' ('FL') is here used for a corresponding semantical system. The system FL applies to a universe of discourse containing a denumerable number of individuals.   Every individual is denoted by an individual constant, and different individual constants denote different individuals.   Therefore the semantical rules (D7-5f) are framed in such a way that '$a_1 = a_2$', for example, is L-false.   [I have also studied other forms of functional logic and corresponding calculi.   In one alternative FL', sentences like '$a_1 = a_2$' are interpreted, not as L-false, but as factual, either true or false.   In this system FL' the state-descriptions must contain = -sentences and are more complicated than in FL (D7-4).   In FL', sentences which state that the cardinal number of the universe of discourse is n, or at least n, or at most n, or the like, are factual, although they are purely general (i.e., without non-logical constants).   If we want a system in which individual descriptions ("the one individual which is . . .") occur and are allowed to be substituted for individual variables, then a form like FL' might be more suitable than the simpler form FL.—Other system forms contain functional variables.]

In the primitive notation, we use only universal quantifiers.   The existential quantifier will occasionally be used as an abbreviation in the customary way; likewise '$x \neq y$' as short for '$\sim(x = y)$'.

The following definitions state the features of the semantical system FL.

D7-1.   The *signs* in FL are the following:

a.   Connectives: '$\sim$', '$\vee$', '$.$'.

b.   Parentheses: '(', ')'.

c.   't'.

d.   An infinite number of individual constants (in): 'a_1', 'a_2', etc. (or 'a', 'b', etc.).

e.   An infinite number of individual variables (i): 'x_1', 'x_2', etc. (or 'x', 'y', etc.).

f.   Any finite or infinite number of predicates (functional constants) of any degree.

g.   The sign of identity: '$=$'.

D7-2.   *Matrices* ($\mathfrak{M}$) in FL are the expressions of the following forms:

a.   Atomic matrix—a predicate of degree n, followed by n individual signs (constants or variables) (e.g., 'Rx_2b_4').

b.   '$=$' preceded and followed by an individual sign.

c. $\sim(\mathfrak{M}_i)$.

d. $(\mathfrak{M}_i) \vee (\mathfrak{M}_j)$.

e. $(\mathfrak{M}_i).(\mathfrak{M}_j)$.

f.   Universal matrix— $(i_k)(\mathfrak{M}_i)$.

D7-3.   *Sentences* ($\mathfrak{S}$) in FL are the closed matrices.   (Atomic sentences are those of form D7-2a; universal sentences, D7-2f; general sentences, those containing variables.)

A state-description is a class of sentences which represents a possible specific state of affairs by giving a complete description of the universe of individuals with respect to all properties and relations designated by predicates in the system.

D7-4. A class of sentences $\mathfrak{R}_i$ is a *state-description* in FL $=_{Df} \mathfrak{R}_i$ contains for every atomic sentence $\mathfrak{S}_i$ either $\mathfrak{S}_i$ itself or $\sim\mathfrak{S}_i$, but not both, and no other sentences.

The class of all state-descriptions is called the *universal range* and denoted by 'V_s'; the null class (of the same type) is called the *null range* and denoted by 'Λ_s'.

We shall now lay down rules which determine for every sentence $\mathfrak{S}_i$ of FL, in which state-descriptions $\mathfrak{S}_i$ holds; in other words, what is the *range* of $\mathfrak{S}_i$, i.e., the class of those state-descriptions in which $\mathfrak{S}_i$ holds. We shall write briefly '$\mathfrak{R}(\mathfrak{S}_i)$' for 'the range of $\mathfrak{S}_i$'. That $\mathfrak{S}_i$ holds in a given state-description means, in non-technical terms, that this state-description entails $\mathfrak{S}_i$; in other words, that $\mathfrak{S}_i$ would be true if this state-description were the description of the actual state of the universe. [I use the term 'true' here only occasionally in informal explanations. I shall not lay down a semantical definition of 'true in FL.' For the present purposes it is sufficient to have the concept of range which is recursively defined by D7-5. With its help the L-concepts can be defined (D7-6).]

D7-5. *Rules of ranges* $(\mathfrak{R})$ for FL.

a. If $\mathfrak{S}_i$ is atomic, $\mathfrak{R}(\mathfrak{S}_i)$ is the class of those state-descriptions to which $\mathfrak{S}_i$ belongs.

b. The range of $\sim\mathfrak{S}_i$ is $V_s - \mathfrak{R}(\mathfrak{S}_i)$.

c. The range of $\mathfrak{S}_i \vee \mathfrak{S}_j$ is the class-sum of $\mathfrak{R}(\mathfrak{S}_i)$ and $\mathfrak{R}(\mathfrak{S}_j)$.

d. The range of $\mathfrak{S}_i.\mathfrak{S}_j$ is the class-product of $\mathfrak{R}(\mathfrak{S}_i)$ and $\mathfrak{R}(\mathfrak{S}_j)$.

e. The range of a $=$-sentence with two occurrences of the same individual constant (e.g., '$a_3 = a_3$') is V_s.

f. The range of a $=$-sentence with two different individual constants (e.g., '$a_3 = a_5$') is Λ_s.

g. The range of $(i_k)(\mathfrak{M}_i)$ is the class-product of the ranges of the instances of $\mathfrak{M}_i$ (i.e., the sentences formed by substituting individual constants for the free occurrences of i_k in $\mathfrak{M}_i$; if $\mathfrak{M}_i$ is closed, $\mathfrak{M}_i$ itself is its only instance).

h. The range of a class $\mathfrak{R}_i$ of sentences is the class-product of the ranges of the sentences of $\mathfrak{R}_i$.

A sentence $\mathfrak{S}_i$ is usually regarded as logically true or logically necessary if it is true in every possible case. Therefore we call $\mathfrak{S}_i$ L-true if it holds in every state-description, in other words, if its range is V_s. Analogously, we use 'L-false' as explicatum for logical falsity or impossibility and define it by the null range. $\mathfrak{S}_j$ follows logically from $\mathfrak{S}_i$ if in every possible case in which $\mathfrak{S}_i$ holds, $\mathfrak{S}_j$ also holds. Therefore we define the explicatum 'L-implication' by the inclusion of the ranges. L-equivalence is meant as mutual L-implication; therefore it is defined by the identity of ranges.[13]

---

[13] More detailed explanations and discussions of the L-concepts are given in [I] §§14 ff. For the definitions of these concepts with the help of 'range,' based on conceptions of

D7-6.

a. $\mathfrak{S}_i$ is L-true (in FL) $=_{\mathrm{Df}} \mathfrak{R}(\mathfrak{S}_i)$ is V_s.

b. $\mathfrak{S}_i$ is L-false $=_{\mathrm{Df}} \mathfrak{R}(\mathfrak{S}_i)$ is Λ_s.

c. $\mathfrak{S}_i$ L-implies $\mathfrak{S}_j =_{\mathrm{Df}} \mathfrak{R}(\mathfrak{S}_i)$ is included in $\mathfrak{R}(\mathfrak{S}_j)$.

d. $\mathfrak{S}_i$ is L-equivalent to $\mathfrak{S}_j =_{\mathrm{Df}} \mathfrak{S}_i$ and $\mathfrak{S}_j$ have the same range.

Since the range of a class of sentences has been defined (D7-5h), these L-concepts can likewise be applied to classes of sentences.

The many well-known theorems concerning FL need not be mentioned here. We shall state, without proofs, only a few theorems which seem less known but are important both here and later in MFL.

The following theorem allows the *variation of an individual constant* under certain conditions. Note that here, and analogously in T7-2, L-implication is not asserted but only the conditional relation with respect to L-truth.

T7-1. Let $\mathfrak{S}_i$ be a sentence in FL, and $\mathfrak{S}_j$ be formed out of $\mathfrak{S}_i$ by replacing all occurrences of in_i by in_j.

a. Let $\mathfrak{S}_i$ not contain in_j (but, in distinction to (b), '=' may occur). If $\mathfrak{S}_i$ is L-true, then $\mathfrak{S}_j$ is L-true, and vice versa.

b. Let $\mathfrak{S}_i$ not contain '=' (but, in distinction to (a), in_j may occur). If $\mathfrak{S}_i$ is L-true then $\mathfrak{S}_j$ is L-true. (The converse does not generally hold.)

That the restrictions in T7-1a and b are necessary is seen from the following counter-example: '$\sim(a = b)$' is L-true, but '$\sim(a = a)$' is not. As a counter-example for the converse of (b), let $\mathfrak{S}_i$ be 'Pa $\supset$ Pb', and $\mathfrak{S}_j$ 'Pa $\supset$ Pa'; then $\mathfrak{S}_j$ is L-true but $\mathfrak{S}_i$ is not.

A theorem analogous to T7-1 holds for the variation of a predicate.

The following theorem allows the *generalization of an individual constant* under certain conditions.

T7-2. Let $\mathfrak{M}_i$ be a matrix in FL with i_k as the only free variable, and $\mathfrak{S}_j$ be formed out of $\mathfrak{M}_i$ by the substitution of in_j for i_k (i.e., for all free occurrences of i_k).

a. Let in_j and '=' not occur in $\mathfrak{M}_i$. If $\mathfrak{S}_j$ is L-true, then the universal sentence $(\mathrm{i}_k)(\mathfrak{M}_i)$ is L-true. (The converse is obvious.) (From T7-1b.)

b. Let $\mathfrak{M}_i$ not contain any individual constant (but '=' may occur). If $\mathfrak{S}_j$ is L-true, then $(\mathrm{i}_k)(\mathfrak{M}_i)$ is L-true. (From T7-1a.)

'$(\ )(\mathfrak{M}_k)$' denotes the *closure* of $\mathfrak{M}_k$, i.e., the sentence formed out of $\mathfrak{M}_k$ by prefixing universal quantifiers for all variables occurring freely in $\mathfrak{M}_k$, in their inverse alphabetical order. (If $\mathfrak{M}_k$ is closed, '$(\ )(\mathfrak{M}_k)$' denotes $\mathfrak{M}_k$ itself.)

T7-3. Let $\mathfrak{M}_i$ be a matrix in FL with n free variables ($n \geq 2$) (i.e., any number of occurrences of n different variable-designs). Let $\mathfrak{S}_j$ be formed out of $\mathfrak{M}_i$ by the substitution of n different individual constants for the variables. Let $\mathfrak{M}_h$ be a disjunction of $=$-matrices, one for any two different ones of the n variables. If $\mathfrak{S}_j$ is L-true, then $(\ )(\mathfrak{M}_h \lor \mathfrak{M}_i)$ is L-true.

T7-2 and 3 say in effect the following. If either no '=' occurs or only one individual constant occurs, simple generalization is allowed (T7-2). If two or

---

Wittgenstein, see [I] §§18 and 19; the method used in the present paper is similar to procedure E in §19, but it can take a simpler form here because FL contains atomic sentences for all atomic propositions.

more individual constants occur, generalization with added =-matrices is allowed. The necessity of the restrictions is seen from the following example. 'a $\neq$ b' is L-true; however, '$(y)(x)[x \neq y]$' is not L-true, but only $(y)(x)[x = y \lor x \neq y]$'.

8. Functional calculus (FC). FC is a formalization for FL. The signs, matrices, and sentences in FC are the same as in FL (D7-1, 2, 3). The only rule of inference is the *rule of implication* (as D4-2e). Further a rule of refutation is laid down (as D4-2f). This rule may be omitted if one wishes the form of the calculus to be more similar to the customary form. We add the rule in order to be able to compare C-falsity in FC with L-falsity in FL.

D8-1. The *primitive sentences* of FC are the sentences of the following forms. [( )($\mathfrak{M}_k$) is the closure of $\mathfrak{M}_k$; see explanation to T7-3.]

a. ( )($\mathfrak{M}_i$), where $\mathfrak{M}_i$ is any matrix whose P-reductum (D3-1) is 't'. (This means in effect that $\mathfrak{M}_i$ has a tautologous form; we could state here instead a sufficient number of particular forms, e.g., the four forms stated by Hilbert and Ackermann as axioms of their propositional calculus.[5])

b. ( )[($(\mathfrak{i}_i)(\mathfrak{M}_j \supset \mathfrak{M}_k) \supset ((\mathfrak{i}_i)(\mathfrak{M}_j) \supset (\mathfrak{i}_i)(\mathfrak{M}_k))$].

c. ( )[$\mathfrak{M}_j \supset (\mathfrak{i}_k)(\mathfrak{M}_j)$], where $\mathfrak{i}_k$ does not occur freely in $\mathfrak{M}_j$.

d. ( )[($(\mathfrak{i}_k)(\mathfrak{M}_k) \supset \mathfrak{M}_j$], where $\mathfrak{M}_j$ is like $\mathfrak{M}_k$ except for containing free occurrences of $\mathfrak{i}_j$ wherever $\mathfrak{M}_k$ contains free occurrences of $\mathfrak{i}_k$.

e. As (d), but here $\mathfrak{M}_j$ is like $\mathfrak{M}_k$ except for containing occurrences of the individual constant $\mathfrak{in}_j$, wherever $\mathfrak{M}_k$ contains free occurrences of $\mathfrak{i}_k$.

f. $(\mathfrak{i}_k)(\mathfrak{i}_k = \mathfrak{i}_k)$.

g. ( )[($\mathfrak{i}_k = \mathfrak{i}_j) \supset (\mathfrak{M}_k \supset \mathfrak{M}_j)$], where $\mathfrak{M}_j$ is like $\mathfrak{M}_k$ except for containing free occurrences of $\mathfrak{i}_j$ wherever $\mathfrak{M}_k$ contains free occurrences of $\mathfrak{i}_k$.

h. $\sim(\mathfrak{in}_i = \mathfrak{in}_j)$, where $\mathfrak{in}_i$ and $\mathfrak{in}_j$ are different individual constants.

D8-1a to d are four of Quine's axiom schemata of quantification.[14] To these we add the schema (e) for the substitution of an individual constant and the schemata (f), (g), (h) for identity.

Thus in our calculus FC, Quine's theorems of quantification[15] hold; that is to say, the sentences of the forms specified by Quine are C-true in FC. Further, all theorems on identity stated by Hilbert and Bernays[16] hold in FC, because they have been proved by these authors on the basis of two axioms corresponding to the schemata D8-1f and g here.

Schema D8-1h serves as a kind of axiom of infinity. With its help, the negations of those sentences can be proved which say that the number of all individuals is at most n, for any finite n (e.g., for $n = 2$, '$\sim(x)(y)(z)[x = y \lor x = z \lor y = z]$' is C-true).

The following theorem says in effect that FC is in accord with FL, as it was intended to be.

---

[14] Quine, op. cit., p. 88, *100, 102, 103, 104. G. Berry has shown (this JOURNAL, vol. 6 (1941), pp. 23–27) that Quine's schema *101 may be omitted if the closure of a matrix is defined by referring to the inverse alphabetical order of the quantifiers. This simplification has here been adopted for FC (and MFC).

[15] Quine, op. cit., §§17–21.

[16] Hilbert and Bernays, op. cit., pp. 179 ff.

T8-1.

a. Every primitive sentence of FC is L-true in FL.

b. Every C-true sentence in FC is L-true in FL. (This follows from (a) together with the fact that for every instance of application of the rule of implication, L-implication holds.)

c. FL is an L-true interpretation of FC. (This follows from (a) and (b) and the fact that the rule of refutation represents an instance of L-falsity.)

It is not clear whether FC is complete in the sense that every sentence which is L-true in FL is C-true in FC. Gödel's theorem of the completeness of the ordinary functional calculus of first order cannot be directly applied to FC because of the following difference. In ordinary functional logic, a sentence is regarded as L-true (or a matrix as universal ("allgemeingültig")) if it is L-true not only in the infinite universe of individuals but, in addition, in every finite (non-null) universe. In FL, on the other hand, more sentences are regarded as L-true, viz. all those which are L-true in the infinite universe. [For example, '$(\exists x)(\exists y)(x \neq y)$' says that there are at least two individuals. Thus this sentence holds in the infinite universe but not in the universe with only one individual. Therefore, it is L-true in FL (D7-5f) (and, moreover, C-true in FC, in virtue of D8-1h), but not L-true in the ordinary functional logic (and not provable in the ordinary functional calculus).] However, the following restricted theorem holds; it would be of interest to investigate the question whether the restriction can be weakened or even eliminated.

T8-2. If $\mathfrak{S}_i$ is an L-true sentence in FL without '$=$', then $\mathfrak{S}_i$ is C-true in FC.

Proof. Let the conditions be fulfilled. Let $\mathfrak{M}_i$ be formed from $\mathfrak{S}_i$ by replacing all individual constants with individual variables not occurring in $\mathfrak{S}_i$ and all predicates by predicate variables (like constants to be replaced by like variables, different by different ones). Then ( )[$\mathfrak{M}_i$] (where the closure applies only to the individual variables) is L-universal (i.e., L-true for all values of the predicate variables) in the infinite universe (compare T7-2a) and also in every non-null finite universe, because '$=$' does not occur. Therefore, according to Gödel's completeness theorem, it is provable in the ordinary functional calculus. Therefore $\mathfrak{S}_i$ is C-true in FC.

9. Modal functional logic (MFL). Now we shall come to our aim, the construction of the semantical and syntactical modal systems with quantification, viz., modal functional logic (MFL) and modal functional calculus (MFC). MFL is built from FL by the addition of the modal sign 'N'; likewise MFC from FC.

D9-1. The *signs* in MFL are the following:

a to g, as D7-1a to g.

h. 'N'.

D9-2. The *matrices* in MFL are the expressions of the following forms:

a to f, as D7-2a to f.

g. $N(\mathfrak{M}_i)$. (N-matrix).

D9-3. *Sentences* in MFL are the closed matrices. (Modal sentences are those containing 'N'.)

D9-4. The *state-descriptions* in MFL are the same as in FL (D7-4).

D9-5. *Rules of ranges* for MFL:

a to h, as D7-5a to h.

i. The range of $N(\mathfrak{S}_i)$ is V_s if the range of $\mathfrak{S}_i$ is V_s, and otherwise Λ_s.

L-truth, L-falsity, L-implication and L-equivalence in MFL are defined as in D7-6.

We shall now state some theorems concerning MFL. The proofs, in most cases, cannot be given here, because they presuppose a number of theorems concerning state-descriptions and ranges which have not been stated here.

T9-1.

a. $N\mathfrak{S}_i$ is L-true if and only if $\mathfrak{S}_i$ is L-true.

b. $N\mathfrak{S}_i$ is L-false if and only if $\mathfrak{S}_i$ is not L-true.

c. $N\mathfrak{S}_i$ is L-true or L-false.   (From D9-5i.)

Exchange of 'N' and universal quantifier:

T9-2. Every sentence of the following form is L-true in MFL:

$N(\ )[(\mathfrak{i}_k)N(\mathfrak{M}_i) \equiv N(\mathfrak{i}_k)(\mathfrak{M}_i)]$.

T9-1 and 2 show that the rules of ranges give to 'N' the interpretation intended (see §1, especially conventions C1-2 and C1-3).

Variation of individual constants under 'N' (analogous to T7-1):

T9-3. Let $\mathfrak{S}_j$ be formed from $\mathfrak{S}_i$ by replacing all occurrences of $\mathfrak{in}_i$ by $\mathfrak{in}_j$.

a. Let $\mathfrak{S}_i$ not contain $\mathfrak{in}_j$ (but it may contain '=' and 'N'). Then $N\mathfrak{S}_i$ is L-equivalent to $N\mathfrak{S}_j$.

b. Let $\mathfrak{S}_i$ contain neither '=' nor 'N' (but it may contain $\mathfrak{in}_j$). Then $N\mathfrak{S}_i$ L-implies $N\mathfrak{S}_j$. (The converse does not generally hold.)

10. Modal functional calculus (MFC).

MFC is constructed as a formalization of MFL. It is built from FC by the addition of 'N'.

The signs, matrices and sentences in MFC are the same as in MFL (D9-1, 2, 3). We shall occasionally use as abbreviations, not belonging to the systems themselves, existential quantifiers and the modal sign of possibility '$\Diamond$'; '$\Diamond p$' is short for '$\sim N\sim p$'.

D10-1. The *primitive sentences* of MFC are those sentences in MFC which have one of the following forms:

a. $N(\ )(\mathfrak{M}_i)$, where $\mathfrak{M}_i$ is any matrix whose P-reductum (D3-1) is 't'. (See remark on D8-1a.)

b. $N(\ )[(\mathfrak{M}_i \supset \mathfrak{M}_j) \supset (N\mathfrak{M}_i \supset N\mathfrak{M}_j)]$.

c. $(\ )[N\mathfrak{M}_i \supset \mathfrak{M}_i]$.

d. $N(\ )[N\mathfrak{M}_i \vee N\sim N\mathfrak{M}_i]$.

e to k. $N\mathfrak{S}_i$, where $\mathfrak{S}_i$ has one of the forms D8-1b to h, respectively (primitive sentences of FC).

l. $N(\ )[(\mathfrak{i}_k)N(\mathfrak{M}_i) \supset N(\mathfrak{i}_k)(\mathfrak{M}_i)]$.

m. $N(\ )[N(\mathfrak{i}_k)(\mathfrak{M}_i) \supset (\mathfrak{i}_k)N(\mathfrak{M}_i)]$.

n. (Assimilation.) $N(\ )[\mathfrak{i}_i = \mathfrak{i}_{k1} \vee \mathfrak{i}_i = \mathfrak{i}_{k2} \vee \cdots \vee \mathfrak{i}_i = \mathfrak{i}_{kn} \vee \sim N\mathfrak{M}_i \vee N\mathfrak{M}_j$; here $\mathfrak{M}_i$ does not contain '=', 'N', any quantifier with $\mathfrak{i}_j$, or any individual constant; it contains free occurrences of $\mathfrak{i}_i$, $\mathfrak{i}_j$, $\mathfrak{i}_{k1}$, $\cdots$, $\mathfrak{i}_{kn}$ but of no other variables; $\mathfrak{M}_j$ is like $\mathfrak{M}_i$ except for containing free occurrences of $\mathfrak{i}_j$ wherever $\mathfrak{M}_i$ contains free occurrences of $\mathfrak{i}_i$.

o. (Variation and Generalization.) $N( )[i_i = i_{k1} \vee i_i = i_{k2} \vee \cdots \vee i_i = i_{kn}$ $\vee \sim N\mathfrak{M}_i \vee i_j = i_{k1} \vee \cdots \vee i_j = i_{kn} \vee N\mathfrak{M}_j]$; here $\mathfrak{M}_i$ contains no individual constant and contains $i_i, i_{k1}, i_{k2}, \cdots, i_{kn}$ as the only free variables, and $\mathfrak{M}_j$ is like $\mathfrak{M}_i$ except for containing free occurrences of i_j wherever $\mathfrak{M}_i$ contains free occurrences of i_i.

p. (Substitution for predicate.) $N( )[N\mathfrak{M}_i \supset N\mathfrak{M}_j]$. Here $\mathfrak{M}_i$ contains a predicate $\mathfrak{pr}_k$ of degree m but no 'N'; $\mathfrak{M}_k$ is any matrix containing the m alphabetically first variables ($i_1, \cdots, i_m$) as the only free variables and not containing a quantifier with any variable occurring in $\mathfrak{M}_i$ ($\mathfrak{M}_k$ may contain 'N' and $\mathfrak{pr}_k$); $\mathfrak{M}_j$ is formed from $\mathfrak{M}_i$ by replacing every atomic matrix containing $\mathfrak{pr}_k$ by the corresponding substitution form of $\mathfrak{M}_k$ (i.e., $\mathfrak{pr}_k a_{k1} \cdots a_{km}$ is replaced by the matrix resulting from $\mathfrak{M}_k$ by substituting a_{k1} for $i_1, \cdots, a_{km}$ for i_m).

The only rule of inference for MFC is the *rule of implication* (as D4-2e). We add further a rule of refutation, as D4-2f (see remark on this rule for FC, §8).

Examples for D10-1. The following examples show either primitive sentences in MFC or sentences easily provable with their help, or cases of simple derivations made possible by the primitive sentences. ('$\cdots x \cdots$' and '$--x--$' are meant to indicate matrices containing 'x' as the only free variable.)

(a) to (d) correspond to D4-2d1 to 4, respectively; hence any primitive sentence of MPC in MFC is a primitive sentence of MFC. However, (a) to (d) admit also sentences like the following examples, which are not primitive sentences of MPC.

a. '$N(x)(Px \vee \sim Px)$'.

b. '$N(x)[(\cdots x \cdots \supset --x--) \supset (N(\cdots x \cdots) \supset N(--x--))]$'.

c. '$(x)[N(\cdots x \cdots) \supset (\cdots x \cdots)]$'. $N\mathfrak{S}_i$ C-implies $\mathfrak{S}_i$.

d. '$(x)[\sim N(\cdots x \cdots) \supset N \sim N(\cdots x \cdots)]$'. $\sim N\mathfrak{S}_i$ C-implies $N \sim N\mathfrak{S}_i$.

(a), (e) to (k) yield, with the help of (c), all the primitive sentences of FC.

(l) and (m), which could of course be combined into a sentence with '$\equiv$', allow the exchange of a universal quantifier and 'N', e.g., the transformation of '$(x)N(\cdots x \cdots)$' into '$N(x)(\cdots x \cdots)$' and vice versa, and hence also any change in the order of a sequence consisting of universal quantifiers and occurrences of 'N'.

n. '$N(z)(y)(x)[x = z \vee \sim N(\cdots x \cdots y \cdots z \cdots) \vee N(\cdots y \cdots y \cdots z \cdots)]$'; hence '$(z)(y)(x)$ $[x \neq z.N(\cdots x \cdots y \cdots z \cdots) \supset N(\cdots y \cdots y \cdots z \cdots)]$'. Thus, '$N(\cdots a \cdots b \cdots c \cdots)$' C-implies '$N(\cdots b \cdots b \cdots c \cdots)$'. In this way, an *assimilation* under 'N' is made possible, i.e., the change of an individual constant ('a') to another individual constant ('b') already occurring.

o. '$N(z)(y)(x)[x = z \vee \sim N(\cdots x \cdots z \cdots) \vee y = z \vee N(\cdots y \cdots z \cdots)]$'; hence '$(z)(y)(x)[x \neq z.N(\cdots x \cdots z \cdots) \supset (y \neq z \supset N(\cdots y \cdots z \cdots))]$'; hence (1) '$N(\cdots a \cdots c \cdots)$ $\supset (y)[y \neq c \supset N(\cdots y \cdots c \cdots)]$', and (2) '$N(\cdots a \cdots c \cdots) \supset N(\cdots b \cdots c \cdots)$'. (1) shows the possibility of a *generalization* under 'N', (2) that of a *variation* under 'N'. (o) permits only the variation into an individual constant otherwise not occurring. The variation into an individual constant already occurring, which we call assimilation, is possible only under the more restricting conditions of (n).

p. '$N(u)(v)[--Rva--Rbu--]$' C-implies '$N(u)(v)[--(\cdots v \cdots a \cdots)--(\cdots b \cdots u \cdots)$ $--]$', e.g., '$N(u)(v)[--(Pv.(\exists w)(Saw))--(Pb.(\exists w)(Suw))--]$'; $\mathfrak{M}_k$ is here '$Px.$ $(\exists w)(Syw)$'.

Some theorems on MFC follow; the proofs are omitted or only briefly indicated.

T10-1.

a. If $\mathfrak{S}_i$ is C-true by PC or by MPC or in FC, both $\mathfrak{S}_i$ and $N\mathfrak{S}_i$ are C-true in MFC.

b. $N\mathfrak{S}_i$ is C-true in MFC if and only if $\mathfrak{S}_i$ is C-true in MFC.

c. *Deduction theorem.* $\mathfrak{S}_i$ C-implies $\mathfrak{S}_j$ (in MFC) if and only if $\mathfrak{S}_i \supset \mathfrak{S}_j$ is C-true.

Proof. Since the rule of implication is the only rule of inference and all sentences are closed, this follows from [II] T6-14b.

d. $\mathfrak{S}_i$ is C-equivalent to $\mathfrak{S}_j$ (in MFC) if and only if $\mathfrak{S}_i \equiv \mathfrak{S}_j$ is C-true. (From (c).)

In what follows, 'ψ' is used to denote any context in which a given expression may occur; thus '$\psi(\mathfrak{M}_i)$' denotes any matrix containing $\mathfrak{M}_i$, and then '$\psi(\mathfrak{M}_j)$' denotes any of those matrices formed from $\psi(\mathfrak{M}_i)$ by replacing one or several (not necessarily all) occurrences of $\mathfrak{M}_i$ by $\mathfrak{M}_j$. $\mathfrak{M}_i$ and $\mathfrak{M}_j$ are called *C-interchangeable* if for any context ψ, $(\ )[\psi(\mathfrak{M}_i) \equiv \psi(\mathfrak{M}_j)]$ is C-true.

T10-2. *Theorems of replacement* in MFC.

a. If $\psi(\mathfrak{M}_i)$ and $\mathfrak{M}_j$ contain no 'N', $(\ )[(\mathfrak{M}_i \equiv \mathfrak{M}_j) \supset (\psi(\mathfrak{M}_i) \equiv \psi(\mathfrak{M}_j))]$ is C-true. (From FC, MPC, T10-1a.)

b. $(\ )[(\mathfrak{M}_i \equiv \mathfrak{M}_j) \supset (\psi(\mathfrak{M}_i) \equiv \psi(\mathfrak{M}_j))]$ is C-true (here 'N' may occur). (From FC, MPC, T10-1a.)

c. If $(\ )[\mathfrak{M}_i \equiv \mathfrak{M}_j]$ is C-true, then $\mathfrak{M}_i$ and $\mathfrak{M}_j$ are C-interchangeable. (From (b).)

d. If $\mathfrak{S}_i \equiv \mathfrak{S}_j$ is C-true, $\mathfrak{S}_i$ and $\mathfrak{S}_j$ are C-interchangeable. (From T10-1b, (c).)

e. If $(\ )(\mathfrak{M}_i)$ is C-true, $\mathfrak{M}_i$ is C-interchangeable with 't'. (From (c).)

f. If $(\ )(\sim\mathfrak{M}_i)$ is C-true, $\mathfrak{M}_i$ is C-interchangeable with '$\sim$t'. (From (e).)

T10-3. Sentences of the following forms are C-true in MFC.

a. $(\ )[(\exists i_k)N(\mathfrak{M}_k) \supset N(\exists i_k)(\mathfrak{M}_k)]$. (The converse implication does not generally hold. 'N' is here analogous to a universal quantifier.) (From FC, T10-1a, D10-1m, D10-1b.)

b. $(\ )[\Diamond (i_k)(\mathfrak{M}_k) \supset (i_k)\Diamond(\mathfrak{M}_k)]$. (The converse implication does not generally hold. '$\Diamond$' is here analogous to an existential quantifier.) (From (a).)

c. $(\ )[(\exists i_i)(i_i \neq i_{k1}.\cdots.i_i \neq i_{kn}.N\mathfrak{M}_i) \supset (i_i)(i_i \neq i_{k1}.\cdots.i_i \neq i_{kn} \supset N\mathfrak{M}_i)]$, where $\mathfrak{M}_i$ contains i_i, i_{k1}, i_{k2}, $\cdots$, i_{kn} as the only free variables and no individual constants. (From D10-1o.)

d. $(\ )[(in_l \neq i_{k1}.\cdots.in_l \neq i_{kn}.N\mathfrak{M}_l) \supset (i_i)(i_i \neq i_{k1}.\cdots.i_i \neq i_{kn} \supset N\mathfrak{M}_i)]$, where $\mathfrak{M}_i$ is as in (c), and $\mathfrak{M}_l$ is formed from $\mathfrak{M}_i$ by substituting in_l for i_i. (From (c).)

e. Generalization and specification under 'N'. $(\ )[in_l = i_{km} \vee \psi_m(in_l = i_{k.m-1} \vee \psi_{m-1}(\cdots \vee \psi_2(in_l = i_{k1} \vee \psi_1[(i_i)(i_i = i_{k1} \vee i_i = i_{k2} \vee \cdots \vee i_i = i_{km} \vee i_i = in_{j1} \vee i_i = in_{j2} \vee \cdots \vee i_i = in_{jn} \vee \mathfrak{M}_h \vee N\mathfrak{M}_i)]))\cdots) \equiv (in_l = i_{km} \vee \psi_m(\cdots \vee \psi_1[(i_i)(i_i = i_{k1} \vee \cdots \vee i_i = i_{km} \vee i_i = in_{j1} \vee \cdots \vee i_i = in_{jn} \vee \mathfrak{M}_h \vee N\mathfrak{M}_l)]))\cdots)]$, (the right side is like the left side except for $\mathfrak{M}_l$ instead of $\mathfrak{M}_i$) where $\mathfrak{M}_i$ contains i_i, i_{k1}, i_{k2}, $\cdots$, i_{km} ($m \geq 0$) as the only free variables and in_{j1}, in_{j2}, $\cdots$, in_{jn} ($n \geq 0$) as the only individual constants, and $\mathfrak{M}_l$ is formed from $\mathfrak{M}_i$ by substituting for i_i in$_l$, which does not belong to the constants mentioned, and $\mathfrak{M}_h$ contains i_i as free variable; $\psi_1, \cdots, \psi_m$ are arbitrary contexts. (From D10-1o, T10-2b.) (For examples of specification under 'N', see the later examples of applications of D11-1y (1).)

f. ( )[$\mathfrak{M}_i \equiv N\mathfrak{M}_i$], where $\mathfrak{M}_i$ is built out of N-matrices and =-matrices with the help of connectives and quantifiers. (From D8-1f, g, MPC, T10-1a, D10-1m, (a).)

g. ( )[N($\mathfrak{M}_i \vee \mathfrak{M}_j$) $\equiv \mathfrak{M}_i \vee N\mathfrak{M}_j$], where $\mathfrak{M}_i$ is as in (f). (From (f).)

h. $\sim N\mathfrak{S}_i$, where $\mathfrak{S}_i$ is a disjunction of n components ($n \geq 1$), each being an atomic sentence or a negation of an atomic sentence, but no atomic sentence occurring together with its negation.

Proof. We form $\mathfrak{S}_j$ from $\mathfrak{S}_i$ by substituting certain matrices for all atomic matrices occurring in the manner explained in D10-1p. The matrices to be substituted are built out of =-matrices with connectives; they can be chosen in such a way that for every disjunctive component $\mathfrak{S}_k$ of $\mathfrak{S}_j$, $\sim\mathfrak{S}_k$ is C-true; hence $\sim\mathfrak{S}_j$ is C-true, hence likewise $N\sim\mathfrak{S}_j$ (T10-2b), and $\sim N\mathfrak{S}_j$ (T4-1k). $N\mathfrak{S}_i \supset N\mathfrak{S}_j$ is C-true (D10-1p), hence also $\sim N\mathfrak{S}_i$.

i. $\sim N\mathfrak{S}_j$, where $\mathfrak{S}_j$ is a conjunction of m components ($m \geq 1$), each having the form described for $\mathfrak{S}_i$ in (h). (From (h), T4-1e.)

11. MF-reduction. We shall describe a method for transforming sentences of MFL or MFC into a normal form which is both L-equivalent in MFL and C-equivalent in MFC to the given sentence. We call this method MF-reduction. It is an extension of MP-reduction. Here likewise, as in the earlier case, the fact that the reduction of a sentence $\mathfrak{S}_i$ leads to 't' is a sufficient condition for $\mathfrak{S}_i$ being both L-true (in MFL) and C-true (in MFC). Here, however, in distinction to the previous method, this result is not a necessary condition. And moreover, it is not possible to construct a method of reduction of such a kind that it leads to 't' if and only if the given sentence is C-true (or L-true); such a method would be a decision method for MFC (and MFL); this, however, is impossible because there is no decision method for FC (or FL), as Church has shown.

Thus MF-reduction has the more modest aim of leading to 't' in many cases of L-true and C-true sentences, and of leading in general to a sentence which shows the logical nature of the given sentence by simplifying its structure. Obviously, this can be done in a more or less thoroughgoing way; to any set of reduction rules of this kind it is always possible to add further rules which bring about a simplification, and in particular also a reduction to 't', in some additional cases. Now our aim will be to effect such a simplification of the structure of the sentence in two important respects: if an expression $N\mathfrak{M}_i$ occurs in the reductum then $\mathfrak{M}_i$ shall always fulfill the following two conditions: (i) $\mathfrak{M}_i$ contains no 'N', (ii) $\mathfrak{M}_i$ is closed, i.e., $\mathfrak{M}_i$ contains no variable which is free in $\mathfrak{M}_i$; in other words, $\mathfrak{M}_i$ is a sentence. (ii) means that an expression of the form '$(x)[\cdots N(\cdots x \cdots)\cdots]$' does not occur; however, quantifiers and hence bound variables may occur in $\mathfrak{M}_i$, e.g., 'N$[\cdots(x)(\cdots x \cdots)\cdots]$'. (i) and (ii) together say that every scope of 'N' is a non-modal sentence, and hence a sentence in FL and FC.

The requirements (i) and (ii) just mentioned are closely connected. It can easily be shown that, if no variables occur, the requirement (i) can be fulfilled very simply. Suppose that N[$\cdots$N$\mathfrak{S}_i\cdots$] occurs where $\mathfrak{S}_i$ contains no 'N' and no variable. Then there is a simple effective method for determining whether $\mathfrak{S}_i$ is L-true or not with the help of P-reduction (T3-1d); we may then replace N$\mathfrak{S}_i$ in the first case by 't', in the second by '$\sim$t'.

If variables occur, the situation is more complicated. Suppose that all rules of MP-reduction (D5-1) which are applicable have been applied. Consider the possible forms resulting, as listed in T5-1; form (c) is either an atomic matrix or a universal matrix. We see that one 'N' can occur under another 'N' (i.e., within its scope) only if the one 'N' stands under a quantifier which in turn stands under the other 'N', e.g., N[$\cdots(x)(\cdots$N$\mathfrak{M}_i\cdots)\cdots$]. Here we have to distinguish two possibilities. 1. Suppose that $\mathfrak{M}_i$ is closed. In this case we can place N$\mathfrak{M}_i$ outside the operand of the quantifier '(x)' and likewise outside the operand of any other quantifier standing between the two 'N' (this is done by rules like D11-1o, p, q below). Then, by applying MP-reduction again, the one 'N' which stands under the other 'N' will disappear. 2. Suppose that 'x' occurs free in $\mathfrak{M}_i$: 'N[$\cdots(x)(\cdots$N$(\cdots x\cdots)\cdots)\cdots$]'. This situation represents the most serious difficulty we have to overcome to reach our aim, the fulfillment of requirement (i). We must find a way of eliminating free variables under 'N'. Consider the two sentences '$(y)[y \neq c \supset$ N$(\cdots y\cdots c\cdots)]$' and 'N$(\cdots a\cdots c\cdots)$', where '$\cdots y\cdots c\cdots$' contains '$y$' as the only free variable and 'c' as the only individual constant and '$\cdots a\cdots c\cdots$' is the result of substituting 'a' for 'y' in '$\cdots y\cdots c\cdots$'. It is clear that the first sentence C-implies the second; the second can be derived from the first by specification (D10-1g corresponding to D8-1d) and the use of 'a $\neq$ c' (D10-1k corresponding to D8-1h). Now the important fact is that, moreover, the second sentence C-implies the first by what we have called generalization under 'N'; see the example for D10-1o. Thus the two sentences are C-equivalent. Therefore we may lay down a rule of reduction which permits the replacement of the first by the second and thereby the elimination of the free variable 'y' under 'N'. This will be done in a more general way by the rule D11-1y(1). Other reduction rules are necessary for bringing the sentence into a form where this rule can be applied. In this way both aims will be reached, first (ii) and then (i), as we shall see (T11-1h).

D11-1. The *MF-reduction* of a matrix $\mathfrak{M}_i$ is its transformation according to the following rules. At any step in the transformation, the first of these rules that can be applied must be applied. (This is especially important in the case of an application of rules (r), (s), (v), and (y).) The replacement applies to any part having the specified form (without any restriction like those in D3-1 and D5-1). The final result to which none of the rules is applicable any more is called the *MF-reductum* of $\mathfrak{M}_i$.

a to k, as D3-1a to k, respectively.

1. Omission of 'N'. N$\mathfrak{M}_k$ is replaced by $\mathfrak{M}_k$ if $\mathfrak{M}_k$ has one of the following forms: (1) 't'; (2) '$\sim$t'; (3) N$\mathfrak{M}_h$ (with any $\mathfrak{M}_h$); (4) any =-matrix; (5) $\mathfrak{M}_k$ is built out of one or several matrices of the forms (3) and (4) with the help of connectives and quantifiers.

m, as D5-1m.

n. (1). A $=$-matrix with two occurrences of the same individual constant or variable is replaced by 't'.

(2). A $=$-matrix with two different individual constants is replaced by '$\sim$t'.

o. Omission of quantifier. $(i_k)(\mathfrak{M}_k)$, where i_k does not occur free in $\mathfrak{M}_k$, is replaced by $\mathfrak{M}_k$.

p. Distribution of quantifier in conjunction. $(i_k)(\mathfrak{M}_{k1}.\mathfrak{M}_{k2}.\cdots.\mathfrak{M}_{kn})$ $(n \geq 2)$ is replaced by $(i_k)(\mathfrak{M}_{k1}).(i_k)(\mathfrak{M}_{k2}).\cdots.(i_k)(\mathfrak{M}_{kn})$.

q. Quantifier with disjunction. Suppose that $(i_k)(\mathfrak{M}_k)$ occurs where $\mathfrak{M}_k$ is a disjunction among whose n components $(n \geq 2)$ there is at least one in which i_k occurs freely and at least one in which it does not. Let $\mathfrak{M}_h$ be the disjunction of the components of the first kind (in their original order) and $\mathfrak{M}_j$ the disjunction of the components of the second kind (in their original order). $(i_k)(\mathfrak{M}_k)$ is replaced by $(i_k)(\mathfrak{M}_h) \vee \mathfrak{M}_j$.

r. Infinity. $(i_k)(\mathfrak{M}_k)$, where $\mathfrak{M}_k$ is either a $=$-matrix or a disjunction of $=$-matrices, is replaced by '$\sim$t'. [Note that after application of (q) and (m), every component of $\mathfrak{M}_k$ contains i_k and, in addition, either a variable different from i_k or an individual constant.]

s. Elimination of $\neq$-matrices.

(1). $(i_k)(\sim\mathfrak{M}_k)$, where $\mathfrak{M}_k$ is a $=$-matrix (which contains i_k, according to (o)), is replaced by '$\sim$t'.

(2). Suppose that $(i_k)(\mathfrak{M}_k)$ occurs where $\mathfrak{M}_k$ is a disjunction among whose n components $(n \geq 2)$ there is at least one of the form $\sim\mathfrak{M}_j$, where $\mathfrak{M}_j$ is a $=$-matrix containing i_k and another individual sign. $(i_k)(\mathfrak{M}_k)$ is replaced by the matrix formed from $\mathfrak{M}_k$ by first omitting the first component of the kind described and then substituting for i_k the other individual sign mentioned.

t. Quantifier under 'N'. Suppose that $N\mathfrak{S}_i$ occurs ($\mathfrak{S}_i$ is a sentence and hence closed!), where either $\mathfrak{S}_i$ itself or the first disjunctive component in $\mathfrak{S}_i$ of universal form is $(i_k)(\mathfrak{M}_k)$. Then this quantifier (i_k) is shifted before the 'N' (that is to say, $N\mathfrak{S}_i$ is replaced by $(i_k)[N\mathfrak{M}_j]$, where $\mathfrak{M}_j$ results from $\mathfrak{S}_i$ by omitting the occurrence in question of the quantifier (i_k)). [After this, rule (w) becomes applicable.]

u. Suppose that a sentence $\mathfrak{S}_i$ (closed!) of the form $(i_{k1})(i_{k2})\cdots(i_{kn})[N\mathfrak{M}_k]$ occurs $(n \geq 2)$. $\mathfrak{S}_i$ is replaced by $(i_{k1})[N(i_{k2})\cdots(i_{kn})(\mathfrak{M}_k)]$. [After this, rule (w) becomes applicable, and then (t).]

v. Factual sentence under 'N'. $N(\mathfrak{S}_i)$ is replaced by '$\sim$t' if $\mathfrak{S}_i$ has one of the following forms: (1) an atomic sentence, (2) the negation of an atomic sentence, (3) a disjunction of sentences of forms (1) and (2). [Note that, because of (a), a disjunction of the form (3) does not contain any atomic sentence together with its negation.]

w. (Some simple cases of elimination of a free variable under 'N'.) Suppose that $\mathfrak{S}_i$ of the form $(i_k)[N\mathfrak{M}_k]$ or $\mathfrak{S}_j$ of the form $(i_k)[\sim N\mathfrak{M}_k]$ occurs. Let the individual constants occurring in $\mathfrak{M}_k$ be (in alphabetical order) in_{k1}, in_{k2}, $\cdots$, in_{kn} $(n \geq 0)$; let in_l be an individual constant (the first in alphabetical order) not occurring in $\mathfrak{M}_k$. Let $\mathfrak{S}_{k1}$, $\mathfrak{S}_{k2}$, $\cdots$, $\mathfrak{S}_{kn}$, and $\mathfrak{S}_l$ be formed from $\mathfrak{M}_k$ by substituting for i_k in_{k1}, in_{k2}, $\cdots$, in_{kn}, and in_l, respectively.

(1). Let $\mathfrak{M}_k$ contain no individual constant. Then $\mathfrak{S}_i$ is replaced by $N\mathfrak{S}_l$, and $\mathfrak{S}_j$ by $\sim N\mathfrak{S}_l$.

(2). Let $\mathfrak{M}_k$ contain neither '=' nor 'N'. Then $\mathfrak{S}_i$ is replaced by $N\mathfrak{S}_l$, and $\mathfrak{S}_j$ by $\sim N\mathfrak{S}_{k1}.\sim N\mathfrak{S}_{k2}.\cdots.\sim N\mathfrak{S}_{kn}$.

(3). (To be applied only if (1) and (2) are not applicable.) $\mathfrak{S}_i$ is replaced by $N\mathfrak{S}_{k1}.N\mathfrak{S}_{k2}.\cdots.N\mathfrak{S}_{kn}.N\mathfrak{S}_l$, and $\mathfrak{S}_j$ by $\sim N\mathfrak{S}_{k1}.\sim N\mathfrak{S}_{k2}.\cdots.\sim N\mathfrak{S}_{kn}.\sim N\mathfrak{S}_l$.

x. Disjunction under 'N'. Suppose that $N\mathfrak{M}_h$ occurs where $\mathfrak{M}_h$ is a disjunction of n components ($n \geqq 2$) of which at least one has one of the forms (3), (4), (5) described in (1). Let $\mathfrak{M}_j$ be the disjunction of all components of this kind (in their original order) and $\mathfrak{M}_l$ be the disjunction of the remaining components (in their original order). [There is at least one remaining component because of (1).] Then $N\mathfrak{M}_h$ is replaced by $\mathfrak{M}_j \vee N\mathfrak{M}_l$.

y. Elimination of a free variable under 'N'. Let $N\mathfrak{M}_k$ be the first occurrence of an N-matrix such that $\mathfrak{M}_k$ contains a free variable but no 'N'. Consider the universal quantifiers binding the free variables in $\mathfrak{M}_k$; let the last one of these quantifiers contain i_j, and let $\mathfrak{R}_k$ be the class of the remaining free variables in $\mathfrak{M}_k$. Let $\mathfrak{M}_j$ be the operand of the quantifier with i_j just mentioned (hence the occurrence of $N\mathfrak{M}_k$ in question is a part of $\mathfrak{M}_j$). Let $\mathfrak{R}_l$ be the class of the individual constants occurring in $\mathfrak{M}_k$, and let in_l be an individual constant (the alphabetically first one) not occurring in $\mathfrak{M}_k$. ($\mathfrak{R}_k$ and $\mathfrak{R}_l$ may be empty.) We examine whether or not the following conditions (α) and (β) are fulfilled. Then one and only one of the subsequent rules (1), (2), (3) is applicable.

(α). $\mathfrak{M}_j$ is a disjunction with two or more components such that for every sign a_i (variable or individual constant) in $\mathfrak{R}_k$ and in $\mathfrak{R}_l$ there is a component $i_j = a_i$ or $a_i = i_j$.

(β). For every variable i_k in $\mathfrak{R}_k$, the operand (which contains the occurrence of $N\mathfrak{M}_k$ in question) of the quantifier with i_k contains as (proper or improper) part a disjunction of which one component is $i_k = in_l$ or $in_l = i_k$ and another component contains the occurrence of $(i_j)(\mathfrak{M}_j)$ in question.

(1). *Specification* under 'N'. Suppose that the conditions (α) and (β) are fulfilled. Then the occurrence of $N\mathfrak{M}_k$ in question is replaced by $N\mathfrak{M}_l$, where $\mathfrak{M}_l$ is formed from $\mathfrak{M}_k$ by substituting in_l for i_j.

(2). Suppose that the condition (α) is not fulfilled for all signs in $\mathfrak{R}_k$ and $\mathfrak{R}_l$. Let the signs for which it is not fulfilled be (in alphabetical order) a_{h1}, a_{h2}, $\cdots$, a_{hn} ($n \geqq 1$). Let $\mathfrak{M}_{h1}$, $\mathfrak{M}_{h2}$, $\cdots$, $\mathfrak{M}_{hn}$ be formed from $\mathfrak{M}_j$ by substituting for i_j a_{h1}, $\cdots$, a_{hn}, respectively. Then the occurrence of $(i_j)(\mathfrak{M}_j)$ in question is replaced by $(i_j)[i_j = a_{h1} \vee i_j = a_{h2} \vee \cdots \vee i_j = a_{hn} \vee \mathfrak{M}_j].[\mathfrak{M}_{h1}.\mathfrak{M}_{h2}.\cdots.\mathfrak{M}_{hn}]$.

(3). Suppose that condition (α) is fulfilled but not (β). Let the last one among those quantifiers which do not fulfill (β) contain the variable i_h and have the operand $\mathfrak{M}_h$. Then this occurrence of $(i_h)(\mathfrak{M}_h)$ is replaced by $(i_h)[i_h = in_l \vee \mathfrak{M}_h].\mathfrak{M}_l$, where $\mathfrak{M}_l$ is formed from $\mathfrak{M}_h$ by substituting in_l for i_h.

T11-1. Let $\mathfrak{S}_j$ be a sentence in MFL and MFC to which none of the rules of MF-reduction (D11-1) is any more applicable. Then every matrix occurring in $\mathfrak{S}_j$, including $\mathfrak{S}_j$ itself, has one of the following forms.

a. Atomic matrix.

b. Negation of (a).

c. =-matrix. This matrix contains either a variable and a constant or two different variables. It is a disjunctive component of an operand whose quantifier contains a variable occurring in this matrix.

d. Disjunction with two or more components. Every component has one of the forms (a), (b), (c), (f), (g), (h). No component is the negation of another component. No two components are alike.

e. Conjunction with two or more components. Every component has one of the forms (a), (b), (d), (f), (g), (h). No component is the negation of another component. No two components are alike. This form occurs only as the whole sentence $\mathfrak{S}_i$ (or as a partial conjunction of the whole conjunction).

f. $(\mathfrak{i}_k)(\mathfrak{M}_k)$. $\mathfrak{M}_k$ has one of the forms (a), (b), (d), (f), (g).

g. $\sim(\mathfrak{i}_k)(\mathfrak{M}_k)$. $\mathfrak{M}_k$ is as in (f).

h. $N\mathfrak{S}_k$ or $\sim N\mathfrak{S}_k$. $\mathfrak{S}_k$ is closed, contains a predicate and a quantifier, but no 'N'. $\mathfrak{S}_k$ has either form (g) or (d); if a disjunction (d), then every component has one of the forms (a), (b), (g), and at least one has (g). The forms (h) do not occur in an operand of a quantifier, but only either (1) as $\mathfrak{S}_i$ itself or (2) as a conjunctive component of $\mathfrak{S}_i$ or (3) as a disjunctive component of $\mathfrak{S}_i$ or of a conjunctive component of $\mathfrak{S}_i$.

i. 't' or '$\sim$t'. This occurs only if the whole sentence $\mathfrak{S}_i$ is 't' or '$\sim$t'.

Examples for rule D11-1y, *elimination of free variables under 'N'.* 1. Let the sentence '$(y)(z)[z = y \vee N(\cdot\cdot y\cdot\cdot z\cdot\cdot)]$' be given, where '$\cdot\cdot y\cdot\cdot z\cdot\cdot$' indicates a matrix containing 'y' and 'z' as the only free variables but containing no individual constants and no 'N', and having such a form that no rule preceding D11-1y is applicable. $\mathfrak{M}_k$ is here '$\cdot\cdot y\cdot\cdot z\cdot\cdot$'; $\mathfrak{i}_j$ is 'z'; $\mathfrak{R}_k$ contains only 'y'; $\mathfrak{R}_l$ is empty; $\mathfrak{in}_l$ is 'a'. (α) is fulfilled because '$z = y$' occurs. (β) is not fulfilled because '$y = a$' does not occur. Therefore rule (y)(3) is applied, with 'y' as $\mathfrak{i}_h$. The result is: '$(y)[y = a \vee (z)[z = y \vee N(\cdot\cdot y\cdot\cdot z\cdot\cdot)]].(z)[z = a \vee N(\cdot\cdot a\cdot\cdot z\cdot\cdot)]$'. Now we apply rule (y) to the first N-matrix. This time (y)(1) is applicable; the result is: '$(y)[y = a \vee (z)[z = y \vee N(\cdot\cdot a\cdot\cdot z\cdot\cdot)]].(z)[z = a \vee N(\cdot\cdot a\cdot\cdot z\cdot\cdot)]$'. We apply rule (y) again to the first N-matrix. (α) is not fulfilled because '$z = a$' does not occur. Therefore rule (y)(2) is applied, with 'a' as a_{h1}: '$(y)[y = a \vee [(z)(z = a \vee z = y \vee N(\cdot\cdot a\cdot\cdot z\cdot\cdot)).(a = y \vee N(\cdot\cdot a\cdot\cdot a\cdot\cdot))]].(z)[z = a \vee N(\cdot\cdot a\cdot\cdot z\cdot\cdot)]$'. Now rule (j) is applied, distribution with respect to the first sign of conjunction; then the quantifier with 'y' is distributed (rule (p)): '$(y)[y = a \vee (z)(z = a \vee z = y \vee N(\cdot\cdot a\cdot\cdot z\cdot\cdot))].(y)[y = a \vee a = y \vee N(\cdot\cdot a\cdot\cdot a\cdot\cdot)].(z)[z = a \vee N(\cdot\cdot a\cdot\cdot z\cdot\cdot)]$'. Now the second N-matrix is moved out of the operand of the quantifier (rule (q)); thereby a disjunctive component of the form '$(y)[y = a \vee a = y]$' is produced, which is then transformed into '$\sim$t' (rule (r)) and then omitted (rule (e)). The result is: '$(y)[y = a \vee (z)(z = a \vee z = y \vee N(\cdot\cdot a\cdot\cdot z\cdot\cdot))].N(\cdot\cdot a\cdot\cdot a\cdot\cdot).(z)[z = a \vee N(\cdot\cdot a\cdot\cdot z\cdot\cdot)]$'. Now (y)(1) can be applied to the first N-matrix, with 'b' as $\mathfrak{in}_l$: '$(y)[y = a \vee (z)(z = a \vee z = y \vee N(\cdot\cdot a\cdot\cdot b\cdot\cdot))].N(\cdot\cdot a\cdot\cdot a\cdot\cdot).(z)[z = a \vee N(\cdot\cdot a\cdot\cdot z\cdot\cdot)]$'. As previously, rules (q), (r), and (e) are applied, first to the first quantifier with 'z' and then to that with 'y', so that both quantifiers disappear; thus the first conjunctive component becomes '$N(\cdot\cdot a\cdot\cdot b\cdot\cdot)$'. Finally we transform the last conjunctive component. According to (y)(1), the matrix '$N(\cdot\cdot a\cdot\cdot z\cdot\cdot)$' is replaced by '$N(\cdot\cdot a\cdot\cdot b\cdot\cdot)$'. Now the rules (q), (r), and (e) are applied again;

thereby the quantifier with 'z' disappears. Thus the component becomes 'N($\cdots$a$\cdots$b$\cdots$)' and disappears (rule (d)). Hence the result of the whole reduction is 'N($\cdots$a$\cdots$b$\cdots$).N($\cdots$a$\cdots$a$\cdots$)'.

2. Let '$- - -$' indicate the original sentence of the first example. Consider 'N(u)N[(x)(Rxu) $\vee$ $---$]'. Here, the first 'N' is of third order. The expression preceding the square bracket becomes '(u)NN' (rule (t)) and then '(u)N' (rule (l)(3)). Then rule (x) yields '(u)[N(x)(Rxu) $\vee$ $- - -$]' because '$- - -$' has form (l)(5). Rule (q) yields '(u)[N(x)(Rxu)] $\vee$ $- - -$'; (w)(1), 'N(x)(Rxa) $\vee$ $- - -$'; (t), '(x)N(Rxa) $\vee$ $- - -$'; (w)(2), 'N(Rba) $\vee$ $- - -$'. The first disjunctive component becomes '$\sim$t' (rule (v)(1)) and disappears (rule (e)). Thus the reductum of the whole is the same as in the first example.

12. Relations between MFC and MFL. The first question concerning the calculus MFC is whether it is not too strong, that is, whether it is indeed a formalization of MFL and hence justified by MFL as an L-true interpretation of it. The affirmative answer to this question is given by T12-1d. Then the second question, whether MFC is strong enough for certain purposes, will be examined.

T12-1.

a. Every primitive sentence of MFC is L-true in MFL. (The proof is based on complicated theorems concerning state-descriptions, which have not been stated here.)

b. Every C-true sentence in MFC is L-true in MFL. (From (a); see T8-1b.)

c. Whenever C-falsity in MFC holds, then L-falsity in MFL holds. Analogously with C-implication and L-implication, and with C-equivalence and L-equivalence. (From (b).)

d. MFL is an *L-true interpretation* of MFC. (From (a) and (b); see T8-1c.)

We shall now show that MFC is strong enough to cover the transformation by MF-reduction (T12-2b). However, the further question whether it is also strong enough to yield all L-true sentences remains open, as we shall see.

T12-2. Let $\mathfrak{S}_i$ be any sentence in MFC and MFL, and $\mathfrak{S}_j$ its *MF-reductum*. Then the following holds:

a. If a sentence is transformed into another sentence by an application of the first rule of MF-reduction which is applicable, then the two sentences are C-equivalent in MFC.

Proof. For the rules (a) to (k) in D11-1 this follows from T6-1e, T10-1a. For rules (l)(1), (2), and (3), from D5-1l(3), (4), (1), T6-1e, T10-1a. For (l)(4), from D10-1i, j (corresponding to D8-1f, g), and m. For (l)(5), from T10-3f. For (m) (= D5-1m), from T4-1e, T10-1a. For (n)(1), from D8-1f, T10-1a, T10-2e. For (n)(2), from D8-1h. For (o), from D8-1c, d. For (p) and (q), from FC, T10-1a. For (r), from D8-1h. For (s)(1), from D8-1f. For (s)(2), from D8-1g. For (t) and (u), from D10-1l, m, and FC. For (v), from T10-3h. For (w), from D10-1n and o. For (x), from T10-3g. For (y), from T10-3e.

b. $\mathfrak{S}_i$ and $\mathfrak{S}_j$ are C-equivalent in MFC. (From (a).)

c. $\mathfrak{S}_i$ and $\mathfrak{S}_j$ are L-equivalent in MFL. (From (b), T12-1c.)

d. If $\mathfrak{S}_j$ is 't', $\mathfrak{S}_i$ is C-true in MFC and L-true in MFL. (From (b), (c).)

e. If $\mathfrak{S}_j$ is '$\sim$t', $\mathfrak{S}_i$ is C-false in MFC and L-false in MFL. (From (b), (c).) —The converses of (d) and (e) do not hold generally. For instance, 'N[$\sim$(x)(Px) $\vee$ Pa]' is L-true and C-true, but is an MF-reductum.

f. If $\mathfrak{S}_i$ contains no predicate, $\mathfrak{S}_j$ is either 't' or '$\sim$t'. (From T11-1.)

g. If $\mathfrak{S}_i$ contains no quantifier under an 'N', then $\mathfrak{S}_j$ contains no 'N'.

Proof. In the reduction procedure, no quantifier comes under 'N' except by rule D11-1u; but these quantifiers are then removed by rules (w) and (t). Therefore, $\mathfrak{S}_j$ contains no quantifier under 'N', and hence no 'N' (T11-1h).

Since there is no general decision method for FL (i.e., no effective method which decides for any given sentence in a finite number of steps whether or not it is L-true in FL), there is, of course, none for MFL. However, the method of MF-reduction makes it possible to apply partial solutions for FL, that is to say, decision methods for restricted classes of sentences in FL, to MFL in the following manner. A sentence is L-true in MFL if and only if its MF-reductum is L-true (T12-2c). The latter has in general the form of a conjunction of disjunctions (T11-1). A conjunction is L-true if and only if each of its components is L-true. Each component is in general a disjunction of the form $\sim N\mathfrak{S}_{i1}$ v $\cdots$ v $\sim N\mathfrak{S}_{im}$ v $N\mathfrak{S}_{j1}$ v $\cdots$ v $N\mathfrak{S}_{jn}$ v $\mathfrak{S}_k$, where $\mathfrak{S}_k$ is a disjunction of p N-free sentences ($m \geqq 0, n \geqq 0, p \geqq 0; m + n + p \geqq 1$). Since the components with '$\sim$N' and 'N' are either L-true or L-false, the whole disjunction is L-true if and only if either one of these components or $\mathfrak{S}_k$ is L-true, in other words, if and only if either one of the sentences $\mathfrak{S}_{i1}$, $\cdots$, $\mathfrak{S}_{im}$ is non-L-true or one of the sentences $\mathfrak{S}_{j1}$, $\cdots$, $\mathfrak{S}_{jn}$, $\mathfrak{S}_k$ is L-true. All these sentences are N-free and hence belong to FL also. Therefore, if all of them belong to classes for which decision methods for FL are available, then their application leads to a decision for the whole disjunction in MFL.

As has been remarked earlier (see the discussion preceding T8-2), it is not known whether FC is complete. Therefore, it is likewise an open question whether or not MFC is complete in the sense that every sentence which is L-true in MFL is C-true in MFC. The following theorem gives a partial answer to this question.

T12-3. Let $\mathfrak{S}_i$ be an MF-reductum in which no '$=$' and no sentence of the form '$\sim N(\cdots)$' occurs. If $\mathfrak{S}_i$ is L-true in MFL, it is C-true in MFC.

Proof. Let the conditions be fulfilled. Then, disregarding simpler cases, in which the theorem is obvious, $\mathfrak{S}_i$ is a conjunction of which all components are L-true. Let $\mathfrak{S}_h$ be such a component. $\mathfrak{S}_h$ has the form $N\mathfrak{S}_{j1}$ v $N\mathfrak{S}_{j2}$ v $\cdots$ v $N\mathfrak{S}_{jn}$ v $\mathfrak{S}_k$, where $\mathfrak{S}_k$ is a disjunction of N-free sentences. At least one of the sentences $\mathfrak{S}_{j1}$, $\cdots$, $\mathfrak{S}_{jn}$, $\mathfrak{S}_k$ must be L-true in MFL and also in FL, since all these sentences are N-free. If $\mathfrak{S}_k$ is L-true in FL, it is C-true in FC (T8-2) and hence in MFC (T10-1a), and hence likewise the whole disjunction. If $\mathfrak{S}_{jr}$ ($r = 1, \cdots, n$) is L-true in FL, it is C-true in FC and in MFC, and hence likewise $N\mathfrak{S}_{jr}$ (T10-1a) and again the whole disjunction. Thus every component of the conjunction $\mathfrak{S}_i$ is C-true in MFC, and hence $\mathfrak{S}_i$ itself.

Whether the two restricting conditions in this theorem can be eliminated remains to be seen. If they can, MFC is complete. The restriction with respect to '$=$' is a problem concerning FC, as earlier discussed. With respect to MFC, the problem remains whether an MF-reductum of the form $\sim N\mathfrak{S}_i$ is C-true if it is L-true in MFL, in other words, if $\mathfrak{S}_i$ is not L-true in MFL. If $\mathfrak{S}_i$ is L-false, the proof of $\sim N\mathfrak{S}_i$ is simple (with T4-1k). If $\mathfrak{S}_i$ is factual, $\sim N\mathfrak{S}_i$ can often be proved in the following way. We construct an L-false sentence $\mathfrak{S}_j$ from $\mathfrak{S}_i$ by suitable substitutions for primitive predicates. Then D10-1p yields $N\mathfrak{S}_i \supset N\mathfrak{S}_j$ and hence $\sim N\mathfrak{S}_j \supset \sim N\mathfrak{S}_i$. Since $\sim N\mathfrak{S}_j$ is C-true, so is $\sim N\mathfrak{S}_i$. As examples, see T10-3h and i, which lead to the following theorem (in which $\mathfrak{S}_k$ is not required to be a reductum).

T12-4. If $\mathfrak{S}_k$ is a factual sentence in MFL without variables, then $\sim N\mathfrak{S}_k$ and $\sim N\sim\mathfrak{S}_k$ (in other symbols, $\Diamond\sim\mathfrak{S}_k$ and $\Diamond\mathfrak{S}_k$) are C-true in MFC. (From T10-3h and i.)

Even if variables occur, the procedure of substitution for predicates yields often a proof for $\sim N\mathfrak{S}_i$ with factual $\mathfrak{S}_i$. The decisive question remains whether this or another procedure will assure C-truth in *all* such cases.

This article has been restricted to an exhibition of some modal systems and to an explanation of some of their technical features. At another place[17] I shall discuss certain fundamental problems connected with modalities which are, so to speak, of a pre-technical nature. These discussions will not presuppose the systems but, on the contrary, try to prepare the ground for the construction of modal systems by clarifying some basic logical and semantical concepts. The problems concern, in particular, the relation of denotation and the nature of denoted entities; further, the concepts of the extension and the intension of linguistic expressions. It is customary to distinguish, for example, between the extension of a predicate (of degree one), which is a class, and its intension, which is a property. Further, we may distinguish between the extension of a sentence, which is a truth-value, and its intension, which is a proposition. I shall try to show that, in an analogous way, we have to distinguish between the extension of an individual expression (e.g., an individual description or a constant abbreviating it), which is an individual (e.g., a physical thing or a space-time point), and its intension, which is a concept of a special kind which we may call an individual concept. The distinction in this case as in the two other cases, becomes essential when the expressions occur in non-extensional contexts, e.g., in modal sentences. It will be shown that this distinction also helps to clarify and to overcome the difficulties which Quine[18] has pointed out for sentences combining modalities and quantification. Without an elimination of these difficulties, no modal functional logic could be constructed. Further, some related problems which have been raised by Church will be discussed.

The approach in the present article, which leaves aside all these fundamental problems, may appear to be uncritical and dogmatic. This appearance, however, is not due to an actual neglect of these problems but merely to the fact that for the sake of brevity this article had to be restricted to the technical aspects of modal systems.

UNIVERSITY OF CHICAGO

---

[17] *Meaning and necessity: A study in semantics and modal logic.* (To appear soon.)

[18] W. V. Quine, *Notes on existence and necessity. The journal of philosophy,* vol. 40 (1943), pp. 113–127.

On Probability and Induction

HANS REICHENBACH

IN A review of my book "Wahrscheinlichkeitslehre",[1] Dr. Ernest Nagel[2] has recently criticized some of my ideas on probability and induction. His review includes a good exposition of my ideas, and I have to thank him for his serious attempts to do justice to my results. He attacks, however, some very essential points of my theory. I may be allowed, therefore, to answer him as frankly and thoroughly as he attacks me.

There are at first some formal objections which I shall discuss briefly. These objections concern the question whether the concept of probability is to be conceived as a semantic term or as a term of "object language".[3] I did not enter into a discussion of this question earlier because it does not possess much relevance for the mathematical and epistemological purposes of my book.

The frequency interpretation opens two ways of interpretation of the probability concept. We may count either *events* or *sentences* concerning events. The result of the first counting is expressed in an object sentence, the result of the second counting in a semantic sentence. According to these two ways of determination we have two ways of

---

[1] Leiden, 1935, Sijthoff.

[2] Mind XLV, No. 180, p. 501.

[3] I use here the terms introduced by Carnap and Tarski. Object language is the ordinary language, concerning things; syntactic language concerns symbols and relations between symbols; semantic language concerns, jointly, symbols and ordinary objects, i.e. qualities of symbols in relation to facts, such as truth. The transition from a symbol to the designation of a symbol is indicated by quotation marks.

writing a probability term.   If we use the predicative conception of probability,[4] these two ways are symbolized as follows:

$$W(\varphi x_i) = p \tag{1}$$

$$W(``(\varphi x_i)") = p \tag{2}$$

(1) is a term of the object language, (2) is a semantic term.   (1) is to be read: "the probability of events x_i of the kind φ, within the series of events x_i, is equal to p".   (2) is to be read: "the probability of the propositional series '(φx_i)' is equal to p".

It is obvious that both conceptions come to the same so far as practical purposes are concerned, and also in respect to the rules of the probability system constructed.   On account of this complete isomorphism the distinction may be dropped.   This is what I did in my book.   But, although the distinction may be suppressed in the formulas we cannot contest its logical correctness.   Strictly speaking, we have to distinguish therefore (1) the *mathematical conception of probability* from (2) the *logical or semantic conception of probability*.   In the last chapter of my book, where I pass from the mathematical conception of probability to the logical conception, it is always the semantic conception which I use.[5]   The dropping of the quotation marks in the formulas of this last chapter is to be interpreted, therefore, as a convention that the function of the quotation marks is transferred to the parentheses of the symbol W( ).   But, I shall never contest the right of adding quotation marks in these formulas, to any one whose logistic inner voice will not submit to this convention.

For the probability implication, I used the form:

$$(i) \quad (\varphi x_i \underset{p}{\rightarrow} \psi y_i) \tag{3}$$

This is to be considered not as a semantic term, but as a term of the object language; it may be conceived, therefore, as a generalization of Russell's formal implication,

$$(i) \quad (\varphi x_i \supset \psi y_i) \tag{4}$$

which is also a term of the object language.   Thus, Dr. Nagel's arguments against this analogy are not tenable.   I have chosen this form for the probability implication, on account of some epistemological

---

[4] cf. Wahrscheinlichkeitslehre, p. 375.

[5] This is made clear on p. 376 where I call the probability statement an analogue of tatements of the form "a is true".

considerations: my construction of the probability implication was expressly intended to fit the application of probability concepts in physics, where it is to take the place of the relation of causality. The relation of causal connection between two events A and B is to be considered as the limiting case of a more general relation of probability connection—this was the program from which my construction started. We talk of a causal connection if A is always followed by B; we talk of a probability connection if A is followed by B sometimes only, but yet in a determinate proportion of the cases. This is the intuitive meaning of the statement that the probability implication is a generalization of the formal implication.

In spite of this there remains a difference between both implications. To the general, or formal implication corresponds the individual, or material, implication; there is, however, no such individual complement to the probability implication. I have pointed this out in a previous paper:[6] it originates from the frequency interpretation according to which there is no individual meaning of probability terms.

The relation between the formal implication and the probability implication is expressed by my axiom II, 1:

$$(i) \ (\varphi x_i \supset \psi \, y_i) \supset [(i) \ (\varphi x_i \xrightarrow{p} \psi \, y_i) \cdot (p = 1)] \tag{5}$$

According to what I said about the object character of both implications, I cannot acknowledge, therefore, Dr. Nagel's attack against this relation. In my interpretation of (5) I write that certainty is a special case of the probability (1). Dr. Nagel will not admit this interpretation since, according to him, certainty is not a concept of the object-language but, "does involve a reference to a state of belief" (p. 503). I do not think that we should use psychological terms like "belief" in the discussion of logical questions. What we should do instead, is to give an analysis of the different definitions available. Now, there are as far as I see, three definitions of "certainty" in dispute. There is first, "certainty" as an object term. If the event A is always followed by the event B, we say: if A occurs, B is certain. Let us call this the "object-sense" of "certainty". Secondly, "certainty" is used as a semantic term. In the given example, we may say: if the sentence "A occurs" is true, the sentence "B occurs" is certain. This may be called the "semantic sense" of "certainty". As both these meanings correspond

[6] Axiomatik der Wahrscheinlichkeitsrechnung, Math. Zeitschr. 34, 1932, p. 572 cf. also the new paper in Erkenntnis VII, 1937, quoted above.

and differ only in respect to the level of the language, they may be united under the common name of "extensional certainty". Thirdly, we use the word "certainty" to denote the relation of deducibility between sentences. If the sentence '*b*' is deducible from the sentence '*a*', i.e. if the syntactic structure of the sentences shows '*b*' to be a consequence of '*a*', we say "if '*a*' is true, '*b*' is certain". This may be called "syntactic certainty."

Now, it is obvious that I use the extensional sense of "certainty" in my theory of probability. In reference to the probability implication, it is the object-sense I use; within the logical conception of probability it is the semantic sense. (On p. 3F8 I use in this sense the word "necessity" which may be concieved as synonomous to "certainty".) I know there are logicians—and after his remark on "belief", I suppose, Dr. Nagel belongs to those who prefer to see the words "certainty" or "necessity" reserved for relations of the kind of the syntactic implication, i.e. for the deducibility relation. But, I do not see any reason to consider their viewpoint as the only correct one. It is true that in conversational language both significations of the term "certainty" occur; but because of this we cannot say that one of these definitions is the only permissible one.

What we can say, however, is that the difference between extensional and syntactic certainty is not relevant for practical purposes. If a physicist uses the term "certainty" in relation to events, all he wants to say is sufficiently expressed in the extensional concept of certainty. Whether the formula combining his events is tautological, or not, i.e. whether he is to apply also the syntactic concept of certainty, depends on nothing but arbitrary decisions which determine the construction of his formulas. This is to say: if the relation of extensional certainty is empirically verified for two events A and B, we are free to coordinate to A and B descriptions such that there is the relation of syntactic certainty between these descriptions. This is all. *We are free* to construct such descriptions, but we may construct also descriptions of another kind which satisfy only the postulate of extensional certainty. The word "certainty", uttered by the physicist, has as much verifiable content as is expressed in the extensional concept—this, I think, cannot be seriously questioned.[7]

With these remarks, I think, the question for the logical position of

---

[7] To avoid misunderstandings: it is, of course, verifiable whether there is the relation of syntactic certainty between two formulas. But the question *whether the events demand* the introduction of syntactic certainty *or* that of extensional certainty, is not verifiable.

the probability concept is answered. Probability may be constructed as an object term, or as a semantic term; but there is no need to introduce purely syntactic terms. We shall choose the object conception of probability for mathematical purposes, the semantic conception for for logical purposes. This distinction, however, has only a formal significance as on account of a complete isomorphism both conceptions come to the same result. Practically speaking, what stands at discussion is the question whether we shall count *events*, or *sentences* concerning events—a difference which, as to all formulas occurring, finds its only expression in the introduction of quotation marks?

This is my answer also to a remark in the review of Dr. Louis O. Kattsoff.[8]

II

I have called my logic of probability an extensional system. A remark concerning this term may be added here. The usual definition of this term reads that a relation between propositions is extensional if it depends on the *truth values* of the propositions only, whereas a relation is intensional if it depends on the *intension* of the proposition. Now, the truth tables of the probability logic constructed by me are different from those of two-valued logic: the probability value of a combination of two propositional series is not determined by the probability values of the propositional series, but is dependent on a third parameter which I have called the "degree of coupling" of the propositional series, and which is identical with the probability of one series *relative to* the other. Instead of this parameter, we may introduce also, as a third independent parameter, the probability value of *one* of the logical combinations of the propositional series, e.g. of their logical product. This peculiarity is a necessary feature of the logic of probability as may easily be shown; it corresponds also to the practical use of probability concepts.[9] A. Tarski[10] however, referring to a remark of S. Mazurkiewicz, has based on this fact the objection that my probability logic is *not extensional*. He is right if he uses the usual definition of the term "extensional" quoted above; but I might add then, that my probability logic is *not intensional* either. It seems to me, however, that this is a very narrow use of the word "extensional", and that we should reason-

[8] Philos. Review, Vol. XLV, 6, p. 625.

[9] Cf. Wahrscheinlichkeitslehre p. 386.

[10] Cf. the report on the discussion of the congress of Prague 1934, in Erkenntnis V, 1935, p. 174.

ably expand the definition of "extensional" in such a way that the dependence of the "truth value" of the combination, on further parameters of the "truth value" type, is included.   This way of an expansion of a definition, in case of occurrence of problems of a more general kind, is used throughout in mathematics.   I refer here to the use of such concepts as "the whole" and "its parts" which, for infinite classes, would lead to contradictions if the older narrower definition of the terms were retained, or to the use of the concept "sum" in vector analysis. If I have used the term "extensional" in my publications, I have of course always referred to such a wider definition.   Maybe the objections of Dr. Nagel against my claims of an extensional character for my probability concept originate from a misunderstanding of this usage of the term.   My wider definition of the term "extensional" is justified by the request to construct a system of logic which corresponds to the rules of the calculus of probability; keeping to the narrower definition would make the postulate of extensionality exclude those very generalizations of two-valued logic which are applicable to the probability concept.   Moreover, all those conditions which make the postulate of extensionality an important epistemological principle, such as the independance of extensional logic of the intension of a proposition, the possibility of basing a theory of meaning on the concept of verifiability, are valid for the wider concept of extensionality as well as for the narrower one.

I turn now to some further objections of Dr. Nagel concerning my probability logic.   As these objections are of a more fundamental character, I want to answer them in a more detailed way.

Dr. Nagel asks why I attach so much importance to the introduction of probability logic (p. 509).   I have shown, in my book that probability logic may be reduced to two-valued logic by the frequency interpretation; Dr. Nagel wants to interpret this fact as a demonstration that probability logic is dispensable, that it is at best "a more convenient mode of statement".   I may add here the remark that some Polish logicians[11] have even drawn the consequence that my probability logic is no genuine multi-valued logic, just because of its reducibility to two-valued logic.   Although these objections may seem convincing at first, they turn out to be untenable under a more profound criticism.

[11] A. Tarski in the discussion of the Congress of Prague 1934; cf. the report in Erkenntnis V, 1935, p. 174, and my answer p. 177.   J. Hosiasson, Actes du Congrès international de Philosophie scientifique, Paris 1936, IV, p. 58; my answer to this is given in my new paper quoted above, appearing in Erkenntnis 7, 1937.

A multi-valued logic is a relational system defined by truth-tables; according to the way that the truth tables are constructed, different systems of multi-valued logic result. Lukasiewicz and Tarski have developed such systems; another one is my probability logic if it is conceived as defined by my truth tables. Given a set of entities which are to be conceived as the elements of the relational system, the question is to be raised whether these entities fulfill the relations of the system. I have shown that if propositional series are chosen as such entities, and if, we consider as their coordinated truth-values, what physicists call the degree of probability, my system of probability logic is fulfilled. The systems of Lukasiewicz and Tarski may be fulfilled by other entities; but their systems, of course, cannot be interpreted by probability concepts, in as much as the calculus of probability prescribes the rules of probability in a determinate way so that only one relational system is compatible with them.

If we introduce now the frequency interpretation of probability, this may be conceived as the dissolution of the original elements into a new set of elements. We may compare this process to the introduction of new variables into a mathematical function. Let us denote a propositional series by "(a_i)," where the parentheses are to be a part of the notation; we may write then the relational system of probability logic in the form[12]

$$L[(a_i), (b_i)...]$$

Introducing the frequency interpretation means reducing the elements (a_i) to new elements a_i; in relation to these new elements we have then another logical system

$$L_o[a_i, b_i,...]$$

The system L is probability logic; the system L_o is two-valued logic.

It would be entirely mistaken, however, to infer from the possibility of such a reduction that L is to be considered as a not-genuine system of logic. Reductions of the given kind may be invented for every other logical system as well, by the construction of a suitable new set of elements; Post[13] e.g. has shown that his multi-valued logical systems are reducible in a similar way to two-valued logic. The same may be possible for the systems of Lukasiewicz and Tarski as well. We may refer

---

[12] In this formula, "(a_i)" is not a propositional series, but the designation of a propositional series.

[13] American Journal of Mathematics, XLIII, 1921, p. 182.

also to the example of non-euclidian geometries which, by the introduction of new elements, are reducible to euclidian geometry. Our system L is a genuine multivalued logic in the same sense as the systems of Bolyai and Lobatschefsky are genuine non-euclidean geometries—in spite of their reducibility to the geometry of Euclid.

This remark allows us to give a first answer to Dr. Nagel's objection. If we conceive propositional series as the elements of science, we are bound to probability logic; *for these elements*, probability logic is not a matter of convenience, but a necessary form. A transition to two-valued logic would involve a transition to other elements.

Now there are, indubitably, elements of science which have the form of propositional series. There are a great many scientific statements which do not concern single events, but classes of events; of this kind are statements containing an all-sign, such as physical laws. We know that the truth claims of statements of this kind are not fully justifiable, that we must be content if the predictions they involve turn out to be true, at least, in a great number of cases. We have, therefore, to replace these all-propositions by propositional series to which a determinate probability-value is coordinated. For this generalized conception, an all-proposition would signify the special case when the relative frequency is equal to 1 (not only for the limit, but for every term of the coordinated frequency series); a case which, however, practically does not occur. All these parts of science find their natural logical form in probability logic.

The question arises what we shall do with single-case statements; i.e. statements concerning a single event, at a definite spatio-temporal position. We might conceive them as two-valued entities, however, it turns out that such a determination remains a mere fiction, as we never are able to determine the truth-value in question. All we can do is make an assumption about the truth-value, based on observations; but we must always be ready to change our original assumption, even after the occurrence of the event.

This is due to the fact that every statement involves predictions. If we say e.g. "On February 22nd, at 12 m., Peter was in the Hagia Sofia at Istanbul", this involves the prediction that I shall not get a reliable report from another person that Peter was in the university at that time; that nobody will see Peter in New York on February 24th, etc. If, in spite of that, I should make such a contradictory observation, I should call the first statement false. It is the fact that every statement of physics, including single-case statements, is to be *controllable*, which leads to this consequence; for to be controllable means to involve predictions. As the class of these predictions is at least practically, infinite, we shall never know the truth value of the statement.

What we have instead is a probability. It may appear strange that I talk here of the probability of a single event, after having so much emphasized the frequency interpretation of probability; I can, however, refer to my book where I have shown that we may, without leaving the frequency interpretation, coordinate a probability to a single case. The single case statement is then to be conceived as a *posit* (I propose this English term for the translation of the German "Setzung"), or as a *wager*; to it belongs a *weight* (I use now this term, in German "Gewicht", instead of my original term "Beurteilung") which is determined by the probability of a propositional series of which the single-case statement in question is an element. As to these concepts, and as to the choice of the corresponding propositional series, I must refer to my book. As an illustration of these concepts, I give the following example: if I want to predict a single throw of a die, the posit "side one will occur" has the weight $\frac{1}{6}$.

We say usually, that *after* the throw has happened, its result is reported by a *true* proposition, and not by a proposition having a probability only. This is however, a schematization; what we have, strictly speaking, is a proposition with a higher probability only or in the terms used above, a posit with a higher weight. This follows from what we said about the possibility of a control, which we must acknowledge for every proposition; the proposition concerning the *observed* throw of the died makes no exception. If we talk, in this case, of a true proposition, we replace a high probability by truth; this is what I call *schematization*.

To understand the logical nature of this schematization, we must enter into a deeper analysis. We spoke of the transition from probability logic to two-valued logic performed by the frequency interpretation; there is, besides, a second transition from probability logic to two-valued logic which we must consider now.

We may divide the scale of probability, reaching from o to 1, into three sections, by introducing "partition values" p_1 and p_2 such that p_1 is a bit superior to o and p_2 a bit inferior to 1. The probability (or weight) of the statement "a" may be denoted by $W(a)$[14]; we apply then the following definitions:

if $W(a) \geqq p_2$, a is called *true*
if $W(a) \leqq p_1$, a is called *false*
if $p_1 < W(a) < p_2$, a is called *indeterminate*

[14] According to my remark above I drop the quotation marks.

We call this transition a *transition by trichotomy*. If we omit proposi-tions of the medium domain, as being indeterminate, the rest of the propositions may then be considered as two-valued entities. The choice of the partition values p_1 and p_2 is, of course, a matter of convention.

This is the second transition from probability logic to two-valued logic we spoke of; but it is obvious that this procedure leads only to an approximate validity of two-valued logic. I have explained this elsewhere.[15] The truth-tables of two-valued logic are then not thor-oughly valid; e.g. if "*a*" and "*b*" are both true propositions, according to the given definition, their conjunction "*a and b*" may be false. This is because, if $W(a) \geqq p_2$ and $W(b) \geqq p_2$, the arithmetical product of both values (which determines the probability of the conjunction) may be $< p_2$. Only the operation of negation remains correct if p_1 and p_2, as chosen, are symmetrical in respect to o and 1.

These are the reasons why the usual conception of science as a system of two-valued propositions is not correct, but valid only in the sense of an approximation. Scientific statements are, either propositional series, or posits having a coordinated weight; their logical system is probability logic.

Jointly with these reflexions, we recognize why the frequency inter-pretation cannot release us from the use of probability logic. The reduction of the degree of probability to a number determined by the frequency of two-valued propositions, is possible only if we *have* such propositions; what we really have, however, are posits. This reduction, therefore, cannot free the resulting two-valued system from an ap-proximative character either. What is actually performed by the fre-quency interpretation is the reduction of the determination of the prob-ability of one statement to the determination of the probability of other statements, with the following advantage: the latter probabilities need not be exactly determined, but are classified within the crude scheme of trichotomy only. In this way, the determination of a degree of probability may be reduced to the application of crude appraisals of other probabilities.

This is a typical approximative method. The analysis of its struc-ture shows that the procedure of science cannot be conceived within the frame of two-valued logic, but demands the wider frame of prob-ability logic.

There is a second reason why the frequency interpretation cannot lead us to an exact validity of two-valued logic. It is connected with

---

[15] Actes du Congrès de Philosophie Scientifique 1935, Paris 1936, IV 26.

the impossibility of knowing the value of the limit of an infinite series, if this series is not intensionally given, as in mathematics, but extensionally given, as physical series are. I have shown in my book, how this problem of the determination of the limit leads into probability logic (§ 68); the statement about the frequency is, as well as the original statements, to be conceived as a posit. Thus, there are two reasons why the transformation of probability logic into two-valued logic by the frequency interpretation cannot be strictly performed: first, the new *elements* introduced are not capable of a strict determination of their truth value, and secondly, the *statement* about the frequency is not capable of a strict determination of its truth-value. The transition from probability logic into two-valued logic by the frequency interpretation is a *theoretical* possibility only; it cannot be *actually* performed, at least not strictly, because the presuppositions of this transition are not fulfilled. Science consists, not of a system of *two-valued propositions*, but of a system of *posits possessing a coordinated weight*: their logic is probability logic. Only approximatively this logical system can be reduced to two-valued logic, by trichotomy and the frequency interpretation; for exact epistemological inquiries, however, the probability character of the logical system cannot be neglected.

A last remark is to be added here. Dr. Nagel reproves me for a "serious mistake" because I "regard 'true' as a special case of 'probable'" (p. 510). He insists that "true" refers to "a relation between a 'proposition' and a 'fact'",[16] and he condemns me because he thinks that my term "probable" cannot be employed in this way. This objection, I think, is settled by my explanation above. Of course, if I call truth a special case of probability, it is the semantic concept of probability I use, not the object-concept; the analogy between truth and probability therefore, can be fully carried through, if I use the predicative conception of probability. Truth is a relation between a proposition and a fact; probability is a relation between a propositional series and a series of facts. If the propositional series is finite, my definitions hold as well; thus if the propositional series reduces to the length n = 1, probability specializes to truth—this is what I explain in § 72 of my book. This is the first interpretation of my saying that truth is a special case of probability.

There is, besides, a second one, which is of greater importance. As

---

[16] I put here the quotation marks beside the terms "proposition" and "fact" because Dr. Nagel does so; I think it however more correct to drop them. We say e.g.: "attraction is a relation between a magnet and a piece of iron"; and not: "attraction is a relation between 'a magnet' and 'a piece of iron'"

I showed above, what we take as truth is, strictly speaking, a high weight. Thus I may say: what we actually use as truth is a special case of probability. This is not to say that the concept of high weight has exactly the same logical structure as the concept of truth, but that it takes its logical place. A weight is a relation between a proposition and a series of events, and therefore has a structure somewhat different from that of truth. But as we never have truth, we use a high weight · as its representative. Thus my statement about scientific truth as a special case of probability may be interpreted: instead of truth, we always use a representative which is a special case of probability.

These are, in short, the reasons why probability logic cannot be dispensed with in science, and why we must consider the truth theory of knowledge as an approximation which, for a more precise analysis, is to be replaced by a probability theory of knowledge.

III

I turn now to the objections made by Dr. Nagel against my application of the concept of probability to scientific theories.

I have developed the idea that the concept "probable" occurring in reference to scientific theories is not different from the concept "probable" in mathematical statistics. I cannot enter here into an exposition of the reasoning which leads me to this conception, but must refer to my book; I will only summarize the main arguments of my theory in the following two statements:

1) The first is that for every proposition containing the term "probable", a coordinated series may be constructed the frequency of which corresponds to the probability of the proposition. If the proposition concerns a single case, the frequency within the coordinated series may be regarded as the probability of the proposition if the latter is conceived as a *posit*, and the probability as its *weight*.

2) The second is that there is no difference, on principle, between an observation proposition and a scientific theory as far as their truth character is concerned; both are claimed to be true in respect to some facts, but can never be strictly verified. There is only a difference of degree.

Dr. Nagel asks me whether my conception is to be "an analysis of the meaning which the phrase 'probability of a theory' does in fact have, or as a proposal to attach a certain meaning to the phrase".

To this my first answer is that every exact definition of a term previously used in an uncritical way, has the character of a proposal. The author who introduces the exact definition wants to propose that his definition be used in the future—no doubt about that.

This does not exclude, however, the possibility that the definition given can be considered equally well as an analysis of the meaning which the term used up to that time "does in fact have". It is only the wording of the last phrase which I must alter a bit, if I want to show this; for the wording given by Dr. Nagel seems to me to suffer from some indistinctness.

What is the meaning which a term "in fact has"? Is it given by the images or representations which a man actually thinks of when he uses the term? The consequence of such a definition would be that the same term would have different meanings for everybody, because everybody combines different representations with it. If we want to construct an objective meaning of a term we must proceed otherwise. We must ask which meaning we are to ascribe to the term, *if the usual usage of the term is to be justifiable*. I may be allowed to illustrate this by an example. Most people use the term "length of a line" without being able to give an exact definition of it; we should certainly not find a good definition of this term if we should take a kind of average of all opinions uttered by a great number of people concerning the meaning of this term. We may however say: if we interpret "length of a line" as the ratio of this line to another line called "unit of length", we obtain a meaning which, if underlaid to the use of the term in the mouth of most people, makes their statements *justifiable*. This, I think, is the only way to construct an "objective meaning".

If Dr. Nagel agrees to this interpretation of his phrase, I claim for my conception of the probability of theories *objective meaning*, or the meaning which this term "does in fact have". For I can show that if my interpretation is adopted, all our attitude in respect to scientific theories can be justified; the way that a theory is constructed, the fact that we make it a basis of actions, the differences of belief we attach to theories of different probabilities, etc. That is all I want to show.

I do not claim to encompass, with my theory, the opinions of all physicists, or even of a greater part of them. Here I may be allowed to intercalate a remark which, for an empiricist, looks almost like sacrilege; but to which a long experience of intercourse with physicists has led me. If we want to analyse the meaning of fundamental concepts in physics, such as the "probability of a theory", I think, we should not ask physicists what *they mean* by this concept; we should ask them what *they do* when they use the concepts.   As to the question of the meaning of the term "probability of a theory", I confess, the opinion of physicists seem to me, in general, not to be better than those of ordinary people as to the meaning of the term "length of a line".   All may be excellent men *in applying* the concept in question; this does not involve, however, that they are able to give a satisfactory *definition* of it.   Clarifying concepts demands another mentality than handling them; this is why this task should be left to the philosophers of science.   I do not pretend, with this, a kind of superiority on the side of philosophers; on the contrary, I know pretty well that it is just this being focussed on the clarification of concepts which prevents philosophers from doing a good positive work in scientific research.   But clarifying concepts is another positive work, and this should be left to the specialists of analysis.   This is why I do not approve of the authority of men of science as far as the meaning of concepts such as "probability of a theory" is concerned.

Dr. Nagel writes: "Eminent men of science repeatedly assert that a theory is found satisfactory by them partly on esthetic grounds, partly because they know no alternative theory, and partly because the consequences of the theory have been tested in accordance with a definite technique".   (p. 508).   This summary of opinions of men of science about their own methods justifies, I think, my sceptical attitude towards epistemological remarks from this side; it shows, I think, that a philosopher should carefully avoid asking a man of science why he believes in his theories.   What we obtain by such an inquiry is a kind of religion for personal use, but not a philosophic argument.   To analyse Dr. Nagel's quotation, let us take up first the esthetical argument which indeed plays a great rôle in the epistemological writings of certain physicists.

It may be true that a physicist believes in his theory because he thinks it to satisfy esthetic standards; but I do not see any reason why *we* should believe in predictions which are based on esthetic arguments; or why a *technician* should do so.   I do not see any relation between esthetic qualities and predictional qualities—and the latter are what a good theory must have.   The beauty and harmony of a theory is a matter of taste; it should be easy to construct theories of an extreme beauty which are obviously false.   Fortunately, scientific theories are better, mostly, than the epistemology combined with them; the esthetic taste of great physicists coincides in an astonishing way, with the postulates of the principle of induction.   I should recommend, therefore, that no attempts be made to dissuade physicists of this type from their esthetic criterion of scientific truth; they may need it for their personal work—as well as, say, a successful statesman may need the belief that he is sent by God to do his work. But I cannot accept the esthetic argument as anything connected with the validity of scientific theories in an objective sense; i.e. as an argument which makes the acceptance of a theory justifiable.

Surely, a theory with esthetic qualities *may*, later, turn out as true; we cannot deny this. But that is not a sufficient reason to accept the theory. If any one foretells, from the grounds of my coffee, that I am to make a great deal of money in the near future, this may later turn out to be true—in spite of this possibility, I shall not trust this prediction, even though the coffee-grounds theory may have admirable qualities from the esthetic point of view. I do not trust it because there is no inductive relation which leads from the coffee grounds to my making money. There are physicists who are not able to offer better reasons for their predictions than the coffee prophet; but if they are good physicists, there *will be* better reasons—although they themselves do not *know* them. It is the task of the philosopher to show, by analytical methods, the inductive relations which justify a good hypothesis in respect to observed facts. If we should not be able to do this and were to remain with an esthetic theory of the construction of scientific theories, science would not be better than coffee-grounds prophecy.

I may be allowed to tell here a little story which shows that physicists turn out to be much better philosophers when, instead of their own theories, those of a colleague are under discussion. When Schrödinger first published his famous formula of wave mechanics, he presented it within a context full of esthetic and even moral "justifications". He referred to an analogy with the transition from geometrical optics to wave optics; he developed an interpretation of his wave function which was dropped soon after his first publications; he deemed his theory necessary if a "complete surrender of arms" (eine vollständige Waffenstreckung) were to be avoided[17]—in short, he expounded all the arguments which, indeed, had had a great influence on his mind during the process of creation of his theory, but which cannot be regarded as the logical justification of the theory. At that time, Einstein—whom Dr. Nagel quotes against me in a remark concerning his own theories—was

[17] Annalen d. Physik 79, 1926, p. 509.

asked what he thought of Schrödinger's publication. "I skipped the novel" answered Einstein; "the formula is excellent". I do not relate this story with the intention of suggesting to the physicists that they drop the psychological motivation in their publications; these parts of their writings have a high psychological and personal interest, especially if they are written in the charming style of Schrödinger's publications. What I want to say is that arguments of this kind should not be forwarded in a philosophic discussion about the probability of theories. Within the frame of analysis of science, we should always "skip the novels".

If we want to construct a philosophy of science, we have to distinguish carefully between two kinds of context in which scientific theories may be considered. The *context of discovery* is to be separated from the *context of justification*;[18] the former belongs to the *psychology of scientific discovery*, the latter alone is to be the object of the *logic of science*. The confusion of the two kinds of context has become the root of many a misinterpretation of the procedure of science. I confess that the remarks of Peirce concerning the construction of scientific theories, quoted against me by Dr. Nagel, seem to me to suffer from the same confusion. I admire Charles Peirce as one of the few men who saw the relations between induction and probability at an early time; but just his remarks concerning what he calls "abduction" suffer from an unfortunate obscurity which I must ascribe to his confounding the psychology of scientific discovery with the logical situation of theories in relation to observed facts.

I may intercalate here the remark that the distinction propounded is not confined to inductive thinking, but occurs as well for purely deductive operations of thought. If we are confronted by a mathematical problem, say the construction of a triangle from three given parameters, the solution (or the class of solutions) is entirely determined by the given problem. If any solution is presented to us, we may decide unambiguously, and with the use of deductive operations alone, whether or not it is correct. The way in which we find the solution, however, remains to a great extent in the darkness of productive thinking, and may be influenced by esthetic considerations, or a feeling of "geometrical harmony". Nobody would here, in spite of this psychological fact, propound as a philosophical theory that the solution of geometrical problems is determined by esthetic points of view. The *objective re-*

---

[18] I have used these terms already in a discussion on the Congress of Prague 1934; cf. Erkenntnis V, 1935, p. 172.

lation from the given entities to the solution, and the *subjective* way of finding it, are clearly separated for problems of a deductive character; we must learn to make the same distinction for the problem of the inductive relation from facts to theories.

It is, therefore, the context of justification which we have to consider when we raise the question of the probability of scientific theories. Returning now to Dr. Nagel's quotation, I find that only the third argument presented there belongs to this context, and this argument is of the inductive type. To accept a theory because its consequences have been confirmed by experience is a procedure based on induction; to it corresponds, in the calculus of probability, the so-called rule of Bayes, which permits calculation of the probability of an assumption as a function of its observed implications. Fortunately this argument plays the decisive rôle in the discussion of physicists about their theories; esthetic arguments disappear with a surprizing quickness if inductions based on experiments are against them.

If the inductive relation from the observed consequences to the theory is once admitted, it cannot be denied that there is likewise an inductive relation which supports the theory *before* the consequences are tested. The situation of the theory in respect to facts *before the experimental test* is not different, in principle, from that *after the experimental test*. In both cases there are facts which do not *verify* the theory, but which confer a determinate probability on it; this probability may be small before the test, and great after it. But even before the test there must be facts on which the theory is based; and there must be, also before the test, a net of inductive relations leading from the facts to the theory—else the theory could not be seriously maintained. The net of relations before the test may be of a very complicated kind, whereas the relations from the observational material of the test to the theory may be of a rather simple kind; but in both cases they must be of the inductive type if the adoption of the theory is to be justifiable.

Dr. Nagel states, as the second argument of the given quotation, the idea that theories are sometimes adopted because the physicists do not know any other theory. If this is to be a quotation from opinions uttered by physicists, it would show once more that it is a rather dangerous matter to believe in epistemological remarks of physicists without further interpretation. The physicist who says that he does not know another theory, lets drop the tacit qualification: "of a kind satisfying inductive rules". It is surely easy to construct a lot of theories for any given set of observations; but most of them are immediately rejected because they do not correspond to inductive rules. The argument in question, therefore, does not apply to the determination of theories in general, but only to the *differential decision*—a problem which we are to deal with later. In any case the argument presupposes the acknowledgement of inductive rules for the establishment of a theory; for without the additional tacit assumption named the argument could not be seriously maintained.

Dr. Nagel may object that he does not talk only of the *opinions* which men of science have about their theories; he wants to maintain besides that the procedure of scientific research is not conformable to a probability theory of hypotheses. He writes: "a careful search of scientific treatises reveals that the probability of theories is not discussed in them, and that in any case no procedures for calculating the probability of theories in the sense specified by Prof. Reichenbach are employed" (p. 508). If this were true, it would be indeed a serious objection against my conception, as this argument is based on the sound intention to derive analysis of science from what physicists *do*, and not from what they *think*. I must answer, however, that just this argument of Dr. Nagel's seems to me to be refuted by every search of scientific treatises which *deserves* the epithet "careful".

For we must be *careful* indeed, in this domain, as the usage of language is frequently built up of concepts which do not openly reveal their logical origin to be in the sphere of probability concepts. The process of schematization which presses knowledge into the frame of two-valued logic, has introduced into this field a great many characteristics which appear remote from a probability theory; we cannot expect, therefore, to find probability considerations other than in a hidden form. If we look, however, with the eyes of the theorist of probability, we shall discover probability relations in all parts of the field of scientific research. We find that hypotheses are to be "simple"; upon a deeper analysis, this concept "simple" turns out to be a typical probability concept, derivable from the inductive principle. We find that hypotheses are not classified as true or false, but are ranked within a scale characterized by such concepts as "likely", "presumably", "surely", etc., thus obviously corresponding to the scale of probability. We find that hypotheses are combined, and that combined hypotheses are considered as supporting each other—a typical form of a probability inference. We find that an increase in the number of confirmations is considered as an increase in the "weight" of the theory—a procedure justifiable only for a probability conception of knowledge. We find that scientific theories are employed as bases of actions, e.g. for technical purposes; a usage which can be justified if the weight of the theory is interpreted as a probability—whereas all other kinds of interpretation return no answer to this question. I could continue this list to a considerable length if it were necessary.[19]

On account of the schematization of two-valued logic, probability inferences appear frequently in a stunted form. Some probabilities occurring in the full logical structure of the inference may be suppressed because they are very near to 1 or 0; in this way, the form of the probability inference may become unrecognizable. The inference of the detective e.g., from observed traces of a murder to the perpetrator, is a stunted form of the rule of Bayes; so is also the inference from physical observations to the validity of a hypothesis, as we mentioned above. I have used, in my book "Wahrscheinlichkeitslehre", a great many examples of the "qualitative" type which may show the occurrence of probability structures in inferences of every day life, or of non-mathematical sciences; they may serve as illustrations of what I mean by the "hidden" probability structure of inferences which do not openly appear in the form of operations of mathematical statistics.

Whereas the "qualitative" treatment of theories contains the probability structure only in a hidden form, this structure becomes obvious at the moment when questions of a greater exactness are discussed: I mean the application of the theory of errors to scientific theories. It has become usual to omit this subject from the philosophical discussion of theories, as a "secondary matter"; this is one of the serious mistakes in the traditional discussion of the problem. There is no sharp limit between a *qualitative* and a *quantitative* statement of a theory; if we pass, however, to statements of the quantitative type, the schematization of two-valued logic is no longer applicable, and the inferences occurring show their full probability structure. If we say: "the velocity of light is 299796 km/sec, with an average error of ± 4, or of $\pm 0.0015\%$",[20] this means according to well-known rules of mathematical statistics:[21] "the probability that the velocity of light lies between 299792 km/sec and 299800 km/sec, is $\frac{2}{3}$". The hypothesis that the velocity of light is a constant, is a statement of a somewhat

---

[19] I have given some examples of this kind at another place. cf. Actes du Congrès International de Philosophie scientifique Paris 1935, IV, 6 and 28.

[20] A. A. Michelson, Astrophys. Journ. 65, 1, 1927.

[21] Cf. the author's Wahrscheinlichkeitslehre, p. 226.

more complicated structure. It is confirmed by the measurement on which our given numbers are based, in so far as these measurements were made at different times, and thus (on account of the motion of the earth) in different spatial directions. (The famous so-called Michelson experiment, though based on measurements different from those used for our number, proves the same point but to a much higher degree of exactness.) The general hypothesis differs from the statement concerning the numerical value in that it states less; it may be true even if the numerical value named is false. It can, however, not be freed from the indication of a numerical width of the constant, as an "absolute constant" is not a meaningful physical concept; what can be meaningfully stated is only that the velocity of light is a constant within determinate narrow limits.[22] Thus these limits enter as an arbitrarily eligible factor into the statement. As the transition from the special value of the constant, to the postulate of constancy in general, involves an increase of probability, and as the influence of other arguments in favour of Einstein's assumption signifies an increase as well, we may infer from the statement above: "the probability of the hypothesis that the velocity of light is a constant within an exactness of 0.0015%, is greater than two-thirds". If Dr. Nagel deems a probability as low as two-thirds incredible for a theory as universally agreed to as Einstein's assumption, (I must infer this from an example given by him where even the value nine tenth seems to him to be too low) he may easily enlarge the probability of Einstein's hypothesis by admitting a somewhat greater width of the constant. A simple calculation making use of the qualities of the Gauss function furnishes the result that the probability of Einstein's hypothesis concerning the constancy of the velocity of light is greater than 99.99%, if a width of 0.0052% is admitted for the "constant". This may serve as an example of a case where at least an inferior limit of the probability of the first level may be numerically calculated.[23]

We see that it is erroneous to say that the methods of scientific

[22] This remark needs some qualification. We could eliminate the numerical width of the constant by passing to a statement about a limit occurring for intervals of a width converging towards zero, and might interpret this as a statement about an absolute constant. As the formulation of this statement, however, presupposes other statements admitting a numerical width of the constant, and as statements of this kind suffice for all purposes of physics, we content ourselves here with a statement of this type.

[23] In Erkenntnis 5, 1935, p. 275. I have used for this the term "Wahrscheinlichkeit erster Form". I must refer to this paper for an explanation of the different types of probabilities occurring for theories.

research do not include methods for the calculation of the probability of theories. Whether we actually discover these latter methods within the net-work of scientific proceedings depends only on our ability to see them.

As to probabilities of the second level, we cannot yet determine numerical values. Dr. Nagel objects that we are here placed before a difficulty of principle because we do not know into which class the theory is to be incorporated if we want to determine its probability in the sense of a frequency; if we want to determine e.g. the probability of the second level of the quantum theory, shall we consider the class of scientific theories in general, or only that of physical theories, or only that of physical theories in modern times? I do not think that this is a serious difficulty as the same question occurs for the determination of the probability of single events; I have indicated in my book (p. 391) the way to proceed in such cases. The narrowest class available is the best; but it must be large enough to afford reliable statistics. If the probability of theories (of the second level) is not yet amenable to a quantitative determination, the reason is to be found, I think, in the fact that in this field we have no statistics on a sufficient number of *uniform* cases. In other words: if we use a class of not too small a number of cases, we may easily indicate a sub-class in which the probability is considerably different. We know this from general reflexions, and thus do not try to secure statistics. In the future, more adequate statistics may perhaps overcome these difficulties, as well as the similar difficulties with meteorological statistics have been overcome. As long as we have no such adequate statistics, crude appraisals will be used in their place—as is the practice in all fields of human knowledge which are not yet accessible to better quantitative determinations. Appraisals of this kind, i.e. concerning the probability of the second level of a theory, may acquire practical importance in cases when we judge a theory by the success obtained with other theories in that domain; if e.g. an astronomer propounds a new theory of the evolution of the universe, we hesitate to trust this theory on account of bad experiences with other theories of that kind.

Dr. Nagel adds a remark which he considers as a *logical* difficulty. If we secure statistics on single events, such as the throw of a die, we are to count true or false statements in order to obtain a probability. Such a procedure however—thus runs Dr. Nagel's objection (p. 507)— is not feasible for a statistics of scientific theories, because we do not know whether the individual theory is true or false; the statistics proposed by me, Dr. Nagel argues, would therefore be logically impossible. This objection reveals one of the erroneous consequences to which a theory of knowledge leads if it does not introduce the concept of probability from the very beginning. Dr. Nagel overlooks the fact that the "truth" of a proposition concerning an event is obtained by a schematization consisting of a transition by trichotomy from a "weight" to "truth". This transition is possible, however, for statements concerning scientific theories as well as for statements concerning a die. All statistics are, therefore, based on counting "posits", and not "true sentences" (I have shown this in §77 of my book); therefore the statistics of theories proposed by me are as feasible as any other statistics. The numerical determination of probabilities by frequencies is not restricted to probabilities of the first level.

I cannot enter here into a more detailed exposition of my conception of the probability of scientific theories. Dr. Nagel asks me to write a more detailed report on my opinions concerning the probability of theories; this desire will be fulfilled as I shall soon publish a new book[24] which will contain, within a comprehensive exposition of the probability theory of knowledge, a closer analysis of this subject. I do not share his opinion, however, that I should have shortened the mathematical parts of my book "Wahrscheinlichkeitslehre". The exact demonstration that the whole of physics can be built up by the inductive rule is contained in the demonstration that the whole calculus of probability (including all applications) can be built up by the inductive rule without any other presuppositions. This is the reason why the mathematical exposition had to be given at such length.

I will briefly summarize my view of the probability of theories. By the application of the rules of probability in a complicated concatenation, observed facts determine coordinated scientific theories with different "weights", or probabilities. The situation *before* an experimental confirmation of a theory is not different, in principle, from the situation *after* such confirmation because in both cases the theory is prescribed by the observations, not unambiguously, but only with a determinate probability; the weight or probability of the theory, however, is increased by a confirmation.

I may add here the discussion of two further questions raised by Dr. Nagel. The first concerns the question of the conventional element in the choice of theories; Dr. Nagel believes that I "overlook the

[24] Experience and Prediction. Chicago University Press, 1937.

large amount of 'convention' (p. 512) which enters into the determination of theories. He then sketches some ideas according to which the deviation from facts, admitted for a theory, is determined by practical purposes.

It is certainly true that this admissible deviation is, logically speaking, a matter of convention; but that does not free us from probability considerations. Whether a deviation observed *will remain* within the boundaries of the deviation admitted, can be determined with probability only; the question, therefore, which is to be answered, from the view point of practical applications of a theory, is in the form: with what probability may we expect that future applications of the theory will furnish results remaining within the practically admissible boundaries of deviation? The convention considered by Dr. Nagel has, therefore, the logical character of an arbitrary parameter without which the question for the probability cannot be formulated; the numerical width to be tolerated for the value of the velocity of light in our example is of this type. The problem of the probability of theories, however, can by no means be eliminated by the consideration of these conventions which concern, not truth, but the formulation of the proposition in question.—The only conventional element in scientific truth is the choice of the partition values p_1 and p_2, occurring in the transition by trichotomy described above; they determine the reliability of truth and may vary according to practical purposes.

The second question raised by Dr. Nagel is whether I believe that *only one* theory is in accordance with observations, as indicated by the inductive rule. The answer is easily to be given if we keep to the principles of the calculus of probability. A set of observations determines not simply *one* theory, but a class of theories, each of them with a specific weight. We may ask whether there is a maximum of the weight; the answer to this, however, cannot be given uniformly. Sometimes there is, and sometimes there is not. There may also be so flat a maximum that we are practically unable to choose between certain theories, whereas others can plainly be crossed out as too improbable. We may call this the case of *differential decision*, analogous to the *differential diagnosis* of physicians (which is, logically considered, a special case of our problem).[25]

It is this case which is usually considered by all those who attack

---

[25] By differential diagnosis the physicians understand a case where the observed symptoms of illness indicate several diseases as their possible origin, but do not permit a decision among the members of this group unless certain new symptoms can be observed.

the idea of an inductive determination of theories. They emphasize the impossibility of an inductive choice among the small group of theories left to us, and do not see the strong inductive prescriptions which exclude all the other ones. We cannot demand, however, that the rules of probability always determine one choice as the best; it may happen that several possibilities are left with equal chances. It would be entirely erroneous, however, to infer that there are no rules of probability at all valid in this game.

The situation of the differential decision opens the way for decisions based on motives other than of the inductive form; for the application of esthetic motives, or economic motives or what else we want. We must not forget, however, that the narrower decision, in such cases, is not justifiable as determining a *better prediction*, as is a decision based on inductive principles. In the situation of Buridan's ass, we may allow the poor animal to choose the bundle of hay with the esthetically preferable form. If two urns are presented to us, one of which is to contain a lump of gold, we may make our choice by counting the buttons of our jacket—we should not pretend, however, that this procedure involves a method of higher intuition. The probability conception of theories leaves open the possibility of *indifference situations*—no doubt as to that. In such a situation, it is better to choose *one* theory, than *none*; if then we need an external impulse to make up our mind, we may use it.

It is this situation of differential decision in which the result of a subsequent test of the consequences involved by the theory may become decisive. The enlargement of the observational material by the subsequent experiment may be of such a kind that the more comprehensive material leads to the indication of one theory as the most probable one, whereas the original material left equal probabilities for several theories. We recognize here the origin of the conception which makes a difference in principle between the situation before and after the test experiment; which denies any determination of the theory by the original facts on which its first establishment is based, but acknowledges a determination by a confirmation of the consequences involved in the theory. This conception is nothing but a schematization of a special situation which finds its natural place in a probability conception of theories; this schematization is, unfortunately, made in such a way that the inductive determination of theories by facts is dropped. It becomes apparent, however, in the moment we consider not only the group of theories for choice in the differential decision, but compare this whole group with a wider class of theories.

This analysis may point out the unfortunate consequences arising from a theory of knowledge which presses knowledge into the frame of two-valued logic. A number of essential features of knowledge are dropped by such a procedure, and the way towards an understanding of the method of scientific prediction is barred. This is the reason why I deem a probability theory of knowledge to be the only solution of the problem of the logical construction of science.

Istanbul, Turkey.

THE TWO CONCEPTS OF PROBABILITY

I. THE PROBLEM OF PROBABILITY

The problem of probability may be regarded as the task of finding an adequate definition of the concept of probability that can provide a basis for a theory of probability. This task is not one of defining a new concept but rather of redefining an old one. Thus we have here an instance of that kind of problem—often important in the development of science and mathematics—where a concept already in use is to be made more exact or, rather, is to be replaced by a more exact new concept. Let us call these problems (in an adaptation of the terminology of Kant and Husserl) problems of *explication;* in each case of an explication, we call the old concept, used in a more or less vague way either in every-day language or in an earlier stage of scientific language, the *explicandum;* the new, more exact concept which is proposed to take the place of the old one the *explicatum.* Thus, for instance, the definition of the cardinal number three by Frege and Russell as the class of all triples was meant as an explication; the explicandum was the ordinary meaning of the word 'three' as it appears in every-day life and in science; the concept of the class of all triples (defined not by means of the word 'triple' but with the help of existential quantifiers and the sign of identity) was proposed as an explicatum for the explicandum mentioned.

Using these terms, we may say that the problem of probability is the problem of finding an adequate explication of the word 'probability' in its ordinary meaning, or in one of its meanings if there are several.

II. THE LOGICAL CONCEPTS OF CONFIRMATION

In the preparation for our subsequent discussion of the problem of probability, let us examine some concepts which are connected with the scientific procedure of confirming or disconfirming hypotheses on the basis of results found by observation.

The procedure of confirmation is a complex one consisting of components of different kinds. In the present discussion, we shall be concerned only with what may be called the logical side of confirmation, namely, with certain logical relations between sentences (or propositions expressed by these sentences). Within the procedure of confirmation, these relations are of interest to the scientist, for instance, in the following situation: He intends to examine a certain hypothesis h; he makes many observations of particular events which he regards as relevant for judging the hypothesis h; he formulates this evidence, the results of all observations made, or as many of them as are relevant, in a report e, which is a long sentence.

513

Then he tries to decide whether and to what degree the hypothesis h is confirmed by the observational evidence e. It is with this decision alone that we shall be concerned. Once the hypothesis is formulated by h and the observational results by e, then this question as to whether and how much h is confirmed by e can be answered merely by a logical analysis of h and e and their relations. Therefore the question is a logical one. It is not a question of fact in the sense that knowledge of empirical fact is required to find the answer. Although the sentences h and e under consideration do themselves certainly refer to facts, nevertheless once h and e are given, the question of confirmation requires only that we are able to understand them, i.e., grasp their meanings, and to discover certain relations which are based upon their meanings. If by semantics[1] we understand the theory of the meanings of expressions, and especially of sentences, in a language then the relations to be studied between h and e may be regarded as semantical.

The question of confirmation in which we are here interested has just been characterized as a logical question. In order to avoid misunderstanding, a qualification should be made. The question at issue does not belong to deductive but to inductive logic. Both branches of logic have this in common: solutions of their problems do not require factual knowledge but only analysis of meaning. Therefore, both parts of logic (if formulated with respect to sentences rather than to propositions) belong to semantics. This similarity makes it possible to explain the logical character of the relations of confirmation by an analogy with a more familiar relation in deductive logic, viz., the relation of logical consequence or its converse, the relation of L-implication (i.e., logical implication or entailment in distinction to material implication). Let i be the sentence 'all men are mortal, and Socrates is a man', and j the sentence 'Socrates is mortal'. Both i and j have factual content. But in order to decide whether i L-implies j, we need no factual knowledge, we need not know whether i is true or false, whether j is true or false, whether anybody believes in i, and if so, on what basis. All that is required is a logical analysis of the meanings of the two sentences. Analogously, to decide to what degree h is confirmed by e—a question in logic, but here in inductive, not in deductive, logic—we need not know whether e is true or false, whether h is true or false, whether anybody believes in e, and, if so, whether on the basis of observation or of imagination or of anything else. All we need is a logical analysis of the meanings of the two sentences. For this reason we call our problem the logical or semantical problem of confirmation,

---

[1] Compare Alfred Tarski, "The Semantic Conception of Truth and the Foundations of Semantics," this journal, vol. IV (1944), pp. 341–376; and R. Carnap, *Introduction to Semantics*, 1942.

in distinction to what might be called the methodological problems of confirmation, e.g., how best to construct and arrange an apparatus for certain experiments in order to test a given hypothesis, how to carry out the experiments, how to observe the results, etc.

We may distinguish three logical concepts of confirmation, concepts which have to do with the logical side only of the problem of confirmation. They are all logical and hence semantical concepts. They apply to two sentences, which we call hypothesis and evidence and which in our example were designated by "h" and "e" respectively. Although the basis is usually an observational report, as in the application sketched above, and the hypothesis a law or a prediction, we shall not restrict our concepts of confirmation to any particular content or form of the two sentences. We distinguish the positive, the comparative, and the metrical concepts of confirmation in the following way.

(i) *The positive concept of confirmation* is that relation between two sentences h and e which is usually expressed by sentences of the following forms:

"h is confirmed by e."

"h is supported by e."

"e gives some (positive) evidence for h."

"e is evidence substantiating (or corroborating) the assumption of h."

Here e is ordinarily, as in the previous example, an observational report, but may also refer to particular states of affairs not yet known but merely assumed, and may even include assumed laws; h is usually a statement about an unknown state of affairs, e.g., a prediction, or it may be a law or any other hypothesis. It is clear that this concept of confirmation is a relation between two sentences, not a property of one of them. Customary formulations which mention only the hypothesis are obviously elliptical; the basis is tacitly understood. For instance, when a physicist says: "This hypothesis is well confirmed," he means ". . . on the evidence of the observational results known today to physicists."

(ii) *The comparative* (or topological) *concept of confirmation* is uusally expressed in sentences of the following forms (a), (b), (c), or similar ones. (a) "h is more strongly confirmed (or supported, substantiated, corrobo rated etc.) by e than h' by e'."

Here we have a tetradic relation between four sentences. In general, the two hypotheses h and h' are different from one another, and likewise the two evidences e and e'. Some scientists will perhaps doubt whether a comparison of this most general form is possible, and may, perhaps, restrict the application of the comparative concept only to those situations where two evidences are compared with respect to the same hypothesis (example (b)), or where two hypotheses are examined with respect to one evidence (example (c)).   In either case the comparative concept is a triadic relation between three sentences.

(b) "The general theory of relativity is more highly confirmed by the results of laboratory experiments and astronomical observations known today than by those known in 1905."

(c) "The optical phenomena available to physicists in the 19th century were more adequately explained by the wave theory of light than by the corpuscular theory; in other words, they gave stronger support to the former theory than to the latter."

(iii) *The metrical* (or quantitative) *concept of confirmation*, the concept of *degree of confirmation*.   Opinion seems divided as to whether or not a concept of this kind ever occurs in the customary talk of scientists, that is to say, whether they ever assign a numerical value to the degree to which a hypothesis is supported by given observational material or whether they use only positive and comparative concepts of confirmation.   For the present discussion, we leave this question open; even if the latter were the case, an attempt to find a metrical explicatum for the comparative explicandum would be worth while.   (This would be analogous to many other cases of scientific explication, to the introduction, for example, of the metrical explicatum "temperature" for the comparative explicandum 'warmer', or of the metrical explicatum 'I.Q.' for the comparative explicandum 'higher intelligence'.)

III. THE TWO CONCEPTS OF PROBABILITY

The history of the theory of probability is the history of attempts to find an explication for the pre-scientific concept of probability.   The number of solutions which have been proposed for this problem in the course of its historical development is rather large.   The differences, though sometimes slight, are in many cases considerable.   To bring some order into the bewildering multiplicity, several attempts have been made to arrange the many solutions into a few groups.   The following is a simple and plausible classification of the various conceptions of probability into three groups[2]: (i) the classical conception, originated by Jacob Bernoulli and Laplace, and represented by their followers in various forms; here, probability is defined as the ratio of the number of favorable cases to the number of all possible cases; (ii) the conception of probability as a certain objective logical relation between propositions (or sentences); the chief representatives of this conception are Keynes[3] and Jeffreys[4]; (iii) the conception of

---

[2] See Ernest Nagel, *Principles of the Theory of Probability*, (International Encyclopedia of Unified Science, Vol. I, 1939, No. 6).

[3] John Maynard Keynes, *A Treatise on Probability*, 1941.

[4] Harold Jeffreys, *Theory of Probability*, 1939.

probability as relative frequency, developed most completely by von Mises[5] and Reichenbach[6].

In this paper, a discussion of these various conceptions is not intended. While the main point of interest both for the authors and for the readers of the various theories of probability is normally the solutions proposed in those theories, we shall inspect the theories from a different point of view. We shall not ask what solutions the authors offer but rather which problems the solutions are intended to solve; in other words, we shall not ask what explicata are proposed but rather which concepts are taken as explicanda.

This question may appear superfluous, and the fact obvious that the explicandum for every theory of probability is the pre-scientific concept of probability, i.e., the meaning in which the word 'probability' is used in the pre-scientific language. Is the assumption correct, however, that there is only one meaning connected with the word 'probability' in its customary use, or at the least that only one meaning has been chosen by the authors as their explicandum? When we look at the formulations which the authors themselves offer in order to make clear which meanings of 'probability' they intend to take as their explicanda, we find phrases as different as "degree of belief," "degree of reasonable expectation," "degree of possibility," "degree of proximity to certainty," "degree of partial truth," "relative frequency," and many others. This multiplicity of phrases shows that any assumption of a unique explicandum common to all authors is untenable. And we might even be tempted to go to the opposite extreme and to conclude that the authors are dealing not with one but with a dozen or more different concepts. However, I believe that this multiplicity is misleading. It seems to me that the number of explicanda in all the various theories of probability is neither just one nor about a dozen, but in all essential respects—leaving aside slight variations—very few, and chiefly two. In the following discussion we shall use subscripts in order to distinguish these two meanings of the term 'probability' from which most of the various theories of probability start; we are, of course, distinguishing between two explicanda and not between the various explicata offered by these theories, whose number is much greater. The two concepts are: (i) $probability_1$ = degree of confirmation; (ii) $probability_2$ = relative frequency in the long run. Strictly speaking, there are two groups of concepts, since both for (i) and for (ii) there is a positive, a comparative, and a metrical concept; however, for our discussion, we may leave aside these distinctions.

Let me emphasize again that the distinction made here refers to two

---

[5] Richard von Mises, *Probability, Statistics, and Truth*, (orig. 1928) 1939.

[6] Hans Reichenbach, *Wahrscheinlichkeitslehre*, 1935.

explicanda, not to two explicata.   That there is more than one explicatum is obvious; and indeed, their number is much larger than two.   But most investigators in the field of probability apparently believe that all the various theories of probability are intended to solve the same problem and hence that any two theories which differ fundamentally from one another are incompatible.   Consequently we find that most representatives of the frequency conception of probability reject all other theories; and, *vice versa*, that the frequency conception is rejected by most of the authors of other theories.   These mutual rejections are often formulated in rather strong terms.   This whole controversy seems to me futile and unnecessary. The two sides start from different explicanda, and both are right in maintaining the scientific importance of the concepts chosen by them as explicanda—a fact which does not, however, imply that on either side all authors have been equally successful in constructing a satisfactory explicatum.   On the other hand, both sides are wrong in most of their polemic assertions against the other side.

A few examples may show how much of the futile controversy between representatives of different conceptions of probability is due to the blindness on both sides with respect to the existence and importance of the probability concept on the other side.   We take as examples a prominent contemporary representative of each conception: von Mises, who constructed the first complete theory based on the frequency conception, and Jeffreys, who constructed the most advanced theory based on probability$_1$. Von Mises[7] seems to believe that probability$_2$ is the only basis of the Calculus of Probability.   To speak of the probability of the death of a certain individual seems to him meaningless.   Any use of the term "probability" in everyday life other than in the statistical sense of probability$_2$ has in his view nothing to do with the Calculus of Probability and cannot take numerical values.   That he regards Keynes' conception of probability as thoroughly subjectivistic[8] indicates clearly his misunderstanding.

On the other hand, we find Jeffreys similarly blind in the other direction. Having laid down certain requirements which every theory of probability (and that means for him probability$_1$) should fulfill, he then rejects all frequency theories, that is, theories of probability$_2$, because they do not fulfill his requirements.   Thus he says[9]: "No 'objective' definition of probability in terms of actual or possible observations ... is admissible," because the results of observations are initially unknown and, consequently, we could not know the fundamental principles of the theory and would have no starting point.   He even goes so far as to say that "in practice,

_____

[7] *Op. cit.*, First Lecture.
[8] *Op. cit.*, Third Lecture.
[9] *Op. cit.*, p. 11.

no statistician ever uses a frequency definition, but that all use the notion of degree of reasonable belief, usually without ever noticing that they are using it."[10] While von Mises's concern with explicating the empirical concept of probability$_2$ by the limit of relative frequency in an infinite sequence has led him to apply the term "probability" only in cases where such a limit exists, Jeffreys misunderstands his procedure completely and accuses the empiricist von Mises of apriorism: "The existence of the limit is taken as a postulate by von Mises .... The postulate is an *a priori* statement about possible experiments and is in itself objectionable."[11] Thus we find this situation: von Mises and Jeffreys both assert that there is only one concept of probability that is of scientific importance and that can be taken as the basis of the Calculus of Probability. The first maintains that this concept is probability$_2$ and certainly not anything like probability$_1$; the second puts it just the other way round; and neither has anything but ironical remarks for the concept proposed by the other.

When we criticize the theory of probability proposed by an author, we must clearly distinguish between a rejection of his explicatum and a rejection of his explicandum. The second by no means follows from the first. Donald Williams, in his paper in this symposium[12], raises serious objections aginst the frequency theory of proability, especially in von Mises's form. The chief objection is that von Mises's explicatum for probability, viz., the limit of the relative frequency in an infinite sequence of events with a random distribution, is not accessible to empirical confirmation—unless it be supplemented by a theory of inductive probability, a procedure explicitly rejected by von Mises. I think Williams is right in this objection. This, however, means merely that the concept proposed by von Mises is not yet an adequate explicatum. On the other hand, I believe the frequentists are right in the assertion that their explicandum, viz., the statistical concept of probability$_2$, plays an important role in all branches of empirical science and especially in modern physics, and that therefore the task of explicating this concept is of great importance for science.

It would likewise be unjustified to reject the concept of probability$_1$ as an explicandum merely because the attempts so far made at an explication are not yet quite satisfactory. It must be admitted that the classical Laplacean definition is untenable. It defines probability as the ratio of the number of favorable cases to the total number of equipossible cases, where equipossibility is determined by the principle of insufficient reason (or indifference). This definition is in certain cases inapplicable,

---

[10] *Op. cit.*, p. 300.

[11] *Op. cit.*, p. 304.

[12] "On the Derivation of Probabilities from Frequencies."

in other cases it yields inadequate values, and in some cases it leads even to contradictions, because for any given proposition there are, in general, several ways of analyzing it as a disjunction of other, logically exclusive, propositions.[13] Modern authors, expecially Keynes, Jeffreys, and Hosiasson[14], proceed more cautiously, but at the price of restricting themselves to axiom systems which are rather weak and hence far from constituting an explicit definition. I have made an attempt to formulate an explicit definition of the concept of degree of confirmation (with numerical values) as an explicatum for probability$_1$, and to construct a system of metrical inductive logic based on that definition[15]. No matter whether this first attempt at an explication with the help of the methods of modern logic and in particular those of semantics will turn out to be satisfactory or not, I think there is no reason for doubting that an adequate explication will be developed in time through further attempts.

The distinction between the two concepts which serve as explicanda is often overlooked on both sides. This is primarily due to the unfortunate fact that both concepts are designated by the same familiar, but ambiguous word 'probability'. Although many languages contain two words (e.g., English 'probable' and 'likely', Latin 'probabilis' and 'verisimilis', French 'probable' and 'vraisemblable'), these words seem in most cases to be used in about the same way or at any rate not to correspond to the two concepts we have distinguished. Some authors (e.g., C. S. Peirce and R. A. Fisher) have suggested utilizing the plurality of available words for the distinction of certain concepts (different from our distinction); however, the proposals were made in an artificial way, without relation to the customary meanings of the words. The same would hold if we were to use the two words for our two concepts; therefore we prefer to use subscripts as indicated above.

Probability$_1$, in other words, the logical concept of confirmation in its different forms (positive, comparative, and metrical), has been explained in the preceding section. A brief explanation may here be given of probability$_2$, merely to make clear its distinction from probability$_1$. A typical example of the use of this concept is the following statement: "The prob-

---

[13] Williams' indications (*op. cit.*, pp. 450 and 469) to the effect that he intends to maintain Laplace's definition even in a simplified form and without the principle of indifference are rather puzzling. We have to wait for the full formulation of his solution, which his present paper does not yet give (*op. cit.*, p. 481), in order to see how it overcomes the well-known difficulties of Laplace's definition.

[14] Janina Hosiasson-Lindenbaum, "On Confirmation," *Journal of Symbolic Logic* Vol. V (1940), pp. 133–148.

[15] A book exhibiting this system is in preparation. The present paper is a modified version of a chapter of the book. The definition is explained and some of the theorems of my system of inductive logic are summarized in the paper "On Inductive Logic," which will appear in *Philosophy of Science*, Vol. XII, 1945.

ability$_2$ of casting an ace with this die is 1/6." Statements of this form refer to two properties (or classes) of events: (i) the reference property M_1, here the property of being a throw with this die; (ii) the specific property M_2, here the property of being a throw with any die resulting in an ace. The statement says that the probability$_2$ of M_2 with respect to M_1 is 1/6. The statement is tested by statistical investigations. A sufficiently long series of, say, n throws of the die in question is made, and the number m of these throws which yield an ace is counted. If the relative frequency m/n of aces in this series is sufficiently close to 1/6, the statement is regarded as confirmed. Thus, the other way round, the statement is understood as predicting that the relative frequency of aces thrown with this die in a sufficiently long series will be about 1/6. This formulation is admittedly inexact; but it intends no more than to indicate the meaning of 'probability$_2$' as an explicandum. To make this concept exact is the task of the explication; our discussion concerns only the two explicanda.

IV. THE LOGICAL NATURE OF THE TWO PROBABILITY CONCEPTS

On the basis of the preceding explanations, let us now characterize the two probability concepts, not with respect to what they mean but merely with respect to their logical nature, more specifically, with respect to the kind of entities to which they are applied and the logical nature of the simplest sentences in which they are used. (Since the pre-scientific use of both concepts is often too vague and incomplete, e.g., because of the omission of the second argument (viz., the evidence or the reference class), we take here into consideration the more careful use by authors on probability. However, we shall be more concerned with their general discussions than with the details of their constructed systems.) For the sake of simplicity, let us consider the two concepts in their metrical forms only. They may be taken also in their comparative and in their positive forms (as explained for probability$_1$, i.e., confirmation, in section II, and these other forms would show analogous differences. Probability$_1$ and probability$_2$, taken as metrical concepts, have the following characteristics in common: each of them is a function of two arguments; their values are real numbers belonging to the interval 0 to 1 (according to the customary convention, which we follow here). Their characteristic differences are as follows:

1. *Probability*$_1$ (degree of confirmation).

(a) The *two arguments* are variously described as events (in the literal sense, see below), states of affairs, circumstances, and the like. Therefore each argument is expressible by a declarative sentence and hence is, in our terminology, a proposition. Another alternative consists in taking as arguments the sentences expressing the propositions, describing the events, etc. If we choose this alternative, probability$_1$ is a semantical concept (as in section II). (Fundamentally it makes no great difference whether propositions or sentences are taken as arguments; but the second method has certain technical advantages, and therefore we use it for our discussion.)

(b) A simple *statement* of probability$_1$, i.e., one attributing to two given arguments a particular number as value of probability$_1$, is either L-true (logically true, analytic) or L-false (logically false, logically self-contradictory), hence in any case L-determinate, not factual (synthetic). Therefore, a statement of this kind is to be established by logical analysis alone, as has been explained earlier (section II). It is independent of the contingency of facts because it does not say anything about facts (although the two arguments do in general refer to facts).

2. *Probability$_2$* (relative frequency).

(a) The *two arguments* are properties, kinds, classes, usually of events or things. [As an alternative, the predicate expressions designating the properties might be taken as arguments; then the concept would become a semantical one. In the present case, however, in distinction to (1), there does not seem to be any advantage in this method. On the contrary, it appears to be more convenient to have the probability$_2$ statements in the object language instead of the metalanguage; and it seems that all authors who deal with probability$_2$ choose this form.]

(b) A simple *statement* of probability$_2$ is factual and empirical, it says something about the facts of nature, and hence must be based upon empirical procedure, the observation of relevant facts. From these simple statements the theorems of a mathematical theory of probability$_2$ must be clearly distinguished. The latter do not state a particular value of probability$_2$ but say something about connections between probability$_2$ values in a general way, usually in a conditional form (for example: "if the values of such and such probabilities$_2$ are q_1 and q_2, then the value of a probability$_2$ related to the original ones in a certain way is such and such a function, say, product or sum, of q_1 and q_2"). These theorems are not factual but L-true (analytic). Thus a theory of probability$_2$, e.g., the system constructed by von Mises or that by Reichenbach, is not of an empirical but of a logico-mathematical nature; it is a branch of mathematics, like arithmetic, fundamentally different from any branch of empirical science, e.g., physics.

It is very important to distinguish clearly between *kinds of events* (war, birth, death, throw of a die, throw of this die, throw of this die yielding an ace, etc.) and *events* (Caesar's death, the throw of this die made yesterday at 10 A.M., the series of all throws of this die past and future). This distinction is doubly important for discussions on probability, because one of the characteristic differences between the two concepts is this: the first concept refers sometimes to two events, the second to two kinds of events (see 1(a) and 2(a)). Many authors of probability use the word 'event' (or the corresponding words 'Ereignis' and 'évènement') when they mean to speak, not about events, but about kinds of events. This usage is of long standing in the literature on probability, but it is very unfortunate. It has only served to reinforce the customary neglect of the fundamental difference between the two probability concepts which arose originally out of the ambiguous use of the word 'probability', and thereby to increase the general confusion in discussions on probability. The authors who use the term 'event' when they mean kinds of events get into trouble, of course, whenever they want to speak about specific events. The traditional solution is to say 'the happenings (or occurrences) of a certain event' instead of 'the events of a certain kind'; sometimes the events are referred to by the term 'single events'. But this phrase is rather misleading; the important difference between events and kinds of events is not the same as the inessential difference between single events (the first throw I made today with this die) and multiple or compound events (the series of all throws made with this die). Keynes, if I interpret him correctly, has noticed the ambiguity of the term 'event'. He says[16] that the customary use of phrases like 'the happening of events' is "vague and unambiguous," which I suppose to be a misprint for "vague and ambiguous"; but he does not specify the ambiguity. He proposes to dispense altogether with the term 'event' and to use instead the term 'proposition'. Subsequent authors dealing with probability$_1$, like Jeffreys, for example, have followed him in this use.

Many authors have made a distinction between two (or sometimes more) kinds of probability, or between two meanings of the word 'probability'. Some of these distinctions are quite different from the distinction made here between probability$_1$ and probability$_2$. For instance, a distinction is sometimes made between mathematical probability and philosophical probability; their characteristic difference appears to be that the first has numerical values, the second not. However, this difference seems hardly essential; we find a concept with numerical values and one without, in other words, both a metrical and a comparative concept on either side of our distinction between the two fundamentally different meanings of 'probability'. Another distinction has been made between subjective and objective probability. However, I believe that practically all authors

---

[16] *Op. cit.*, p. 5.

really have an objective concept of probability in mind, and that the appearance of subjectivist conceptions is in most cases caused only by occasional unfortunate formulations; this will soon be discussed.

Other distinctions which have been made are more or less similar to our distinction between probability₁ and probability₂. For instance, Ramsey[17] says: ". . . the general difference of opinion between statisticians who for the most part adopt the frequency theory of probability and logicians who mostly reject it renders it likely that the two schools are really discussing different things, and that the word 'probability' is used by logicians in one sense and by statisticians in another."

It seems that many authors have taken either probability₁ or probability₂ as their explicandum. I believe moreover that practically all authors on probability have intended one of these two concepts as their explicandum, despite the fact that their various explanations appear to refer to a number of quite different concepts.

For one group of authors, the question of their explicandum is easily answered. In the case of all those who support a frequency theory of probability, i.e., who define their explicata in terms of relative frequency (as a limit or in some other way), there can be no doubt that their explicandum is probability₂. Their formulations are, in general, presented in clear and unambiguous terms. Often they state explicitly that their explicandum is relative frequency. And even in the cases where this is not done, the discussion of their explicata leaves no doubt as to what is meant as explicandum.

This, however, covers only one of the various conceptions, i.e., explicata proposed, and only one of the many different explanations of explicanda which have been given and of which some examples were mentioned earlier. It seems clear that the other explanations do not refer to the statistical, empirical concept of relative frequency; and I believe that practically all of them, in spite of their apparent dissimilarity, are intended to refer to probability₁. Unfortunately, many of the phrases used are more misleading than helpful in our efforts to find out what their authors actually meant as explicandum. There is, in particular, one point on which many authors in discussions on probability₁, or on logical problems in general, commit a certain typical confusion or adopt incautiously other authors' formulations which are infected by this confusion. I am referring to what is sometimes called psychologism in logic.

Many authors in their general remarks about the nature of (deductive) logic say that it has to do with ways and forms of thinking or, in more cautious formulations, with forms of correct⸴ or rational thinking. In spite of these subjectivistic formulations, we find that in practice these

[17] F. P. Ramsey, *The Foundations of Mathematics*, 1931; see p. 157.

authors use an objectivistic method in solving any particular logical problem. For instance, in order to find out whether a certain conclusion follows from given premises, they do not in fact make psychological experiments about the thinking habits of people but rather analyze the given sentences and show their conceptual relations. In inductive logic or, in other words, the theory of probability₁, we often find a similar psychologism. Some authors, from Laplace and other representatives of the classical theory of probability down to contemporary authors like Keynes and Jeffreys, use subjectivistic formulations when trying to explain what they take as their explicandum; they say that it is probability in the sense of degree of belief or, if they are somewhat more cautious, degree of reasonable or justified belief. However, an analysis of the work of these authors comes to quite different results if we pay more attention to the methods the authors actually use in solving problems of probability than to the general remarks in which they try to characterize their own aims and methods. Such an analysis, which cannot be carried out within this paper, shows that most and perhaps all of these authors use objectivistic rather than subjectivistic methods. They do not try to measure degrees of belief by actual, psychological experiments, but rather carry out a logical analysis of the concepts and propositions involved. It appears, therefore, that the psychologism in inductive logic is, just like that in deductive logic, merely a superficial feature of certain marginal formulations, while the core of the theories remains thoroughly objectivistic. And, further, it seems to me that for most of those authors who do not maintain a frequency theory, from the classical period to our time, the objective concept which they take as their explicandum is probability₁, i.e., degree of confirmation.

<h3>V. EMPIRICISM AND THE LOGICAL CONCEPT OF PROBABILITY</h3>

Many empiricist authors have rejected the logical concept of probability₁ as distinguished from probability₂ because they believe that its use violates the principle of empiricism and that, therefore, probability₂ is the only concept admissible for empiricism and hence for science. We shall now examine some of the reasons given for this view.

The concept of probability₁ is applied also in cases in which the hypothesis h is a prediction concerning a particular "single event," e.g., the prediction that it will rain tomorrow or that the next throw of this die will yield an ace. Some philosophers believe that an application of this kind violates the principle of verifiability (or confirmability). They might say, for example: "How can the statement 'the probability of rain tomorrow on the evidence of the given meteorological observations is one-fifth' be verified? We shall observe either rain or not-rain tomorrow, but we shall not observe anything that can verify the value one-fifth." This objection, however, is based on a misconception concerning the nature of the probability$_1$ statement. This statement does not ascribe the probability$_1$ value 1/5 to tomorrow's rain but rather to a certain logical relation between the prediction of rain and the meteorological report. Since the relation is logical, the statement is, if true, L-true; therefore it is not in need of verification by observation of tomorrow's weather or of any other facts.

It must be admitted that earlier authors on probability have sometimes made inferences which are inadmissible from the point of view of empiricism. They calculated the value of a logical probability and then inferred from it a frequency, hence making an inadvertent transition from probability$_1$ to probability$_2$. Their reasoning might be somewhat like this: "On the basis of the symmetry of this die the probability of an ace is 1/6; therefore, one-sixth of the throws of this die will result in an ace." Later authors have correctly criticized inferences of this kind. It is clear that from a probability$_1$ statement a statement on frequency can never be inferred, because the former is purely logical while the latter is factual. Thus the source of the mistake was the confusion of probability$_1$ with probability$_2$. The use of probability$_1$ statements cannot in itself violate the principle of empiricism so long as we remain aware of the fact that those statements are purely logical and hence do not allow the derivation of factual conclusions.

The situation with respect to both objections just discussed may be clarified by a comparison with deductive logic. Let h be the sentence 'there will be rain tomorrow' and j the sentence 'there will be rain and wind tomorrow'. Suppose somebody makes the statement in deductive logic: "h follows logically from j." Certainly nobody will accuse him of apriorism either for making the statement or for claiming that for its verification no factual knowledge is required. The statement "the probability$_1$ of h on the evidence e is 1/5" has the same general character as the former statement; therefore it cannot violate empiricism any more than the first. Both statements express a purely logical relation between two sentences. The difference between the two statements is merely this: while the first states a complete logical implication, the second states only, so to speak, a partial logical implication; hence, while the first belongs to deductive logic, the second belongs to inductive logic. Generally speaking, the assertion of purely logical sentences, whether in deductive or in inductive logic, can never violate empiricism; if they are false, they violate the rules of logic. The principle of empiricism can be violated only by the assertion of a factual (synthetic) sentence without a sufficient empirical foundation, or by the thesis of apriorism when it contends that for knowledge with respect to certain factual sentences no empirical foundation is required.

According to Reichenbach's view[18], the concept of logical probability or weight, in order to be in accord with empiricism, must be identified with the statistical concept of probability. If we formulate his view with the help of our terms with subscripts, it says that probability$_1$ is identical with probability$_2$, or, rather, with a special kind of application of it. He argues for this "identity conception" against any "disparity conception," like the one presented in this paper, which regards the two uses of 'probability' as essentially different. Reichenbach tries to prove the identity conception by showing how the concept which we call probability$_1$, even when applied to a "single event," leads back to a relative frequency. I agree that in certain cases there is a close relationship between probability$_1$ and relative frequency. The decisive question is, however, the nature of this relationship. Let us consider a simple example. Let the evidence e say that among 30 observed things with the property M_1 20 have been found to have the property M_2, and hence that the relative frequency of M_2 with respect to M_1 in the observed sample is $2/3$; let e say, in addition, that a certain individual b not belonging to the sample is M_1. Let h be the prediction that b is M_2. If the degree of confirmation c is defined in a suitable way as an explicatum for probability$_1$, $c(h,e)$ will be equal or close to $2/3$; let us assume for the sake of simplicity that $c= 2/3$.[19] However, the fact that, in this case, the value of c or probability$_1$ is equal to a certain relative frequency by no means implies that probability$_1$ is here the same as probability$_2$; these two concepts remain fundamentally different even in this case. This becomes clear by the following considerations (i) to (iv).

(i) The c-statement '$c(h,e) = 2/3$' does not itself state a relative frequency although the value of c which it states is calculated on the basis of a known relative frequency and, under our assumptions, is in this case exactly equal to it. A temperature is sometimes determined by the volume of a certain body of mercury and is, under certain conditions, equal to it; this, however, does not mean that temperature and volume are the same concept. The c-statement, being a purely logical statement, cannot possibly state a relative frequency for two empirical properties like M_1 and M_2. Such a relative frequency can be stated only by a factual sentence; in the example, it is stated by a part of the factual sentence e. The c-statement

[18] Hans Reichenbach, *Experience and Prediction*, 1938, see §§ 32–34.

[19] According to Reichenbach's inductive logic, in the case described $c = 2/3$. According to my inductive logic, c is close to but not exactly equal to $2/3$. My reason for regarding a value of the latter kind as more adequate has been briefly indicated in the paper mentioned above "On Inductive Logic," § 10. For our present discussion, we may leave aside this question.

does not imply either e or the part of e just mentioned; it rather speaks about e, stating a logical relation between e and h. It seems to me that Reichenbach does not realize this fact sufficiently clearly. He feels, correctly, that the c-value 2/3 stated in the c-statement is in some way based upon our empirical knowledge of the observed relative frequency. This leads him to the conception, which I regard as incorrect, that the c-statement must be interpreted as stating the relative frequency and hence as being itself a factual, empirical statement. In my conception, the factual content concerning the observed relative frequency must be ascribed, not to the c-statement, but to the evidence e referred to in the c-statement.

(ii) The relative frequency 2/3, which is stated in e and on which the value of c is based, is not at all a probability$_2$. The probability$_2$ of M_2 with respect to M_1 is the relative frequency of M_2 with respect to M_1 in the whole sequence of relevant events. The relative frequency stated by e, on the other hand, is the relative frequency observed within the given sample. It is true that our estimate of the value of probability$_2$ will be based on the observed relative frequency in the sample. However, observations of several samples may yield different values for the observed relative frequency. Therefore we cannot identify observed relative frequency with probability$_2$, since the latter has only one value, which is unknown. (I am using here the customary realistic language as it is used in everyday life and in science; this use does not imply acceptance of realism as a metaphysical thesis but only of what Feigl calls "empirical realism."[20])

(iii) As mentioned, an estimate of the probability$_2$, the relative frequency in the whole sequence, is based upon the observed relative frequency in the sample. I think that, in a sense, the statement '$c(h,e) = 2/3$' itself may be interpreted as stating such an estimate; it says the same as: "The best estimate on the evidence e of the probability$_2$ of M_2 with respect to M_1 is 2/3." If somebody should like to call this a frequency interpretation of probability$_1$, I should raise no objection. It need, however, be noticed clearly that this interpretation identifies probability$_1$ not with probability$_2$ but with the best estimate of probability$_2$ on the evidence e; and this is something quite different. The best estimate may have different values for different evidences; probability$_2$ has only one value. A statement of the best estimate on a given evidence is purely logical; a statement of probability$_2$ is empirical. The reformulation of the statement on probability$_1$ or c in terms of the best estimate of probability$_2$ may be helpful in showing the close connection between the two probability concepts. This formulation must, however, not be regarded as eliminating

[20] Herbert Feigl, "Logical Empiricism," in *Twentieth Century Philosophy*, ed. D. Runes, 1943, pp. 373–416; see pp. 390 f.

probability$_1$. The latter concept is still implicitly contained in the phrase "the best estimate," which means nothing else but "the most probable estimate," that is, "the estimate with the highest probability$_1$." Generally speaking, any estimation of the value of a physical magnitude (length, temperature, probability$_2$, etc.) on the evidence of certain observations or measurements is an inductive procedure and hence necessarily involves probability$_1$, either in its metrical or in its comparative form.

(iv) The fundamental difference between probability$_1$ and probability$_2$ may be further elucidated by analyzing the sense of the customary references to *unknown probabilities*. As we have seen under (ii), the value of a certain probability$_2$ may be unknown to us at a certain time in the sense that we do not possess sufficient factual information for its calculation. On the other hand, the value of a probability$_1$ for two given sentences cannot be unknown in the same sense. (It may, of course, be unknown in the sense that a certain logico-mathematical procedure has not yet been accomplished, that is, in the same sense in which we say that the solution of a certain arithmetical problem is at present unknown to us.) In this respect also, a confusion of the two concepts of probability has sometimes been made in formulations of the classical theory. This theory deals, on the whole, with probability$_1$; and the principle of indifference, one of the cornerstones of the theory, is indeed valid to a certain limited extent for this concept. However, this principle is absurd for probability$_2$, as has often been pointed out. Yet the classical authors sometimes refer to unknown probabilities or to the probability (or chance) of certain probability values, e.g., in formulations of Bayes' theorem. This would not be admissible for probability$_1$, and I believe that here the authors inadvertently go over to probability$_2$. Since a probability$_2$ value is a physical property like a temperature, we may very well inquire into the probability$_1$, on a given evidence, of a certain probability$_2$ (as in the earlier example, at the end of (iii)). However, a question about the probability$_1$ of a probability$_1$ statement has no more point than a question about the probability$_1$ of the statement that $2 + 2 = 4$ or that $2 + 2 = 5$, because a probability$_1$ statement is, like an arithmetical statement, either L-true or L-false; therefore its probability$_1$, with respect to any evidence, is either 1 or 0.

VI. PROBABILITY AND TRUTH

It is important to distinguish clearly between a concept characterizing a thing independently of the state of our knowledge (e.g., the concept 'hard') and the related concept characterizing our state of knowledge with respect to the thing (e.g., the concept 'known to be hard'). It is true that a person will, as a rule, attribute the predicate 'hard' to a thing b only if he knows it to be hard, hence only if he is prepared to attribute to it also the predicate 'known to be hard'. Nevertheless, the sentences 'b is hard' and 'b is known to be hard' are obviously far from meaning the same. One point of difference becomes evident when we look at the sentences in their complete form; the second sentence, in distinction to the first (if we regard hardness as a permanent property), must be supplemented by references to a person and a time point: 'b is known to X at the time t to be hard'. The distinction between the two sentences becomes more conspicuous if they occur within certain larger contexts. For example, the difference between the sentences 'b is not hard' and 'b is not known to X at the time t to be hard' is clear from the fact that we can easily imagine a situation where we would be prepared to assert the second but not the first.

The distinction just explained may appear as obvious beyond any need of emphasis. However, a distinction of the same general form, where 'true' is substituted for 'hard', is nevertheless often neglected by philosophers. A person will, in general, attribute the predicate 'true' to a given sentence (or proposition) only if he knows it to be true, hence only if he is prepared to attribute to it also the predicate 'known to be true' or 'established as true' or 'verified'. Nevertheless 'true' and 'verified (by the person X at the time t)' mean quite different things; and so do 'false' and 'falsified' (in the sense of 'known to be false', 'established as false'). A given sentence is often neither verified nor falsified; nevertheless it is either true or false, whether anybody knows it or not. (Some empiricists shy away from the latter formulation because they believe it to involve an anti-empiricist absolutism. This, however, is not the case. Empiricism admits as meaningful any statement about unknown fact and hence also about unknown truth, provided only the fact or the truth is know*able*, or confirmable.) In this way an inadvertent confusion of 'true' and 'verified' may lead to doubts about the validity of the principle of excluded middle. The question of whether and to what extent a confusion of this kind has actually contributed to the origin of some contemporary philosophical doctrines rejecting that principle is hard to decide and will not be investigated here.

A statement like 'this thing is made of iron' can never be verified in the strictest sense, i.e., definitively established as true so that no possibility remains of refuting it by future experience. The statement can only be more or less confirmed. If it is highly confirmed, that is to say, if strong evidence for it is found, then it is often said to be verified; but this is a weakened, non-absolutistic sense of the term. I think it is fair to say that most philosophers, and at least all empiricists, agree today that the concept 'verified' in its strict sense is not applicable to statements about physical things. Some philosophers, however, go further; they say that, because we can never reach absolutely certain knowledge about things, we ought to abandon the concept of truth. It seems to me that this view is due again to an unconscious confusion of 'true' and 'verified'.[21] Some of these philosophers say that, in order to avoid absolutism, we should not ask whether a given statement is true but only whether it has been confirmed, corroborated, or accepted by a certain person at a certain time.[22] Others think that 'true' should be abandoned in favor of 'highly confirmed' or 'highly probable'. Reichenbach[23] has been led by considerations of this kind to the view that the values of probability (the logical concept of probability$_1$) ought to take the place of the two truth-values, truth and falsity, of ordinary logic, or, in other words, that probability logic is a multivalued logic superseding the customary two-valued logic. I agree with Reichenbach that here a concept referring to an absolute and unobtainable maximum should be replaced by a concept referring to a high degree in a continuous scale. However, what is superseded by 'highly probable' or 'confirmed to a high degree' is the concept 'confirmed to the maximum degree' or 'verified', and not the concept 'true'.

Values of probability$_1$ are fundamentally different from truth-values. Therefore inductive logic, although it introduces the continuous scale of probability$_1$ values, remains like deductive logic two-valued. While it is true that to the multiplicity of probability$_1$ values in inductive logic only a dichotomy corresponds in deductive logic, nevertheless, this dichotomy is not between truth and falsity of a sentence but between L-implication and non-L-implication for two sentences. If, to take our previous example, $c(h,e) = 2/3$, then h is still either true or false and does not have an intermediate truth-value of $2/3$.

It has been the chief purpose of this paper to explain and discuss the two concepts of probability in their role as explicanda for theories of probability. I think that in the present situation clarification of the explicanda is the most urgent task. When every author has not only a clear understanding of his own explicandum but also some insight into the existence, the importance, and the meaning of the explicandum on the other side, then it will be possible for each side to concentrate entirely on

[21] I have given earlier warnings against this confusion in "Wahrheit und Bewährung," *Actes du Congrès International de Philosophic Scientifique*, Paris, 1936, Vol. IV, pp. 1–6; and in *Introduction to Semantics*, p. 28.

[22] See, e.g., Otto Neurath, "Universal Jargon and Terminology," *Proceedings Aristotelian Society*, 1940–1941, pp. 127–148, see esp. pp. 138 f.

[23] *Op. cit. (Experience)*, §§ 22, 35.

the positive task of constructing an explication and a theory of the chosen explicatum without wasting energy in futile polemics against the explicandum of the other side.

RUDOLF CARNAP.

UNIVERSITY OF CHICAGO.

EXTRACTO

Las diversas teorías de la probabilidad tienen por fin substituir el concepto pre-científico de probabilidad por un concepto exacto y científico. Sin embargo, hay *dos* sentidos del término "probabilidad" empleados generalmente; ambos están conectados el uno con el otro, aunque son fundamentalmente diferentes. Primero, hay el concepto lógico de probabilidad, o sea el grado de confirmación de una hipótesis sobre la base de las pruebas aducidas. La proposición que aplica este concepto no se basa en la observación de los hechos, sino en el análisis lógico. Segundo, hay el concepto estadístico de probabilidad, o sea la relativa frecuencia. La proposición que emplea este concepto es sintética y empírica. Ambos conceptos son importantes para la ciencia. El primero constituye la base de la lógica inductiva. El segundo es útil en las investigaciones estadísticas. Muchos autores que se ocupan de sistematizar uno de estos dos conceptos, no se percatan de la importancia y aun de la existencia del otro concepto. Cuando el uno y el otro sean claramente reconocidos, se evitarán muchas controversias inútiles.

PHILOSOPHICAL FOUNDATIONS OF PROBABILITY

HANS REICHENBACH

UNIVERSITY OF CALIFORNIA, LOS ANGELES

I

In sections I–V we deal with the formal structure of probability; in sections VI–XI we investigate the meaning and the assertability of probability statements.[1]

The concept of probability refers to a relation. If we cast a die, the probability of a certain face is 1/6; the condition introduced by "if" is necessary for this instance of a probability relation as well as for all others. When the condition is omitted the statement must be regarded as elliptic; such omission is possible if it is obvious from the context what condition is understood. We therefore regard probability as having the logical form of an implication, which we call the *probability implication*.

This implication, however, holds not between individuals but between classes. Thus the phrase "cast the die" defines a class A of events, and similarly the phrase "face 6 turns up" defines a class B of events. The class A is called the *reference class*, the class B is named the *attribute class*. Furthermore, the events x_i and y_i belonging respectively to these classes are regarded as given in a certain order and in such a way that a one-to-one correspondence between the elements of the sequences is known, which we express by the use of the same subscript "i". Since the probability statement refers to all events belonging to the classes A and B, it can be written in the form of an all-statement:

$$(i) \quad (x_i \epsilon A \underset{p}{\rightarrow} y_i \epsilon B) \tag{1}$$

The symbol "(i)" is the all-operator of logistics; the symbol "ϵ", as usual, denotes the relation of class membership. The real number p is the degree of probability.

Instead of the *implicational notation* presented in (1) it is convenient to introduce a *mathematical notation*, or *functor notation*. We write

$$P(A,B) = p \tag{2}$$

The symbol "$P(\quad)$" is a functor, meaning "the probability of". Expression (2) has the same meaning as (1) and can be regarded as an abbreviation in which the reference to the sequences x_i and y_i is not expressed. We read (2) in the form "the probability from A to B is p".

---

[1] For a detailed account of the following ideas we refer the reader to the author's *Wahrscheinlichkeitslehre* (Leiden, 1935), A. W. Sijthoff. A summary in the French language was published under the title "Les fondements logiques du calcul des probabilités," *Annales de l'Institut Henri Poincaré*, t. VII, fasc. v (Paris, 1937), pp. 267–348. The general ideas of secs. VI–XI are presented in chap. v of the author's *Experience and Prediction* (Chicago, 1938).

For operations inside the P-symbol we apply the rules of logistics, using the signs " $\vee$ " for the inclusive "or", "." for the "and", and a line on top of the letter for the negation. Thus "$P(A,\bar{B})$" means the probability from A to non-B; "$P(A,B \vee C)$" means the probability from A to B or C. For every expression it is permissible to substitute tautologically equivalent expressions; this rule allows for manipulations of the kind used in logistics. The P-symbol as a whole is regarded as a mathematical variable used in equations like (2) or of a more general form.

Like all mathematical systems the calculus of the P-symbol is employed in two conceptions. In the *formal conception* we do not give any meaning to the P-symbol, but set up formal relations connecting various forms of expressions. In other words, we define the P-symbol implicitly by a set of axioms. In the *material conception* or *interpretation* we introduce a meaning for the P-symbol in terms of other mathematical or physical concepts. We then have to show that the *coördinative definition* so introduced satisfies the set of axioms. The latter condition restricts the class of admissible interpretations, but it does not single out one interpretation as the only admissible one.

The axioms which we introduce are the following ones:

α. Normalization:

     1. $P(A,A \vee B) = 1$

     2. $P(A,B.\bar{B}) = 0$

     3. $0 \leqq P(A,B)$

β. Addition:

$$P(A,B \vee C) = P(A,B) + P(A,C) - P(A,B.C)$$

γ. Multiplication:

$$P(A,B.C) = P(A,B) \cdot P(A.B,C)$$

The variables expressed by small letters are real numbers; a restriction to the values 0 and 1 limits included follows from axiom α, 3. To this axiom, however, we add the condition (not expressed in the symbolic notation) that it holds only in case the class A is not empty. The general restriction, therefore, is bound to the same condition.

We do not claim that between any two classes A and B a probability exists, but regard it as an empirical question whether there is such a probability. On the other hand, we set up the rule that a probability exists if it is numerically determined by given probabilities in terms of the axioms or derivable theorems (*rule of existence*). This rule permits us to solve a probability equation for any of its variables.

Axiom β is called the *general theorem of addition*. For exclusive events B and C the term $P(A,B.C)$ drops out, and the formula then becomes the *special theorem of addition.* Similarly we call axiom γ the *general theorem of multiplication.* For independent events we have

$$P(A.B,C) = P(A,C) \tag{3a}$$

The theorem then assumes the form

$$P(A,B.C) = P(A,B) \cdot P(A,C) \tag{3b}$$

which we call the *special theorem of multiplication.*

The frequency interpretation can be written in the form

$$P(A,B) = \lim_{n \to \infty} \frac{N^n(A.B)}{N^n(A)} \tag{4}$$

where the symbol "$N^n(X)$" means the number of those elements of the sequence up to the nth element which belong to the class X. It can be shown that if the frequency interpretation (4) is used the axioms follow tautologically. The frequency interpretation is therefore admissible. For the formal manipulations within the calculus of probability, however, the interpretation (4) is not necessary. All theorems of the calculus, for instance the rule of Bayes, can be formally derived from the axioms. The calculus defined by the axioms $\alpha - \gamma$ we call the *elementary calculus of probability.*

II

We have introduced probability as a property of sequences; if we use the frequency interpretation this property is identified with the limit of the frequency. The only condition a sequence must have in order to be a probability sequence, therefore, is that it has a probability, or limit of the frequency.

This definition leaves open the structure of probability sequences. We include among probability sequences the special forms that are usually called random sequences, but we do not hesitate likewise to include such strictly ordered forms as a sequence in which B and $\bar{B}$ alternate regularly. It cannot be the task of the mathematician to deny the name "probability sequence" to certain types of sequences; it should rather be regarded as his task to define various types of order and to derive the laws holding for them. This task is undertaken by *the theory of the order of probability sequences,* which represents the second chapter of the calculus of probability.

A definition of types of order is achieved by certain means of structural characterization. They are based on the study of the probability in subsequences resulting from the original sequence by means of selections. This means of structural characterization was introduced by v. Mises; we apply it, however, in a generalized form since we do not always require that the probabilities of the selected subsequences are equal to that of the main sequence. For every sequence there will exist a number of selections resulting in subsequences of the same probability as the main sequence; the class of selections satisfying this (and a further) condition is called the *domain of invariance*. A certain type of sequence will be characterized by rules which set up minimum requirements for the domain of invariance. This method of minimum requirements has the advantage that for the derivation of theorems we do not introduce more postulates than are necessary for the derivation.

We extend our symbolism by using *phase superscripts*. Assume, for instance, that in a sequence of events B and $\overline{B}$ we select all events preceded by a B. This probability will be written in the implicational notation

$$(i) \quad (x_i \epsilon A \,.\, y_i \epsilon B \vec{p} y_{i+1} \epsilon B) \tag{5}$$

In the functor notation we use phase superscripts, and write instead of (5)

$$P(A.B,B^1) \tag{6}$$

The condition that a selection by the predecessor leads to the same probability as the main sequence is then expressed by the equation

$$P(A.B,B^1) = P(A,B) \tag{7}$$

By a sequence *free from aftereffect* we understand the condition

$$P(A,B_{i_1}{}^1 \ldots B_{i_{\nu-1}}{}^{\nu-1},B_{i_\nu}{}^\nu) = P(A,B_{i_\nu}) \qquad (\nu = 1, 2, 3, \ldots) \tag{8}$$

Another form of selection is given by the arithmetical progressions. We say that an element x_i belongs to the selection $S_{\lambda\kappa}$ when

$$i = \kappa + (m-1) \cdot \lambda \qquad \begin{array}{l} m = 1, 2, 3, \ldots \\ \kappa = 1, 2, \ldots \lambda \end{array} \tag{9}$$

A selection $S_{\lambda\kappa}$ is called a *regular division*.

The *domain of invariance* is defined as the class of all selections that leave unchanged not only the probability of the major sequence but also all phase probabilities. We define *normal sequences* by the condition that the regular divisions and the selections by predecessors belong to the domain of invariance. The first condition is expressed by the following equations:

$$P(A.S_{\lambda\kappa},B_i) = P(A,B_i) \tag{10}$$

$$P(A.S_{\lambda\kappa}{}^a.B_{i_1}{}^1 \ldots B_{i_{\nu-1}}{}^{\nu-1},B_{i_\nu}{}^\nu) = P(A.B_{i_1}{}^1 \ldots B_{i_{\nu-1}}{}^{\nu-1},B_{i_\nu}{}^\nu)$$

$$\kappa = 1, 2, \ldots, \lambda; a = 1, 2, \ldots, \lambda; \nu = 1, 2, \ldots, \lambda; \lambda = 1, 2, 3, \ldots)$$

The second condition is given by (8); it is identical with the condition that the sequence be free from aftereffect. It can be shown that for normal sequences so defined the special theorem of multiplication holds for any succession of consecutive elements and that they therefore satisfy the Bernoulli theorem. The normal sequences of our definition are identical with the admissible numbers introduced by A. Copeland.[1] The postulates introduced by R. v. Mises for his collectives are much stronger than our postulates; every collective is a normal sequence, but not vice versa.

A randomness which is not restricted to an arbitrary class of selections, and which makes it impossible to select deviating subsequences by arithmetical methods that do not refer to the attribute, may be called a *logical randomness*. The plan of defining a logical randomness was set forth by R. v. Mises[2] and has been continued by A. Wald[3] and A. Church.[4] The results of Church, in particular, are interesting for the logician. But it should not be forgotten that the random sequences represent only a special case of probability sequences and that therefore the logical problems of probability are independent of the definition of randomness.

Among the sequences with aftereffect we meet with an interesting type in which the aftereffect depends only on the immediate predecessor. Such sequences were studied by Markoff. We speak here of transfer of probability. This condition is expressed by the following equations:

$$P(A.B_{i_1}{}^1 \ldots B_{i_{.1}}.{}^{r-1},B_{i_r}{}') = P(A.B_{i_{.1}}.{}^{r-1},B_{i_r}{}') \tag{11}$$

The types of order so far considered refer to individual sequences. In many cases, however, we are concerned with sequences of sequences, i.e., with a *sequence lattice*. The probabilities of horizontal and vertical sequences must here be studied individually; they are logically independent of each other. If a dependence—for instance, equality of all limits—is assumed, it must be introduced by a special postulate defining a certain type of sequence lattice. An important type of this kind is given by *normal sequences in the narrower sense*. Another type, which combines properties of probability transfer with certain properties of normal sequences, is important for the analysis of such physical phenomena as occur in the mixture of liquids.

The definition of types of order is not limited; it will be advisable, however, to construct the definitions in such a way that they lend themselves to practical applications.

III

The probability sequences so far considered possess only a finite number of attributes, in the simplest case only B and $\overline{B}$. A more general form is given by sequences possessing an infinite number of attributes. Translating a term introduced by v. Mises, we speak here of an *attribute space*. Such sequences may be called *primitive probability sequences*. A division of the attribute space into areas will produce probability sequences of a finite number of attributes.

[1] A. H. Copeland, "Admissible numbers in the theory of probability," *Amer. Jour. Math.*, vol. 50, no. 4 (1928), p. 535.
[2] R. v. Mises, *Wahrscheinlichkeitsrechnung* (Leipzig, 1931), and earlier publications.
[3] A. Wald, *Die Widerspruchsfreiheit des Kollektivbegriffs, Ergebnisse eines mathematischen Kolloquiums*, Heft 8 (Wien, 1937).
[4] Alonzo Church, *Bull. Amer. Math. Soc.*, vol. 46 (1940), p. 130.

That probability can be defined as a ratio of areas in the attribute space is made possible by the fact that the formal system of axioms admits of a geometrical interpretation. For many mathematical purposes it is convenient to use only the geometrical interpretation; and some theories of probability therefore read like chapters of set theory. It should not be forgotten, however, that the probability concept of applications is always the frequency concept, not the geometrical concept. That it is permissible to apply the results of geometrical probabilities to practical statistics derives from the isomorphism between the two interpretations.

For the treatment of the continuous attribute space a condition of complete additivity is required which states that the probability of the limit of an infinite set of classes is equal to the limit of the probabilities of these classes.

IV

The conception of probability developed in the preceding considerations may be called the *mathematical conception*. In it probability is conceived as a property of sequences of events or of other physical objects. We now turn to the *logical conception* of probability, which regards probability as a property of sentences, comparable to truth. The first to see this duality of interpretation was G. Boole.[6]

In order to make probability the analogue of truth we shall omit the general reference class A in the notation, regarding this class as understood. In the frequency interpretation it is given by the subscript of the symbols of the elements. We thus write "$P(B)$" instead of "$P(A,B)$".

The frequency interpretation can be transferred to the logical conception by the device of counting sentences about events instead of counting events. Since the number of true sentences of the form "$x_i \epsilon B$" corresponds to the number of events B, the two interpretations are isomorphous. The logical interpretation, however, offers certain advantages since it is required for the understanding of linguistic forms in which probability is used as the analogue of truth. Thus we say it is probable that it will rain tomorrow; or we speak of the probability of a certain hypothetical assumption. For the present we shall study only the logical structure of such forms, postponing the question of interpretation.

Two-valued logic is based on the truth tables (12). In addition to the symbols introduced above we employ here the signs "$\supset$" for implication and "$\equiv$" for equivalence.

TRUTH TABLES OF TWO-VALUED LOGIC

a	$\bar{a}$		a	b	$a \vee b$	$a.b$	$a \supset b$	$a \equiv b$
T	F		T	T	T	T	T	T
F	T		T	F	T	F	F	F
			F	T	T	F	T	F
			F	F	F	F	T	T

(12)

[6] G. Boole, *An Investigation of the Laws of Thought* (London, 1854), pp. 247–248.

The truth tables can be read in two directions. The first direction goes from right to left, i.e., from the compound proposition to the elementary propositions. Thus if "$a \lor b$" is true we know that "$a.b$" is true or that "$\bar{a}.b$" is true or that "$a.\bar{b}$" is true. The second direction goes from left to right, i.e., from the elementary propositions to the compound proposition. Thus if "$a.b$" is true we regard the statement "$a \supset b$" as verified. When we use only the first direction we follow a *connective interpretation*; when we use both directions we apply an *adjunctive interpretation*. In conversational language the implication is usually interpreted connectively; the adjunctive implication, therefore, is often regarded as "unreasonable." The "or" is used in both interpretations; an adjunctive "or" therefore appears "reasonable." For logical purposes the adjunctive operations appear preferable; the definition of connective operations can then be achieved by means of the metalanguage.[7] For "adjunctive" the word "extensional" has been used; we prefer the use of "adjunctive" because the word "extensional" has been used in several meanings.

The truth tables of probability logic can be derived from the calculus of probability; they are given by theorems concerning the probability of certain compound classes. We present them in the table (13).[8] The symbols "b" and "c" used in these tables may be regarded as standing for individual sentences or for sequences of sentences, depending on the interpretation employed.

TRUTH TABLES OF PROBABILITY LOGIC

$P(b)$	$P(\bar{b})$
p	$1-p$

(13 a)

$P(b)$	$P(c)$	$P(b.c)$	$P(b \lor c)$	$P(b.c)$	$P(b \supset c)$	$P(b \equiv c)$	$P(c.b)$
p	q	u	$p+q-p \cdot u$	$p \cdot u$	$1-p+p \cdot u$	$1-p-q+2p \cdot u$	$\dfrac{p \cdot u}{q}$

(13 b)

The main difference between the tables (13) and (12) is that in table (13) we need a third argument column. The probability of the compound sentence is not determined by the probabilities of the individual sentences; we need a third probability, the relative probability from "b" to "c", which may be regarded as a degree of coupling between two sentences. Some logicians have objected that for this reason probability logic is not extensional. This is true only for a very narrow meaning of the word "extensional." The truth tables of probability logic are certainly adjunctive in the sense defined above. That the probability of a compound sentence is a function of three parameters, and not

---

[7] Cf. the author's *Elements of Symbolic Logic* (New York, 1947), § 7 and § 9, and chap. viii.

[8] These truth tables were first published by the author in *Sitzungsber. d. Preuss. Akad., Phys.-math. Kl.* (Berlin, 1932, p. 476). The truth tables (20) of the quantitative negation were added in the present publication.

óf two, expresses a generalization of a kind well known in mathematics. Thus in Euclidean geometry the third angle of a triangle is determined by the two other angles, whereas in a non-Euclidean geometry of a given constant curvature this determination requires a third parameter, the area.

Incidentally, we could use in probability logic, instead of the relative probability from "b" to "c", the probability of the compound "b.c" (or of another compound) as the third independent parameter; the probabilities of the other compounds would then be determined. We shall regard the expression "a,b" also as a compound proposition; it has properties similar to those of the other compounds. The expression "P(a,b)" then is of the same form as the expression "P(a.b)" or "P(a ∨ b)".

In addition to the truth tables we have for the three independent parameters the inequality

$$\frac{p+q-1}{p} \leqq u \leqq \frac{q}{p} \tag{14}$$

This relation is derived from the postulate that not only the fundamental probabilities but also all probabilities derivable from them are subject to the normalization condition of being numbers between 0 and 1 limits included. For the value $p = 1$ we derive from (14) that $u = q$. In this case, therefore, the third parameter becomes a function of the two other ones. It can be shown that by the use of this condition the truth tables (12) of two-valued logic can be derived from the truth tables (13) of probability logic. Two-valued logic appears therefore as a special case of probability logic; it is even a degenerate case in which the general dependence on three parameters is eliminated and replaced by a dependence on two parameters only.

A problem that requires particular investigation is the problem of assertion in probability logic. In two-valued logic we follow the rule that only true sentences can be asserted. If a sentence "a" is false, however, we need not resort to the metalinguistic statement " "a" is false", but can express the falsehood by asserting the negation "ā". In fact it is one of the major functions of the negation that it allows us to express falsehood in the object language.

In a logic of the three truth values t_1, t_2, t_3, assertability is restricted to one of these values, for which we may choose t_3. In order to express the other two truth values in the object language we introduce a cyclical negation by the tables (15)

a	$\sim a$
t_3	t_2
t_2	t_1
t_1	t_3

$$\tag{15}$$

When we wish to assert that a sentence "a" has the truth value t_1 we assert the sentence

$$\sim a \tag{16}$$

When the sentence "a" has the truth value t_2 we assert the sentence

$$\sim\sim a \tag{17}$$

We see that the subscript of the truth value corresponds to the number of negations in the assertable sentence. This holds also for the value t_3 since the true sentence can also be written in the form

$$\sim\sim\sim a \tag{18}$$

This procedure can be transferred to probability logic. Usually the probability p of a sentence "a" is expressed by the metalinguistic statement

$$P(a) = p \tag{19}$$

In order to construct an object equivalent of this metalinguistic sentence we introduce a *quantitative negation*. We negate a sentence "a" to the degree w by putting the symbol "$[w]$" before the sentence; we thus obtain the sentence "$[w]a$". The quantitative negation is defined by the truth tables (20):

$P(a)$	$P([w]a)$
p	$p - w + \delta_{p-w}$ $$\delta_{p-w} = \begin{cases} +1 & \text{for } p - w \leqq 0 \\ 0 & \text{for } 0 < p - w < 1 \\ -1 & \text{for } p - w = 1 \end{cases}$$

$$\tag{20}$$

w is a real number between 0 and 1 limits included. We see from the table (20) that if we negate the sentence "a" to the degree p, i.e., if $w = p$, the resulting sentence has the probability 1. Following the rule that only sentences of the probability 1 can be asserted, we can express the degree p of a statement "a" by asserting the statement

$$[p]a \tag{21}$$

This sentence of the object language takes the place of the metalinguistic statement (19). It is easily seen from the truth tables (20) that the statement (21) has the probability 1 if and only if p is the probability of "a". We see, furthermore, that a negation to the degree $w = 1$ leaves the truth value of the statement unchanged. A negation to the degree $w = 0$ in general leaves also the truth value unchanged, except for the case $p = 0$ or $p = 1$, where this negation reverses the truth value.

By means of the quantitative negation we can write derivations in the object language. Thus we have the inferential schema:

$$\begin{array}{l} [p]a \\ [u]\,(a,b) \\ [v]\,(a,b) \\ \hline [p{\cdot}u + (1 - p){\cdot}v]\,b \end{array} \tag{22}$$

For the case $p = 1$ the schema assumes the simple form

$$\frac{\overset{a}{\lceil u \rceil (a,b)}}{\lceil u \rceil b} \tag{23}$$

The two schemas (22) and (23) can be regarded as generalizations of the modus ponens of two-valued logic.

The concept of tautology can be transferred to probability logic by the definition that tautologies are formulas which have the probability 1 for all probability values of their components. It can be shown that all tautologies of two-valued logic remain tautologies within probability logic. In addition, however, we can construct in probability logic tautologies that cannot be written in two-valued logic. This is achieved by means of the quantitative negation. Thus the schema (22) furnishes the tautology

$$\lceil p \rceil a \cdot \lceil u \rceil (a,b) \cdot \lceil v \rceil (\bar{a},b) \rightarrow \lceil p \cdot u + (1 - p) \cdot v \rceil b \tag{24}$$

The arrow indicates a particular form of implication which we do not specify here.

These considerations show that probability logic is a generalization of two-valued logic containing the latter as a special case. Probability logic, therefore, will be applicable to those forms of knowledge in which truth is replaced by probability.

V

Let us summarize the two major results of our analysis.

1. We constructed the calculus of probability in axiomatic form. The calculus so constructed is of a general kind including all types of sequences; the random sequences represent only a special type. The axioms of this general calculus of probability follow from the frequency interpretation.

2. We transformed the calculus of probability into a probability logic by means of truth tables that appear as a generalization of those of two-valued logic.

VI

After presenting the formal system of probability we now turn to the analysis of the application of the formal system to physical reality.

The problem of application may be divided into two parts. In the first part we analyze the meaning of the concept of probability underlying applications to physical reality. In the second part we inquire with what title probability statements about physical objects can be asserted. The two problems, that of *meaning* and that of *assertability*, are closely connected since the assertability will depend largely on the meaning assumed for the probability statement.

Turning to the problem of meaning, let us repeat that the formal system does not prescribe one interpretation as the only admissible one. We said that a set of axioms can be regarded as a set of implicit definitions; they delimit the fundamental concepts only to a certain extent, leaving open a class of admissible interpretations. The meaning of the applied concept of probability is therefore not determined by the formal system. This system will rule out some interpretations as inadmissible; but which of the admissible interpretations is to be used for the applied concept must be determined by considerations outside the formal system.

The problem of meaning may be subdivided into two parts, depending on the type of object to which the concept of probability is applied. The first type of interpretation is given when the object to which the concept of probability is applied consists in a *sequence*, which may be either a sequence of physical events or of propositions about events (propositional sequence). The second type is given by an interpretation which refers to individual events or individual propositions. Let us discuss the two types of interpretation in this order.

VII

As long as sequences are regarded as the objects of probability statements the question of interpretation is not controversial. Practically all logicians are agreed that for such objects only the limit of the relative frequency supplies an adequate interpretation. Let us, therefore, restrict our discussion of the first type of interpretation to the frequency interpretation and study its advantages and difficulties.

The problem of meaning is here identical with the question of whether a statement about the limit of the frequency can be regarded as meaningful. Now it is obvious that there is no difficulty with limit statements if they refer to an *intensionally given* sequence, i.e., a sequence given by a mathematical rule. For instance, to use an example suggested by Poincaré, that the relative frequency of odd and even numbers in the last digit of a table of logarithms converges to the limit $1/2$ is mathematically demonstrable; the meaning of the statement is therefore not controversial. Now it is an advantage of our construction of the calculus of probability that intensionally given sequences are admissible interpretations; even normal sequences can be constructed by means of mathematical rules, as has been shown by Arthur Copeland.[*] But we know that in physical applications we are usually concerned with *extensionally given* sequences, i.e., sequences given element by element. Since it is impossible to enumerate an infinite sequence, it has been questioned whether a statement about the limit in such a sequence is meaningful. We cannot determine, for such sequences, the relation between the width ϵ of convergence and the number n of the element, i.e., we cannot say for which n a convergence within ϵ is reached.

This difficulty, however, does not seem to me so serious as it has appeared to other logicians. That in applied statistics we speak of infinite sequences must be regarded as an idealization which we use for the sake of convenience although we know that all applications are restricted to finite sequences. The idealization may be compared to the one used in geometry, where we speak of

[*] As cited in note 2 above.

lines and points without width although we know that physical objects do not strictly satisfy this requirement. It would be possible to replace idealized geometry by a geometry in which points are small areas and lines have a narrow width; in this geometry the usual theorems would hold only approximately. Similarly, we could construct a *finitized* calculus of probability which deals with sequences of a finite length possessing some sort of convergence toward a limit within an interval ε which could be defined by suitable methods. While the theorems of the elementary calculus of probability would even hold strictly for that calculus, the theorems of the theory of order would hold only approximately and would have to be carefully worded in a way that would include reference to the interval of convergence and the length of the subsequences. There is no doubt that by the construction of such a calculus all postulates of the *finitist* could be strictly satisfied. When we prefer to speak of infinite sequences we do so because of the great simplification introduced by this idealization. If, however, the question of meaning is under consideration, we may always refer to the fact that for all purposes of application finite sequences will suffice. In fact, the "limit" of the applied calculus is meant to be a *practical limit*. A sequence has a practical limit if, for a finite number of elements, large enough and yet accessible to human experience, it shows properties of convergence. This definition excludes sequences that converge so late that for all human observations they behave like nonconvergent series; on the other hand, it admits sequences that diverge after the section of practical convergence and have no limit when continued to infinity. The concept of practical limit will answer the question of meaning for the probability sequences of applications.

Let us now turn to the problem of assertability. It may be subdivided into the problem of the assertability of probability laws and the problem of the ascertainment of the degree of probability. The first problem is easily answered: since the axioms of the calculus follow from the frequency interpretation, the laws of probability are guaranteed, for this interpretation, by deductive logic. This is a great advantage of the frequency interpretation; it is made possible by the generalized form of the calculus of probability, in which the inner structure of probability sequences is regarded as unessential. The definition of various types of order is given within a special chapter, the theory of order. We saw that all these definitions are constructed in the form of postulates stating that certain probabilities are equal. The determination of a type of order, therefore, requires no more than the ascertainment of a degree of probability.

It is the second problem, the ascertainment of the degree of probability, that leads into difficulties for extensionally given sequences. While for an intensionally given sequence the value of the limit of the frequency can be derived from the defining rule, there exists for extensionally given sequences only the method of enumeration: we have to count the whole sequence in order to know the limit. For infinite sequences this is certainly impossible. We should realize, however, that the difficulties of this problem are not removed by a finitization. Although a finite sequence can be enumerated, in principle, this possibility does not help us because, in all practical applications, we wish to know the value of the limit before the total sequence is observed. We resort here to the method of counting an initial section of the sequence and then assuming that the observed frequency will persist on further prolongation of the sequence. This procedure represents the *inductive inference*. The frequency interpretation of probability, whether employed for infinite or for finite sequences, is therefore burdened with the problem of induction. A satisfactory solution of the problem of applicability can be given only when it is possible to solve the problem of induction.

It is here that we meet with a specific difficulty of the calculus of probability. For other axiomatic systems there exists no problem of application; the axioms can be applied if the physical objects under consideration have the properties required in the axioms, and instead of a problem of applicability we have the requirement of a *suitable choice of the objects*. With respect to probability sequences this method breaks down because it is the very question whether the physical objects possess the necessary properties that cannot be answered. If we could wait until the total probability sequence is observed we could apply the method of the suitable choice of the object, rejecting as unsuitable any sequence that does not have the required property. In practical statistics, however, we cannot wait until the whole sequence is observed for the reason that we wish to use probability values for predictions. It is the predictive nature of the applied calculus of probabilities that leads into the difficulties of the problem of induction.

VIII

Let us now turn to the investigation of applications in which the concept of probability refers to an individual event or to an individual proposition. It is on this ground that the frequency interpretation has been questioned. Some logicians have argued that we are concerned here with a different notion of probability not reducible to frequencies. Let us inquire whether the contention of the existence of two disparate notions of probability is tenable.

At first sight, indeed, it appears as though a probability applied to a single case has nothing to do with a frequency. We say, "it is probable that it will rain tomorrow"; "it is improbable that Julius Caesar was in Great Britain"; etc.; and thus refer probability to a single event, or in the logical conception to a single proposition. What does it help us to know that in a certain percentage of days of certain meteorological conditions it will rain, when we wish to know the probability for rain on one individual day? Similarly the example of Julius Caesar's stay in Britain has often been quoted as denying a frequency interpretation. Let us analyze the various meanings that can be suggested for such a second concept of probability.

In the first interpretation, the degree of probability is regarded as a measure of the intensity of expectation with which we anticipate a future event. This interpretation, however, leads into difficulties because the feeling of expectation varies from person to person; we rather use probabilities as a standard of what the intensity of expectation *should be*, but not as a measure of what it *is*.

Thus the optimist is controlled by too high an expectation if the event expected is desirable; the pessimist, on the other hand, suffers from too low an expectation. But if probability is a standard of what expectation should be it cannot be identified with the intensity of a psychological status. Furthermore, the validity of the laws of probability is by no means warranted by such an interpretation.

The second and third interpretations to be considered derive from a problem which historically speaking constituted the first philosophical issue by which the theory of probability was confronted. The historical origin of the calculus of probability from the study of the games of chance has led to the conception that a degree of probability can be ascertained by means of an a priori method that is attached to a disjunction of equally probable cases. Throwing a die, for instance, we frequently argue that the six possible cases should be equally probable because we have no reason to prefer one face to the other. The principle of this inference was called the *principle of indifference*, or of *no reason to the contrary*. It was introduced by Laplace; and although mathematicians have long since abandoned this principle, it has haunted the field of philosophical inquiry into the nature of probabilities. Up to our day it is defended fervently by some philosophers who hope by such means to secure for themselves a fenced-off area, a reservation, safe from the precision of mathematical methods.

That the principle of indifference is logically untenable has been sufficiently demonstrated. Maybe we have no reason to prefer one face of the die to the other; but then we have no reason to assume that the faces are equally probable, either. To transform the absence of a reason into a positive reason represents a feat of oratorical art worthy of an attorney of the defense but not permissible in the court of logic. Moreover, it has been demonstrated clearly enough that the principle of indifference leads to contradictions when applied to geometrical probabilities connected by a nonlinear measure transformation. I should like to classify the principle of indifference as a *fallacy of incomplete schematization*. It leads to a true conclusion in the case of a die possessing geometrical symmetry, but to false conclusions in other cases. Where it leads to true conclusions it can do so only because more is known than is stated in the principle. In the case of the die, for instance, we can derive the equiprobability of the faces by a more complicated schema of inference that contains in its premises a very general empirical statement about probability functions.[10] Unfortunately, the incomplete schematization formulated in the principle of indifference has misled some philosophers into construing the concept of probability in such a way that the principle can be maintained.

The first attempt to save the principle of indifference is based on the logical *principle of retrogression*. This principle plays a part in the theory of meaning; it states that the meaning of a sentence is given by the method of its verification. Since, according to the adherents of the a priori conception of probability, we determine a probability by counting the terms of an exclusive disjunction in which we have no reason to prefer one term, the principle of retrogression

---

[10] Cf. the author's *Wahrscheinlichkeitslehre* (Leiden, 1935), § 65.

furnishes the result that the meaning of probability is given by reference to such a disjunction. We thus arrive at a *retrogressive interpretation* of probability. For instance, that the probability of obtaining face 6 with a die is 1/6 means, according to this interpretation, that face 6 is a term of a disjunction of six terms and that we have no reason to prefer one of the terms.[11] It is obvious that with this interpretation the use of the principle of indifference is justified since then the probability statement states no more than what is assumed as the premise of that principle; but it is equally obvious that with this interpretation the probability statement has lost its predictional value. Why should we bet on the occurrence of the event "non-six" rather than on the occurrence of "six"? The retrogressive interpretation narrows down the meaning of probability statements in such a way that the assertion of the statement is justified; but in the transition from probability statements to bets, or advices to action, there reappears the very problem that the retrogressive interpretation was intended to evade and that the principle of indifference cannot solve.

The retrogressive interpretation has been used in a second form to cover another fallacious application of probability inferences, called the *inference by confirmation.*[12] It is claimed by some logicians that the probability of hypothetical assumptions is derived by means of the following inference: When a certain consequence of the assumption is verified, such observation is regarded as conferring a certain degree of probability to the assumption. This means an inference of the form

$$\text{``}a \supset b\text{'' is proved}$$
$$\text{``}b\text{'' is verified}$$

$$\overline{\text{``}a\text{'' is probable}}$$

The theory of probability knows no such inference. Whenever such inference is successfully applied, we can show that it represents an incomplete schematization, that much more is known than is expressed in the inference. In fact, the inferences made in the theory of indirect evidence and in the verification of scientific theories must be construed as inferences in terms of the rule of Bayes. They can be made only when the values of some other probabilities are known or, at least, can be roughly estimated.

In order to save the inference by confirmation an attempt has been made to construct a retrogressive interpretation of probability in such a way that the meaning of probability is given by the premises of the inference. This kind of probability is called *degree of confirmation.* Although, of course, such an interpretation would justify the inference, it is clear that the degree of confirmation so defined possesses no predictional value and cannot account for the reliability of hypothetical assumptions.

[11] As far as I know, the first to use this interpretation was K. Stumpff in a paper published in *Berichte d. Bayer. Akad., Philos. Kl.* (Munich, 1892). The interpretation was taken up in our day by some logicians under the influence of ideas of L. Wittgenstein, *Tractatus Logico-philosophicus* (London, 1922), p. 113; thus by A. Waisman, *Erkenntnis,* V, 1 (1930), 229.

[12] Cf. R. Carnap, "Testability and meaning," *Philosophy of Science,* III (1937), 420. Carnap's ideas on confirmation were carried on chiefly by E. Nagel and C. Hempel.

Finally we may mention here a retrogressive interpretation of the principle of induction. According to this interpretation, the meaning of the conclusion of the inductive inference is given by the statement of the premises. Thus when we infer from past observations that the sun will rise tomorrow we mean by this conclusion, according to the interpretation, that we have seen the sun rising in past observations. This interpretation is so absurd that it has scarcely ever been maintained. It is obvious that with it the principle of induction would lose its predictional value.

In view of the deficiencies of the retrogressive interpretation, the adherents of the a priori determination of probability degrees have introduced another interpretation, which abandons the principle of retrogression and regards probability as a *primitive concept* not capable of further definition. According to this conception, the statement that the probability of an expected event is 1/6 has a meaning of its own, comparable to the meaning of the primitive notions of logic; and we cannot interpret this meaning as a frequency or a report about terms in a disjunction. This conception is sometimes stated in the form that probability is a *rational belief*, that the laws of probability constitute a *quantitative logic* based on a self-evidence comparable to those of ordinary logic.[13] As far as I see, the primitive-concept interpretation is not always clearly distinguished from the retrogressive interpretation by its adherents; it appears that some logicians vacillate between the two interpretations and use sometimes the one, sometimes the other, depending upon what they wish to prove.

The difficulties of the primitive-concept interpretation appear to me so overwhelming that I cannot see how a logician can commit himself to this interpretation. First, the degree of probability remains unverifiable. When the event expected with a probability 5/6 is observed, does this observation verify the probability statement? Obviously not, since the nonoccurrence of the event is also compatible with the probability statement. A numerical value of a probability cannot be ascertained by one observation. We do not escape this predicament by restricting probability statements to relations of order stating that a probability is higher or lower than another; such relations cannot be verified by one observation, either. It is sometimes argued that the verification of the degree of probability is obtained not by the observation of the event but by other methods such as used in the principle of indifference. But such procedure can be regarded as a verification only if a retrogressive interpretation of probability is adopted. With this turn, however, the primitive-concept interpretation is abandoned and the interpretation loses its predictional value.

Another difficulty of the primitive-concept interpretation is the justification of the laws of probability. In fact, the whole calculus of probability and its application to physical objects appears here as a system based on *synthetic self-evidence*. A philosophy of probability that would commit itself to interpret

---

[13] This conception is represented by the ideas of J. M. Keynes, *Treatise on Probability* (London, 1921), and was continued by H. Jeffreys, *Theory of Probability* (Oxford, 1939). I could not say to what extent it is present also in the ideas of Carnap and others about confirmation. If the so-called degree of confirmation is meant to be a measure of reliability and an advice to action, it falls under this category.

probability as a primitive notion would lead logic back to rationalism, to a *synthetic a priori;* in other words, to a metaphysics that claims an intrinsic correspondence between reason and physical reality.

IX

The analysis of meaning has suffered from too close an attachment to psychological considerations. The meaning of a sentence has been identified with the pictures and representations associated with the utterance of the sentence. Such conception will lead to meanings varying from person to person; and it will not help us to find out the meaning a man would adopt if he had a clear insight into the implications of his words. Logic is interested not in what a man means but in what he *should* mean; that is, in that meaning which, if assumed for his words, would make his words compatible with his actions. When we analyze the meaning of probability statements about single events by the use of this objective criterion, we find that the frequency interpretation can be applied to this case too, and that we need not resort to one of the questionable interpretations based on the reconstruction of subjective psychological intentions, discussed above.

Assume that the frequency of an event B in a sequence is $= 5/6$. Confronted by the question of whether an individual event B will happen, we shall prefer to answer in the affirmative because, if we do so repeatedly, we shall be right in $5/6$ of the cases. We shall not claim that the individual assertion is true; we shall assert it in the sense of a *posit*, i.e., in the attitude of a man who lays a bet. A posit is a statement with which we deal as true although we have no knowledge about its truth. The greater a probability in the frequency sense, the more favorable will it be to posit the individual statement because on repetition we shall have a greater number of successful predictions. The probability appears as a *rating* of the posit, which we call its *weight.*

According to this conception, the probability of an individual statement appears as a fictitious property resulting from a *transfer of meaning from the general to the particular case.* Strictly speaking, it has no meaning that in an individual case the probability of casting "non-six" with a die is $= 5/6$; but when we coördinate to this statement a fictitious meaning it will lead to a behavior that, in repeated applications, will be the most successful one. The frequency interpretation allows us to construct a fictitious meaning for the probability of individual events, or propositions, of such a kind that it makes our words compatible with our actions.

It is not necessary for this conception that the events to be repeated are all of the same kind. Events of various kinds and probabilities may be connected into a sequence such that always positing the more probable event will lead to the greater number of successes. The sequences of insignificant events of everyday life furnish large enough numbers admitting this application of the frequency interpretation.

An apparent difficulty results from the fact that, given an individual event, we often do not know in what reference class we should incorporate it. Conversational language is none too precise in this respect; we speak of the probability of the death of a certain person, of the probability of an expected political event, etc., without explicitly stating a reference class for which the probability is to be constructed. In such cases the statement may be understood to mean: the probability of the event with respect to the best reference class available. This reference class may be defined as the narrowest reference class for which we have reliable statistics. Thus a physician, when asked for the probability of the death of a certain person, will know into which reference class the case should be incorporated. That a suitable choice of the reference class for political events is so difficult indicates that the statistical laws of politics are none too well known.

The logical form of a language that deals with probabilities as truth values of individual sentences is the probability logic presented above. Since conversational as well as scientific language, to a great extent, is of this type, probability logic constitutes the form of a great part of actual language. In fact, the use of two-valued logic in statements about physical reality must be regarded as a degenerate form of probability logic, in which only high and low probabilities are employed, while intermediate values are omitted. That the meaning of the term "probable" in statements such as "Peter will probably come," or "the enemy will probably accept the ultimatum," can be assumed to be the same as the one used in mathematical statistics, is guaranteed by the fact that the truth tables of probability logic are derivable from the calculus of probability.

X

The concept of weight may also be applied on a higher level, i.e., with reference not to propositions about facts but to propositions stating the probabilities of other propositions. From the calculus of probability it is known that probabilities for the occurrence of certain frequencies can be derived; thus the Bernoulli theorem furnishes probabilities for the convergence of the frequency toward a certain limit. If the limit of the frequency under consideration is regarded as a probability of the first level, the Bernoulli probability is a probability of the second level. In those cases where probabilities of the second level can be computed we can employ them as weights determining a rating for an inductive inference; they tell us the reliability of the assumption that the observed frequency will persist. In fact, most inductive inferences are made, not as isolated inferences, but within a network of other inductions. The theory of indirect evidence can be constructed in terms of such a network; the inferences connecting the individual inductions are covered by the theorems of the calculus of probability. The concatenation of inductions thus achieved improves greatly the reliability of inductive methods; on the other hand, the analysis of such network inferences allows us to account for more complicated forms of induction which do not directly possess the form of induction by enumeration.

In view of the fact that all axioms of the calculus of probability are derivable from the frequency interpretation, we can now formulate the two following results:

1. All probability inferences are reducible to induction by enumeration with the addition of deductive inferences.

2. Although some inductive inferences can be given a rating expressed by probabilities, there will always remain other inductive inferences whose weight is unknown. The general theory of induction does not constitute a chapter of probability theory, but must be given without the use of probability considerations.

XI

We now turn to the discussion of the one remaining difficulty connected with the frequency interpretation—of the general problem of induction.

The inference of induction has found its critic in David Hume. Although Hume attached his famous criticism to the particular case that the relative frequency of the event is = 1, his results hold likewise for the general case of the statistical inference in which the persistence of an observed percentage of events is assumed. What Hume[14] has shown is:

1) The inductive inference is not logically a priori, i.e., the conclusion is not a necessary consequence of the premises.

2) The inductive inference is not logically a posteriori. Any attempt to explain the inductive inference as a result of past experiences in which the inference was used successfully is circular reasoning because the inductive inference would be used for its own justification.

The truth of these results is unquestionable. Does it follow, however, that no justification of induction can be given?

Hume believed that this consequence is inescapable. His theory that induction is a habit is not meant to be a justification; and it is no way out of the difficulty, because it refers to a psychological fact that is logically irrelevant. There are good habits and bad habits; the logical problem is whether induction is a good habit.

I think that it is possible to give a positive answer to this question. The inductive inference can be justified; induction can be shown to be a good habit. This proof, however, requires a reinterpretation of language; scientific and other statements must be regarded, not as *assertions* claimed to be *true*, but as *posits* claimed to be *the best posits* we can make.

If this revision of the claims of language is accepted, the justification of induction is rather easily given. We speak of a justification if it can be shown that for the pursuit of a certain aim it is advisable to apply a certain means; a justification always concerns a means with respect to an end. Let us formulate the aim of making predictions, which is common to science and everyday life, in the form that we wish to find limits of the relative frequency in sequences. When we now regard the inductive inference as a rule for constructing posits, to be applied repeatedly in the sense of a trial subject to later correction, we can show that, if the sequence has a limit, the inductive inference will lead to this limit within an interval ϵ of exactness in a finite number of steps. This result follows from the definition of the limit. If, on the other hand, the se-

[14] David Hume, *Enquiry concerning Human Understanding* (1748).

quence has no limit of the frequency, the inductive rule will not find it—but then, no other method will find it, either. The use of the rule of induction, therefore, can be regarded as the fulfillment of a necessary condition of success in a situation in which a sufficient condition is unknown to us. To speak in such a case of a justified use of a rule appears in agreement with linguistic usage concerning the word "justification." Thus we call Magellan's enterprise justified because, if he wanted to find a thoroughfare through the Americas, he had to sail along the coast until he found one—that there was a thoroughfare was by no means guaranteed. He could act only on the basis of necessary conditions of success; that they would turn out sufficient was unknowable to him.

This simple consideration solves the problem of induction. It removes the last difficulty connected with the frequency interpretation of probability. It requires, on the other hand, a renunciation of a rationalistic attitude toward knowledge; we have to give up the quest for certainty if we wish to account for the use of probability methods. Such reconversion of emotional attitudes is not always easy; but once it is achieved it offers us the greatest reward that philosophical analysis can ever find: it supplies a proof that our methods of knowledge are the best instrument of finding predictions, if predictions can be found. I could not think of a better justification of scientific method than the proof that to apply such method is the best we can do.

The propensity interpretation of the calculus of probability, and the quantum theory

by

K. R. POPPER

In this paper, I propose briefly to put forth and to explain the following theses, and to indicate the manner of their defence.

(1) The solution of the problem of interpreting probability theory is fundamental for the interpretation of quantum theory; for quantum theory is a probabilistic theory.

(2) The idea of a statistical interpretation is correct, but is lacking in clarity.

(3) As a consequence of this lack of clarity, the usual interpretation of probability in physics *oscillates* between two extremes: an *objective* purely statistical interpretation and a *subjective* interpretation in terms of our incomplete knowledge, or of the available information.

(4) In the orthodox Copenhagen interpretation of quantum theory we find the same oscillation between an objective and a subjective interpretation: *the famous intrusion of the observer into physics.*

(5) As opposed to all this, a revised or reformed statistical interpretation is here proposed. It is called the *propensity interpretation of probability.*

(6) The propensity interpretation is a purely objective interpretation. It eliminates the oscillations between objective and subjective interpretation, and with it the intrusion of the subject into physics.

(7) The idea of propensities is 'metaphysical', in exactly the same sense as forces or fields of forces are metaphysical.

(8) It is also 'metaphysical' in another sense: in the sense of providing a coherent programme for physical research.

These are my theses. I begin by explaining what I call the propensity interpretation of probability theory.[1]

Section 1. Objective and Subjective Interpretations of Probability.

Let us assume that we have two dice: one is a *regular* die of homogeneous material, the other is *loaded*, in such a way that in long sequences of throws the side marked '6' comes uppermost in about 1/4 of the throws. We say, in this case, that the probability of throwing a 6 is 1/4.

Now the following line of arguing seems attractive.

We ask what we *mean* by saying that the probability is 1/4; and we may arrive at the answer: What we *mean*, precisely, is that the relative frequency, or the statistical

---

[1] I have explained the propensity interpretation of probability and of quantum theory very briefly in my paper 'Three Views Concerning Human Knowledge', in *Contemporary British Philosophy*, edited by H. D. Lewis, 1956, p. 388. A full treatment of the propensity interpretation and of its repercussions upon quantum theory will be found in the *Postscript: After Twenty Years* to my *Logic of Scientific Discovery*, 1957.

135

frequency, of the results in long sequences is 1/4. Thus probability is relative frequency in the long run. This is the statistical interpretation.

The statistical interpretation has been often criticized because of the difficulties of the phrase 'in the long run'. I will *not* discuss this question. Instead I will discuss the question of *the probability of a* SINGLE EVENT. This question is of importance in connexion with quantum theory because the ψ-function determines the probability of a *single electron* to take up a certain state, under certain conditions.

Thus we ask ourselves now what it *means* to say 'The probability of throwing 6 *with the next throw* of this loaded die is 1/4.'

From the point of view of the *statistical interpretation*, this can only mean one thing: 'The next throw is a *member of a sequence* of throws, and the relative frequency within this sequence is 1/4.'

At first sight, this answer seems satisfactory. But we can ask the following awkward question:

What if the sequence consists of throws of a *loaded* die, with one or two throws of a *regular* die occurring in between the others? Clearly, we shall say about the throws with the regular die that their probability is different from 1/4, in spite of the fact that these throws are members of a sequence of throws with the frequency 1/4.

This simple objection is of fundamental importance. It can be answered in various ways. I shall mention two of these answers, one leading to a *subjective interpretation*, the other to the *propensity interpretation*.

The first or subjective answer is this. 'You have assumed in your question', the subjectivist may address me, 'that *we know* that the one die is loaded, the other regular, and also that *we know* whether the one or the other is used at a certain place in the sequence of throws. In view of this information, we shall of course attribute the proper probabilities to the various single throws. For probability, as your own objection shows, is not simply a frequency in a sequence. Admittedly, observed frequencies are important as providing us with valuable *information*. But we must use *all* our information. The probability is our assessment, in the light *of all we know*, of reasonable betting odds. It is a measure which depends essentially upon our incomplete information, and *it is a measure of the incompleteness of our information*: if our information about the conditions under which the die will be thrown were sufficiently precise, then there would be no difficulty in predicting the result with certainty.'

This is the subjectivist's answer, and I shall take it as a characterization of the subjectivist position which I shall not discuss further in this paper, although I shall mention it in various places.[2]

Now what will the defender of an objective interpretation say to our fundamental objection? Most likely he will say (as I myself used to say for a long time) the following:

'To make a statement about probability is to propose a *hypothesis*. It is a hypothesis about frequencies in a sequence of events. In proposing this hypothesis, we can make use of all sorts of things—of past experience, or of inspiration: it does not matter *how we get* it; all that matters is *how we test it*. Now in the case mentioned, we all agree on the frequency hypothesis, and we all agree that the frequency of 1/4

---

[2] I have discussed and criticized the subjectivist position very fully elsewhere (see the preceding footnote). The subjectivist interpretation of probability is a necessary consequence of determinism. Its retention within the quantum theory is a residue of a not yet fully eliminated determinist position.

66

will not be affected by having one or two throws with a regular die in between the throws with a loaded die. As to the regular throws, *if* we consider them merely as belonging to this sequence, we have to attribute to them, strange as it may sound, the probability of $1/4$, even though they are throws with a regular die. And if, on the other hand, we attribute to them the probability of $1/6$, then we do so oecause of the hypothesis that in *another* sequence—one of throws with the regular die—the frequency will be $1/6$.'

This is the objectivist's defence of the purely statistical interpretation, or of the frequency interpretation, and *as far as it goes* I still agree with it.

But I now think it strange that I did not press my question further. For it seems clear to me now that this answer of mine, or of the objectivist's, implies the following. In attributing probabilities to sequences, we consider as decisive the *conditions under which the sequence is produced.* In assuming that a sequence of throws of a loaded die will be different from a sequence of throws of a regular die, we attribute the probability to the *experimental conditions.* But this leads to the following result.

Even though probabilities may be said to be frequencies, we believe that these *frequencies will depend on the experimental arrangement.*

But with this, we come to a new version of the objective interpretation. It is as follows.

Every experimental arrangement is *liable to produce,* if we repeat the experiment very often, a sequence with frequencies which depend upon this particular experimental arrangement. These virtual frequencies may be called probabilities. But since the probabilities turn out to depend upon the experimental arrangement, they may be looked upon as *properties of this arrangement. They characterize the disposition, or the propensity,* of the experimental arrangement to give rise to certain characteristic frequencies *when the experiment is often repeated.*

Section 2. *The Propensity Interpretation*

We thus arrive at the propensity interpretation of probability.[3] It differs from the

---

[3] What we interpret is not a word, 'probability', and its 'meaning', but formal systems—the probability calculus (especially in its measure-theoretical form), and the formalism of quantum theory.

A formalized set of axioms for relative (or conditional) probability is the following. (The theory of real numbers is assumed.)

(A) $a, b \in K \to : p(a, b)$ is a real number, and

     (a) $((c)(c \in K \to p(a, c) = p(b, c))) \to (d)(d \in K \to p(d, a) = p(d, b))$

     (a') $p(a, a) = 1$

(B) $a, b \in K \to : ab \in K$; and $b, c, bc \in K \to p(a, bc)p(b, c) = p(ab, c) \leqslant p(a, c)$

(C) $a \in K \to : \,'a \in K$, and $b, c \in K \to . \, p(b, c) \neq 1 \to p(a, c) + p('a, c) = 1$

(D) $(Ea)(Eb)(a, b \in K,$ and $p(a, b) \neq 1)$

(ab is here, of course, the meet or conjunction of a and b, and $'a$ the complement of a.)

This set is equivalent to a more concise set in which no numerical constant such as '1' occurs. It is obtained by retaining (a), omitting (a'), and replacing the second (operational) lines of (B) and (C), as well as (D), respectively, by the following lines ((B) is now made 'organic'):

     (b) $d \in K \to ((p(a, bc)p(b, c) = p(d, c) \to p(a, c) < p(d, c)) \to p(ab, c) < p(d, c))$

     (c) $d \in K \to (p(a, a) \neq p(b, c) \to p(a, c) + p('a, c) = p(d, d))$

     (d) $(Ea)(Eb)(a, b \in K,$ and $p(a, b) = p(b, b))$

This axiom system has the following properties: (1) if we define absolute probability by

$$p(a) = p(a, \,'(a' \, a)),$$

then we can say that $p(a, b)$ is defined in our system even if $p(b) = 0$. In this respect, the system is a generalization of the systems known so far (except my systems published in *B.J.P.S.*, 6, 1955, pp. 56f., and in *British Philosophy in the Mid-Century*, ed. C. A. Mace, 1957, p. 191). (2) It is not, as in other systems, tacitly assumed that the elements of K satisfy the postulates of Boolean algebra (this fact, on the contrary, can be proved). This is of advantage in connexion with the problem of interpretation, such as the problem whether propensities satisfy Boolean algebra: it becomes superfluous to postulate that they do.

67

137

purely statistical or frequency interpretation only in this—that it considers the probability as a characteristic property of the experimental arrangement rather than as a property of a sequence.

The main point of this change is that we now take as fundamental *the probability of the result of a single experiment*, with respect to its *conditions*, rather than the frequency of results in a sequence of experiments. Admittedly, if we wish to *test* a probability statement, we have to test an experimental sequence. But now the probability statement is not a statement *about* this sequence: it is a statement *about* certain properties of the experimental conditions, of the experimental set-up. (Mathematically, the change corresponds to the transition from the frequency-theory to the measure-theoretical approach.)

A statement about propensities may be compared with a statement about the strength of an electric field. We can test this statement only if we introduce a test body and measure the effect of the field upon this body. But the statement which we test speaks about the field rather than about the body. It speaks about certain *dispositional properties* of the field. And just as we can consider the field as physically real, so we can consider the propensities as physically real. They are *relational* properties of the experimental set-up. For example, the propensity 1/4 *is not a property of our loaded die*. This can be seen at once if we consider that in a very weak gravitational field, the load will have little effect—the propensity of throwing a 6 may decrease from 1/4 to very nearly 1/6. In a strong gravitational field, the load will be more effective and the same die will exhibit a propensity of 1/3 or 1/2. The tendency or disposition or propensity is therefore, as a relational property of the experimental set-up, something more abstract than, say, a Newtonian force with its simple rules of vectorial addition. *The propensity distribution attributes weights to all possible results of the experiment.* Clearly, it can be represented by a vector in the *space of possibilities*.

Section 3. Propensity and Quantum Theory

The main thing about the propensity interpretation is that *it takes the mystery out of quantum theory, while leaving probability and indeterminism in it*. It does so by pointing out that all the apparent mysteries would also involve thrown dice, or tossed pennies—*exactly* as they do electrons. In other words, it shows that quantum theory is a probability theory just as any theory of any other game of chance, such as the bagatelle board (pin board).

In our interpretation, Schrödinger's ψ-function determines the propensities of the states of the electron. We therefore have no 'dualism' of particles and waves. The electron is a particle, but its wave theory is a propensity theory which attributes weights to the electron's possible states. The waves in configuration space are waves of weights, or waves of propensities.

Let us consider Dirac's example of a photon and a polarizer. According to Dirac, we have to say that the photon is in both possible states at once, half in each; even although it is indivisible, and although we can find it, or observe it, in only one of its possible states.

We can translate this as follows. The theory describes, and gives weight to, all the possible states—in our case, two. The photon will be in one state only. The

68

138

situation is exactly the same as with a tossed penny. Assume that we have tossed the penny, and that we are shortsighted and have to bend down before we can observe which side is upmost. The probability formalism tells us then that each of the possible states has a probability of $1/2$. So we can say that the penny is half in one state, and half in the other. And when we bend down to observe it, the Copenhagen spirit will inspire the penny to make a quantum jump into one of its two *Eigen*-states. For nowadays a quantum jump is said by Heisenberg to be the same as a reduction of the wave packet. And by 'observing' the penny, we induce exactly what in Copenhagen is called a 'reduction of the wave packet'.

The famous two-slit experiment allows exactly the same analysis. If we shut one slit, we interfere with the possibilities, and therefore get a different ψ-function, and a different probability distribution of the possible results. *Every change in the experimental arrangement such as the shutting of a slit, will lead to a different distribution of weights to the possibilities* (just as will the shifting of a pin on a pin board). That is, we obtain a different ψ-function, determining a different distribution of the propensities.

There is nothing peculiar about the role of the observer: he does not come in at all. What 'interferes' with the ψ-function are only changes of experimental arrangements.

The opposite impression is due to an oscillation between an objective and a subjective interpretation of probability. It is the subjective interpretation which drags in our knowledge, and its changes, while we ought to speak only of experimental arrangements, and the results of experiments.

Section 4. Metaphysical Considerations

I have stressed that the propensities are not only as objective as the experimental arrangements but also *physically real*—in the sense in which forces, and fields of forces, are *physically real*. Nevertheless they are *not* pilot-waves in ordinary space, but weight functions of possibilities, that is to say, vectors in possibility space. (Bohm's 'quantum-mechanical potential' would become here a propensity to accelerate, rather than an accelerating force. This would give full weight to the Pauli-Einstein criticism of the pilot-wave theory of de Broglie and Bohm.) We are quite used to the fact that such abstract things as, for example, degrees of freedom, have a very real influence on our results, and are in so far something physically real. Or consider the fact that, compared with the mass of the sun, the masses of the planets are negligible, and that, compared with the masses of the planets, those of their moons are also negligible. This is an abstract, a relational fact, not attributable to any planet or to any point in space, but a relational property of the whole solar system. Nevertheless, there is every reason to believe that it is one of the 'causes' of the stability of the solar system. Thus abstract relational facts can be 'causes', and in that sense physically real.

It seems to me that by stressing that the ψ-function describes physical realities, we may be able to bridge the gap between those who rightly stress the statistical character of modern physics and those who, like Einstein and Schrödinger, insist that physics has to describe an objective physical reality. The two points of view are incompatible on the subjectivist assumption that statistical laws describe our own imperfect state of knowledge. They become compatible if only we realize that these

69

statistical laws describe propensities, that is to say, objective relational properties of the physical world.

Beyond this, the propensity interpretation seems to offer a new metaphysical interpretation of physics (and incidentally also of biology and psychology). For we can say that all physical (and psychological) properties are dispositional. That a surface is coloured red means that it has the disposition to reflect light of a certain wave length. That a beam of light has a certain wave length means that it is disposed to behave in a certain manner if surfaces of various colours, or prisms, or spectographs, or slotted screens, etc., are put in its way.

Aristotle put the propensities as potentialities *into* the things. Newton's was the first *relational* theory of physical dispositions and his gravitational theory led, almost inevitably, to a theory of fields of forces. I believe that the propensity interpretation of probability may take this development one step further.

The Propensity Interpretation of the Calculus of Probability, and the Quantum Theory, in *Observation and Interpretation, Proceedings of the Ninth Symposium of the Colston Research Society, University of Bristol*, edited by S. Körner in collaboration with M.H.L. Pryce; Volume 9 of the *Colston Papers*, Butterworths, London, 1957, pp.65–70 and 88–89. For *Errata* see *British Journal for the Philosophy of Science*, 8, 1958, No.32, p.301, Note 1. [See also 1962 (o).]

ON THE JUSTIFICATION OF INDUCTION

MISS CREED'S paper gives me the welcome opportunity to explain more fully what I understand by a justification of induction. We both agree that the question of justification can be raised only if a decision on a certain aim has been made; i.e., justification concerns the question of the appropriateness of a certain means in respect to a chosen aim, not the question of the choice of the aim itself. The question of appropriateness, however, includes the question of the attainability of the aim; thus we have to distinguish different kinds of justification according to differences in what we know about the attainability of the aim.   I say "differences in *what we know*," for a man's actions can be guided only by his knowledge of the world, and not by unknown features of the world; if we have to decide whether his actions are reasonable we have to ask whether they are reasonable in relation to what he knows.

Now as to our knowledge of the attainability of the aim I would distinguish the following cases:

1. We know something about the objective possibility of reaching the aim A by applying the means M. We can introduce here the subdivisions:

α we know that by applying the means M we shall certainly reach the aim A;

β we know the probability p that A will occur if M is applied;

γ we know at least that $p > 0$;

δ we know that although p may be $= 0$ the aim is possible.

The difference between γ and δ is here not very great; we have to distinguish these cases only because the probability $p = 0$ is not equivalent to impossibility (i.e., the limit of the frequency may be 0 although the event sometimes happens). For the same reason case α is slightly different from the case $p = 1$. But these differences are not very important for our problem.

2. We do not know whether or not application of the means M will lead to the aim A.

3. We know that by applying the means M we shall never reach the aim A.

Now we certainly agree that in case 3 a justification of the procedure M can not be given. But it seems to me that in all the other cases it can be given. Although we know only in case 1α that the means will lead to success we should call the procedure justified also in the cases 1β, γ, δ, and 2, if the means M can be shown to be the best means we know in respect to A. Miss Creed would agree with me as far as 1α, β, γ are concerned; her condition b is equivalent to my case 1γ. The difference then would be that a justification is not considered possible in the cases 1δ and 2.

Instead Miss Creed considers another case which is called b' and defined by the clause "that it be assumed or believed that success is practically possible, though there be no evidence to support the assumption or belief." I confess that this seems to me a rather strange case. Belief can be the *motive* of an action, but belief as such can never *justify* an action; only a *justified* belief can do that. But if according to the definition of case b', the belief is unjustified— how can it lead to a justification of induction?

This is a point in which, it seems to me, Miss Creed's paper reveals the tracks of David Hume, and unfortunately of that part of his theory of induction which has brought so much confusion into the empiricist camp. It seems to me that after his brilliant criticism of induction, the merits of which can not be overestimated, Hume ran the problem into a side track by his defense of inductive belief as a habit. Even today the consideration of this habit, and belief, is in the foreground of the philosophic discussion of the inductive problem, and eclipses the logical problem behind this psychological fact. Miss Creed speaks in the title of her paper of the justification of the *habit* of induction. I do not think she would ever have written a paper on the habit of the syllogism. Although syllogistic inference is a habit also, as well as inductive inference, nobody would mention this fact within a logical analysis. Unfortunately, ever since David Hume's turning of the problem of the inference into the problem of a habit logicians have shared his escape from logic into psychology. But unwarranted belief, even if it is a habit, can never confer any title of justification to the scientific procedure of induction.

So what we have in case *b'* is, logically speaking, nothing but the clause added, that there is no evidence against the possibility of success. But this is exactly my case 2, the very case which corresponds to the logical situation of the problem of induction. The fact that even in this case a justification can be given, though overlooked in the traditional exposition of the problem, seems to me to open the way to a solution of the inductive problem.

The situation is obscured by some ambiguity in the word "possible." We have to consider first *objective possibility;* by this term we mean cases in which an event sometimes happens. For instance, we say it is possible that it will rain tomorrow, and mean by this that there are summer days on which it rains. Besides this, there is what we might call *epistemic possibility*, i.e., the case that we know the objective possibility. Whereas the terms "objective possibility" and "objective impossibility" form a complete disjunction, the epistemic terms lead to a trichotomy: either we know the possibility of an event, or we know the impossibility, or we do not know anything about the possibility or impossibility. Thus "not impossible" in the epistemic sense is not equivalent to "possible"; it would include the case of indeterminacy. Sometimes, however, we understand by epistemic "possibility" the same as the epistemic "not impossible," that is, we include the case of indeterminacy into the term "possible." I think that if somebody insists that a justification should include a demonstration of the possibility of success he means "possibility" in this sense. If this is meant I would, of course, agree with such a postulate. It is clear that in this sense success of the inductive procedure is possible.

I turn now to a consideration of condition *c*. Miss Creed is certainly right in saying that if we speak of the justification of a means, or procedure, this justification is conditional in so far as it refers to a certain aim. If somebody does not want the aim in question an application of the means by him would not be justified. But it would mean misunderstanding my theory of induction if it is said that for a justification I should have to prove condition *c*, or *c'*, both of which certainly are synthetic propositions of a psychological content. It is not my task to prove these propositions; they stand in the implicans of my theorem, and all I want to maintain, and need to maintain for a justification, is an implication of the form: if a man fulfills condition *c*, or *c'*, i.e., if he wants to predict, he should make inductions.

But apart from this easily corrigible misunderstanding I readily consent to Miss Creed's very pertinent distinction between *c* and *c'*, between the case of wanting the aim in general, and wanting it in face of the more or less problematic chance of success. It is quite right that my implication presupposes *c'*, and that *c* is not sufficient. I am glad Miss Creed pointed this out so clearly; my writings are perhaps not clear enough on this point although I never wished to maintain anything else.

It seems, however, Miss Creed considers condition *c'* unsatisfactory. She would prefer to have a method which permits us to deduce *c'* from *c*. This refers to the methods using the concept of "mathematical expectation" which determine whether or not a bet is acceptable. If *a* is the amount of pleasure attached to the aim *A*, *m* the amount of displeasure combined with the application of the means *M*, *p* the probability that *A* will occur if *M* is applied, then the mathematical theory furnishes the result that the bet is acceptable if the "valuational balance"

$$p \cdot a - m$$

is greater than 0. This condition is easily deducible if *a* means the amount of money to be gained in a game, *m* the amount of money to be paid in advance, i.e., the individual stake. We sometimes extend this condition to cases in which *a* and *m* mean emotional values, assuming that there is a possibility of measuring them. Only such a calculation of the valuational balance, so far as I see, would be considered by Miss Creed as a perfect justification of induction; if she denies to my theory of induction the predicate of justification it seems to be because I do not give an analysis of this kind. That is, I am invited to prove that the pleasure of successful prediction multiplied by the chance of success exceeds the displeasure of the troublesome procedure of induction.

Now I will not deny that in some cases of volitional decisions we engage in considerations which at least qualitatively may be represented by the mathematical calculation described. However, such an analysis of the volitional decision to act is not always possible; and even if it is possible it mostly is not relevant. It is not possible if p is not known; thus already in case 1γ it is not possible unless we are ready to consider a as an infinite value which seems to me rather absurd. And if it is possible it is not relevant because it does not free the volitional decision from a subjective element. That is, instead of having in the implicans the subjective decision to act, we have there now the subjective constants a and m; and if a man rejects our suggestion to act he may always defend himself by saying that his emotional constants do not furnish a positive valuational balance.

Moreover, it is very dubitable whether we are entitled to assume the existence of such emotional constants as a and m. Emotional values vary widely with the situation. The means M may appear very unpleasant to us if considered in isolation but if we know it is the means to reach the coveted aim A our dislike for M may turn into a liking for it. Any intellectual calculation of the valuational balance does violence to the actual psychological processes by which the volitional decision is made. What actually happens in most cases is that when confronted by the situation, we perform a decision *en bloc*, i.e., we perform an act not analyzable into the components p, a, m. Thus Miss Creed's condition c' is a much better representation of the actual situation than would be a deduction of c' from c.

I may add here the remark that even if such a calculation of the valuational balance could be given, a positive value for this balance would not involve an obligation to act. Our man might object that he accepts a bet only if the valuational balance is greater than a certain valuational threshold t. This is, for instance, the practice of insurance companies which calculate the premiums so that statistical laws guarantee to them a profit. Or our man might object that he accepts a bet only if the probability p is greater than a certain safety quantity s; if p is smaller than s he considers the event as practically impossible. The case then would be better represented by case 3 in which the action is not justifiable. I think this is a principle which we all more or less apply because "impossible" never means for us more than "practically impossible." If, for instance—to make use of Miss Creed's interesting example—we refuse to recognize an obligation to comply with the rites of a certain church in spite of there being a slight chance of truth in the promise that such a behavior would lead us to eternal happiness, this negative decision I think is best explained by the introduction of the principle that a probability of so small a degree is considered as a practical impossibility. (It follows, by the way, that cases 1γ, δ, and 2 will permit a better justification than case 1β, if in the last case p is known to be small; knowing nothing about p is a less unfavorable situation than knowing that p is small.) With these considerations two further subjective constants t and s are introduced, which make the relation between wanting the aim in general, and deciding under given conditions for an action in its favor, more flexible and questionable.

On the other hand there are cases in which the calculation of the valuational balance is unnecessary because its positive value can immediately be asserted. This is the case if m is not a displeasure but a pleasure; the sign of the m-term in the valuational balance then would be positive. There are many examples of this kind. Imagine the father saying to his son: If you want to have a car of your own take the old car in the garage and repair it. The boy may be glad to have, not only a car, but also the opportunity to repair a car. This case, I suppose, corresponds best to our actual behavior in face of the problem of induction. I think most of us would not envy the lazy man who waits until the fried pigeons fly into his mouth; we like to work, to try success, to gamble—at least as long as we like to live. I do not say this with the intention of giving an exhaustive psychological analysis of the decision to attempt to predict. I should have no objection if one should speak here of the great contribution which habit adds to our decision to act, for this is a point where we have left the logical field and are concerned with the psychological motives of our actions. Habit, though never a permissible argument for the choice of the *means* of prediction, may well be accepted as the motive which determines the *aim*. Miss Creed speaks so charmingly of the "gentle force of the habit of induction" which leads us, instead of "an intellectual calculation of values"; let us consider this idea together with the joy of gambling as the psychological motives of our decision to act. But let us carefully distinguish between the psychological question of the motives of our decision, and the logical question of the best means we can choose in order to reach the aim of our decision.

This logical question remains unchanged whatever may be the result of the psychological analysis as to the motives of our pursuit of the aim. Even if the application of the means M is combined, not with displeasure, but with pleasure, the question whether it is an appropriate means for the aim A has to be decided independently of this psychological fact. M involves pleasure for us only because it is the way towards A.

I shall summarize. The problem of the justification of induction includes both the question of the decision to attempt predictions and the question of the choice of the best means of making them. The decision on the attempt is justified if the aim is not proved to be unattainable. To demand more, to show that the aim will be reached with a probability p and that the valuational balance is positive would be equivalent to an analysis of the volitional decision which seems inappropriate from a psychological viewpoint and which would not eliminate the subjective element reappearing in the emotional constants a, m, l, s. I therefore can not consider such an analysis as necessary for a justification; a decision *en bloc* is permissible in the implicans of our proof. The word "justification" is chosen with the very intention of indicating the subjective element; otherwise we should speak of a demonstration of an *obligation* to pursue the aim. Only when we turn to the second point are we confronted by a problem free from subjective elements, as the question of the choice of the best means is purely logical. Justification of the choice of the means is therefore equivalent to the logical demonstration that the means is the best we have. If this proof is given the problem of justification is solved, since it is obvious that the first condition, the lack of a disproof of the possibility of success, is fulfilled.

HANS REICHENBACH.

UNIVERSITY OF CALIFORNIA AT LOS ANGELES.

147

ON INDUCTIVE LOGIC

RUDOLF CARNAP

§1. INDUCTIVE LOGIC

Among the various meanings in which the word 'probability' is used in every-day language, in the discussion of scientists, and in the theories of probability, there are especially two which must be clearly distinguished. We shall use for them the terms 'probability$_1$' and 'probability$_2$'. Probability$_1$ is a logical concept, a certain logical relation between two sentences (or, alternatively, between two propositions); it is the same as the concept of degree of confirmation. I shall write briefly "c" for "degree of confirmation," and "c(h, e)" for "the degree of confirmation of the hypothesis h on the evidence e"; the evidence is usually a report on the results of our observations. On the other hand, probability$_2$ is an empirical concept; it is the relative frequency in the long run of one property with respect to another. The controversy between the so-called logical conception of probability, as represented e.g. by Keynes[1], and Jeffreys[2], and others, and the frequency conception, maintained e.g. by v. Mises[3] and Reichenbach[4], seems to me futile. These two theories deal with two different probability concepts which are both of great importance for science. Therefore, the theories are not incompatible, but rather supplement each other.[5]

In a certain sense we might regard deductive logic as the theory of L-implication (logical implication, entailment). And inductive logic may be construed as the theory of degree of confirmation, which is, so to speak, partial L-implication. "e L-implies h" says that h is implicitly given with e, in other words, that the whole logical content of h is contained in e. On the other hand, "c(h, e) = 3/4" says that h is not entirely given with e but that the assumption of h is supported to the degree 3/4 by the observational evidence expressed in e.

In the course of the last years, I have constructed a new system of inductive logic by laying down a definition for degree of confirmation and developing a theory based on this definition. A book containing this theory is in preparation. The purpose of the present paper is to indicate briefly and informally the definition and a few of the results found; for lack of space, the reasons for the choice of this definition and the proofs for the results cannot be given here. The book will, of course, provide a better basis than the present informal summary for a critical evaluation of the theory and of the fundamental conception on which it is based.[6]

---

[1] J. M. Keynes, *A Treatise on Probability*, 1921.

[2] H. Jeffreys, *Theory of Probability*, 1939.

[3] R. v. Mises, *Probability, Statistics, and Truth*, (orig. 1928) 1939.

[4] H. Reichenbach, *Wahrscheinlichkeitslehre*, 1935.

[5] The distinction briefly indicated here, is discussed more in detail in my paper "The Two Concepts of Probability," which will appear in *Philos. and Phenom. Research*, 1945.

[6] In an article by C. G. Hempel and Paul Oppenheim in the present issue of this journal, a new concept of degree of confirmation is proposed, which was developed by the two authors and Olaf Helmer in research independent of my own.

148

§2. SOME SEMANTICAL CONCEPTS

Inductive logic is, like deductive logic, in my conception a branch of semantics. However, I shall try to formulate the present outline in such a way that it does not presuppose knowledge of semantics.

Let us begin with explanations of some semantical concepts which are important both for deductive logic and for inductive logic.[7]

The system of inductive logic to be outlined applies to an infinite sequence of finite language systems L_N (N = 1, 2, 3, etc.) and an infinite language system L_∞. L_∞ refers to an infinite universe of individuals, designated by the individual constants 'a_1', 'a_2', etc. (or 'a', 'b', etc.), while L_N refers to a finite universe containing only N individuals designated by 'a_1', 'a_2', $\cdots$ 'a_N'. Individual variables 'x_1', 'x_2', etc. (or 'x', 'y', etc.) are the only variables occurring in these languages. The languages contain a finite number of predicates of any degree (number of arguments), designating properties of the individuals or relations between them. There are, furthermore, the customary connectives of negation ('$\sim$', corresponding to "not"), disjunction ('$\vee$', "or"), conjunction ('$\cdot$', "and"); universal and existential quantifiers ("for every x," "there is an x"); the sign of identity between individuals '$=$', and 't' as an abbreviation for an arbitrarily chosen tautological sentence. (Thus the languages are certain forms of what is technically known as the lower functional logic with identity.) (The connectives will be used in this paper in three ways, as is customary: (1) between sentences, (2) between predicates (§8), (3) between names (or variables) of sentences (so that, if 'i' and 'j' refer to two sentences, '$i \vee j$' is meant to refer to their disjunction).)

A sentence consisting of a predicate of degree n with n individual constants is called an *atomic sentence* (e.g. 'Pa_1', i.e. 'a_1 has the property P', or 'Ra_3a_5', i.e. 'the relation R holds between a_3 and a_5'). The conjunction of all atomic sentences in a finite language L_N describes one of the possible states of the domain of the N individuals with respect to the properties and relations expressible in the language L_N. If we replace in this conjunction some of the atomic sentences by their negations, we obtain the description of another possible state. All the conjunctions which we can form in this way, including the original one, are called *state-descriptions* in L_N. Analogously, a state-description in L_∞ is a class containing some atomic sentences and the negations of the remaining atomic sentences; since this class is infinite, it cannot be transformed into a conjunction.

In the actual construction of the language systems, which cannot be given here, semantical rules are laid down determining for any given sentence j and any state-description i whether j holds in i, that is to say whether j would be true if i described the actual state among all possible states. The class of those state-descriptions in a language system L (either one of the systems L_N or L_∞) in which j holds is called the *range* of j in L.

The concept of range is fundamental both for deductive and for inductive logic; this has already been pointed out by Wittgenstein. If the range of a

---

[7] For more detailed explanations of some of these concepts see my *Introduction to Semantics*, 1942.

sentence j in the language system L is universal, i.e. if j holds in every state-description (in L), j must necessarily be true independently of the facts; there-fore we call j (in L) in this case *L-true* (logically true, analytic). (The prefix 'L-' stands for "logical"; it is not meant to refer to the system L.) Analogously, if the range of j is null, we call j *L-false* (logically false, self-contradictory). If j is neither L-true nor L-false, we call it *factual* (synthetic, contingent). Sup-pose that the range of e is included in that of h. Then in every possible case in which e would be true, h would likewise be true. Therefore we say in this case that e *L-implies* (logically implies, entails) h. If two sentences have the same range, we call them *L-equivalent*; in this case, they are merely different formula-tions for the same content.

The L-concepts just explained are fundamental for deductive logic and there-fore also for inductive logic. Inductive logic is constructed out of deductive logic by the introduction of the concept of degree of confirmation. This intro-duction will here be carried out in three steps: (1) the definition of regular c-functions (§3), (2) the definition of symmetrical c-functions (§5), (3) the defini-tion of the degree of confirmation c^* (§6).

§3. REGULAR C-FUNCTIONS

A numerical function m ascribing real numbers of the interval 0 to 1 to the sentences of a finite language L_N is called a regular m-function if it is constructed according to the following rules:

(1) We assign to the state-descriptions in L_N as values of m any positive real numbers whose sum is 1.

(2) For every other sentence j in L_N, the value $m(j)$ is determined as follows:
  (a) If j is not L-false, $m(j)$ is the sum of the m-values of those state-descrip-tions which belong to the range of j.
  (b) If j is L-false and hence its range is null, $m(j) = 0$.

(The choice of the rule (2)(a) is motivated by the fact that j is L-equivalent to the disjunction of those state-descriptions which belong to the range of j and that these state-descriptions logically exclude each other.)

If any regular m-function m is given, we define a corresponding function c as follows:

(3) For any pair of sentences e, h in L_N, where e is not L-false, $c(h, e) = \dfrac{m(e \cdot h)}{m(e)}$.

$m(j)$ may be regarded as a measure ascribed to the range of j; thus the func-tion m constitutes a metric for the ranges. Since the range of the conjunction $e \cdot h$ is the common part of the ranges of e and of h, the quotient in (3) indicates, so to speak, how large a part of the range of e is included in the range of h. The numerical value of this ratio, however, depends on what particular m-function has been chosen. We saw earlier that a statement in deductive logic of the form "e L-implies h" says that the range of e is entirely included in that of h. Now we see that a statement in inductive logic of the form "$c(h, e) = 3/4$" says that a certain part—in the example, three fourths—of the range of e is included in the range of h.[8] Here, in order to express the partial inclusion numerically, it is necessary to choose a regular m-function for measuring the ranges. Any m chosen leads to a particular c as defined above. All functions c obtained in this way are called *regular c-functions*.

One might perhaps have the feeling that the metric m should not be chosen once for all but should rather be changed according to the accumulating experiences.[9] This feeling is correct in a certain sense. However, it is to be satisfied not by the function m used in the definition (3) but by another function m_e dependent upon e and leading to an alternative definition (5) for the corresponding c. If a regular m is chosen according to (1) and (2), then a corresponding function m_e is defined for the state-descriptions in L_N as follows:

(4) Let i be a state-description in L_N, and e a non-L-false sentence in L_N.
   (a) If e does not hold in i, $m_e(i) = 0$.

   (b) If e holds in i, $m_e(i) = \dfrac{m(i)}{m(e)}$.

Thus m_e represents a metric for the state-descriptions which changes with the changing evidence e. Now $m_e(j)$ for any other sentence j in L_N is defined in analogy to (2) (a) and (b). Then we define the function c corresponding to m as follows:

(5) For any pair of sentences e, h in L_N, where e is not L-false, $c(h, e) = m_e(h)$.

It can easily be shown that this alternative definition (5) yields the same values as the original definition (3).

Suppose that a sequence of regular m-functions is given, one for each of the finite languages L_N ($N = 1, 2$, etc.). Then we define a corresponding m-function for the infinite language as follows:

(6) $m(j)$ in L_∞ is the limit of the values $m(j)$ in L_N for $N \to \infty$.

c-functions for the finite languages are based on the given m-functions according to (3). We define a corresponding c-function for the infinite language as follows:

(7) $c(h, e)$ in L_∞ is the limit of the values $c(h, e)$ in L_N for $N \to \infty$.

The definitions (6) and (7) are applicable only in those cases where the specified limits exist.

We shall later see how to select a particular sub-class of regular c-functions (§5) and finally one particular c-function c^* as the basis of a complete system of inductive logic (§6). For the moment, let us pause at our first step, the definition of regular c-functions just given, in order to see what results this definition alone can yield, before we add further definitions. The theory of regular c-functions, i.e. the totality of those theorems which are founded on the definition

[8] See F. Waismann, "Logische Analyse des Wahrscheinlichkeitsbegriffs," *Erkenntnis*, vol. 1, 1930, pp. 228–248.
[9] See Waismann, op. cit., p. 242.

stated, is the first and fundamental part of inductive logic. It turns out that we find here many of the fundamental theorems of the classical theory of probability, e.g. those known as the theorem (or principle) of multiplication, the general and the special theorems of addition, the theorem of division and, based upon it, Bayes' theorem.

One of the cornerstones of the classical theory of probability is the principle of indifference (or principle of insufficient reason). It says that, if our evidence e does not give us any sufficient reason for regarding one of two hypotheses h and h' as more probable than the other, then we must take their probabilities$_1$ as equal: $c(h, e) = c(h', e)$. Modern authors, especially Keynes, have correctly pointed out that this principle has often been used beyond the limits of its original meaning and has then led to quite absurd results. Moreover, it can easily be shown that, even in its original meaning, the principle is by far too general and leads to contradictions. Therefore the principle must be abandoned. If it is and we consider only those theorems of the classical theory which are provable without the help of this principle, then we find that these theorems hold for all regular c-functions. The same is true for those modern theories of probability$_1$ (e.g. that by Jeffreys, op.cit.) which make use of the principle of indifference. Most authors of modern axiom systems of probability$_1$ (e.g. Keynes (op.cit.), Waismann (op.cit.), Mazurkiewicz[10], Hosiasson[11], v. Wright[12]) are cautious enough not to accept that principle. An examination of these systems shows that their axioms and hence their theorems hold for all regular c-functions. Thus these systems restrict themselves to the first part of inductive logic, which, although fundamental and important, constitutes only a very small and weak section of the whole of inductive logic. The weakness of this part shows itself in the fact that it does not determine the value of c for any pair h, e except in some special cases where the value is 0 or 1. The theorems of this part tell us merely how to calculate further values of c if some values are given. Thus it is clear that this part alone is quite useless for application and must be supplemented by additional rules. (It may be remarked incidentally, that this point marks a fundamental difference between the theories of probability$_1$ and of probability$_2$ which otherwise are analogous in many respects. The theorems concerning probability$_2$ which are analogous to the theorems concerning regular c-functions constitute not only the first part but the whole of the logico-mathematical theory of probability$_2$. The task of determining the value of probability$_2$ for a given case is—in contradistinction to the corresponding task for probability$_1$—an empirical one and hence lies outside the scope of the logical theory of probability$_2$.)

[10] St. Mazurkiewicz, "Zur Axiomatik der Wahrscheinlichkeitsrechnung," *C. R. Soc. Science Varsovie*, Cl. III, vol. 25, 1932, pp. 1–4.

[11] Janina Hosiasson-Lindenbaum, "On Confirmation," *Journal Symbolic Logic*, vol. 5, 1940, pp. 133–148.

[12] G. H. von Wright, *The Logical Problem of Induction*, (Acta Phil. Fennica, 1941, Fasc. III). See also C. D. Broad, *Mind*, vol. 53, 1944.

§4. THE COMPARATIVE CONCEPT OF CONFIRMATION

Some authors believe that a metrical (or quantitative) concept of degree of confirmation, that is, one with numerical values, can be applied, if at all, only in certain cases of a special kind and that in general we can make only a comparison in terms of higher or lower confirmation without ascribing numerical values. Whether these authors are right or not, the introduction of a merely comparative (or topological) concept of confirmation not presupposing a metrical concept is, in any case, of interest. We shall now discuss a way of defining a concept of this kind.

For technical reasons, we do not take the concept "more confirmed" but "more or equally confirmed." The following discussion refers to the sentences of any finite language L_N. We write, for brevity, "$MC(h, e, h', e')$" for "h is confirmed on the evidence e more highly or just as highly as h' on the evidence e'".

Although the definition of the comparative concept MC at which we aim will not make use of any metrical concept of degree of confirmation, let us now consider, for heuristic purposes, the relation between MC and the metrical concepts, i.e. the regular c-functions. Suppose we have chosen some concept of degree of confirmation, in other words, a regular c-function c, and further a comparative relation MC; then we shall say that MC is in accord with c if the following holds:

(1) For any sentences h, e, h', e', if $MC(h, e, h', e')$ then $c(h, e) \geq c(h', e')$.

However, we shall not proceed by selecting one c-function and then choosing a relation MC which is in accord with it. This would not fulfill our intention. Our aim is to find a comparative relation MC which grasps those logical relations between sentences which are, so to speak, prior to the introduction of any particular m-metric for the ranges and of any particular c-function; in other words, those logical relations with respect to which all the various regular c-functions agree. Therefore we lay down the following requirement:

(2) The relation MC is to be defined in such a way that it is in accord with *all* regular c-functions; in other words, if $MC(h, e, h', e')$, then for every regular c, $c(h, e) \geq c(h', e')$.

It is not difficult to find relations which fulfill this requirement (2). First let us see whether we can find quadruples of sentences h, e, h', e' which satisfy the following condition occurring in (2):

(3) For every regular c, $c(h, e) \geq c(h', e')$.

It is easy to find various kinds of such quadruples. (For instance, if e and e' are any non-L-false sentences, then the condition (3) is satisfied in all cases where e L-implies h, because here $c(h, e) = 1$; further in all cases where e' L-implies $\sim h'$, because here $c(h', e') = 0$; and in many other cases.) We could, of course, define a relation MC by taking some cases where we know that the condition (3) is satisfied and restricting the relation to these cases. Then the relation would fulfill the requirement (2); however, as long as there are cases which satisfy the condition (3) but which we have not included in the relation, the relation is unnecessarily restricted. Therefore we lay down the following as a second requirement for MC:

(4) MC is to be defined in such a way that it holds in all cases which satisfy the condition (3); in such a way, in other words, that it is the most comprehensive relation which fulfills the first requirement (2).

These two requirements (2) and (4) together stipulate that $MC(h, e, h', e')$ is to hold if and only if the condition (3) is satisfied; thus the requirements determine uniquely one relation MC. However, because they refer to the c-functions, we do not take these requirements as a definition for MC, for we intend to give a purely comparative definition for MC, a definition which does not make use of any metrical concepts but which leads nevertheless to a relation MC which fulfills the requirements (2) and (4) referring to c-functions. This aim is reached by the following definition (where ' $=_{Df}$' is used as sign of definition).

(5) $MC(h, e, h', e') =_{Df}$ the sentences h, e, h', e' (in L_N) are such that e and e' are not L-false and at least one of the following three conditions is fulfilled:
(a) e L-implies h,
(b) e' L-implies $\sim h'$,
(c) $e' \cdot h'$ L-implies $e \cdot h$ and simultaneously e L-implies $h \vee e'$.

((a) and (b) are the two kinds of rather trivial cases earlier mentioned; (c) comprehends the interesting cases; an explanation and discussion of them cannot be given here.)

The following theorem can then be proved concerning the relation MC defined by (5). It shows that this relation fulfills the two requirements (2) and (4).

(6) For any sentences h, e, h', e' in L_N the following holds:
(a) If $MC(h, e, h', e')$, then, for every regular c, $c(h, e) \geqq c(h', e')$.
(b) If, for every regular c, $c(h, e) \geqq c(h', e')$, then $MC(h, e, h', e')$.

(With respect to L_∞, the analogue of (6)(a) holds for all sentences, and that of (6)(b) for all sentences without variables.)

§5. SYMMETRICAL C-FUNCTIONS

The next step in the construction of our system of inductive logic consists in selecting a narrow sub-class of the comprehensive class of all regular c-functions. The guiding idea for this step will be the principle that inductive logic should treat all individuals on a par. The same principle holds for deductive logic; for instance, if '$\cdot \cdot a \cdot \cdot b \cdot \cdot$' L-implies '$-b-c-$' (where the first expression in quotation marks is meant to indicate some sentence containing 'a' and 'b', and the second another sentence containing 'b' and 'c'), then L-implication holds likewise between corresponding sentences with other individual constants, e.g. between '$\cdot \cdot d \cdot \cdot c \cdot \cdot$' and '$-c-a-$'. Now we require that this should hold also for inductive logic, e.g. that $c('-b-c-', '\cdot \cdot a \cdot \cdot b \cdot \cdot') = c('-c-a-', '\cdot \cdot d \cdot \cdot c \cdot \cdot')$. It seems that all authors on probability₁ have assumed this principle—although it has seldom, if ever, been stated explicitly—by formulating theorems in the following or similar terms: "On the basis of observations of s things of which s_1 were found to have the property M and s_2 not to have this property, the probability that another thing has this property is such and such." The fact that these theorems refer only to the number of things observed and do not mention particular things shows implicitly that it does not matter which things are involved; thus it is assumed, e.g., that $c('Pd', 'Pa \cdot Pb \cdot \sim Pc') = c('Pc', 'Pa \cdot Pd \cdot \sim Pb')$.

The principle could also be formulated as follows. Inductive logic should, like deductive logic, make no discrimination among individuals. In other words, the value of c should be influenced only by those differences between individuals which are expressed in the two sentences involved; no differences between particular individuals should be stipulated by the rules of either deductive or inductive logic.

It can be shown that this principle of non-discrimination is fulfilled if c belongs to the class of symmetrical c-functions which will now be defined. Two state-descriptions in a language L_N are said to be *isomorphic* or to have the same structure if one is formed from the other by replacements of the following kind: we take any one-one relation R such that both its domain and its converse domain is the class of all individual constants in L_N, and then replace every individual constant in the given state-description by the one correlated with it by R. If a regular m-function (for L_N) assigns to any two isomorphic state-descriptions (in L_N) equal values, it is called a symmetrical m-function; and a c-function based upon such an m-function in the way explained earlier (see (3) in §3) is then called a *symmetrical c-function*.

§6. THE DEGREE OF CONFIRMATION c^*

Let i be a state-description in L_N. Suppose there are n_i state-descriptions in L_N isomorphic to i (including i itself), say i, i', i'', etc. These n_i state-descriptions exhibit one and the same structure of the universe of L_N with respect to all the properties and relations designated by the primitive predicates in L_N. This concept of structure is an extension of the concept of structure or relation-number (Russell) usually applied to one dyadic relation. The common structure of the isomorphic state-descriptions i, i', i'', etc. can be described by their disjunction $i \lor i' \lor i'' \lor \cdots$. Therefore we call this disjunction, say j, a *structure-description* in L_N. It can be shown that the range of j contains only the isomorphic state-descriptions i, i', i'', etc. Therefore (see (2)(a) in §3) $m(j)$ is the sum of the m-values for these state-descriptions. If m is symmetrical, then these values are equal, and hence

(1) $m(j) = n_i \times m(i)$.

And, conversely, if $m(j)$ is known to be q, then

(2) $m(i) = m(i') = m(i'') = \cdots = q/n_i$.

This shows that what remains to be decided, is merely the distribution of m-values among the structure-descriptions in L_N. We decide to give them equal m-values. This decision constitutes the third step in the construction of our inductive logic. This step leads to one particular m-function m* and to the c-function c* based upon m*. According to the preceding discussion, m* is characterized by the following two stipulations:

(3)   (a)   m* is a symmetrical m-function;
       (b)   m* has the same value for all structure-descriptions (in L_N).

We shall see that these two stipulations characterize just one function. Every state-description (in L_N) belongs to the range of just one structure-description. Therefore, the sum of the m*-values for all structure-descriptions in L_N must be the same as for all state-descriptions, hence 1 (according to (1) in §3). Thus, if the number of structure-descriptions in L_N is m, then, according to (3)(b),

(4) for every structure-description j in L_N, $m^*(j) = \dfrac{1}{m}$.

Therefore, if i is any state-description in L_N and n_i is the number of state-descriptions isomorphic to i, then, according to (3)(a) and (2),

(5) $$m^*(i) = \frac{1}{mn_i}.$$

(5) constitutes a definition of m* as applied to the state-descriptions in L_N. On this basis, further definitions are laid down as explained above (see (2) and (3) in §3): first a definition of m* as applied to all sentences in L_N, and then a definition of c* on the basis of m*. Our inductive logic is the theory of this particular function c* as our concept of degree of confirmation.

It seems to me that there are good and even compelling reasons for the stipulation (3)(a), i.e. the choice of a symmetrical function. The proposal of any non-symmetrical c-function as degree of confirmation could hardly be regarded as acceptable. The same can not be said, however, for the stipulation (3)(b). No doubt, to the way of thinking which was customary in the classical period of the theory of probability, (3)(b) would appear as validated, like (3)(a), by the principle of indifference. However, to modern, more critical thought, this mode of reasoning appears as invalid because the structure-descriptions (in contra-distinction to the individual constants) are by no means alike in their logical features but show very conspicuous differences. The definition of c* shows a great simplicity in comparison with other concepts which may be taken into consideration. Although this fact may influence our decision to choose c*, it cannot, of course, be regarded as a sufficient reason for this choice. It seems to me that the choice of c* cannot be justified by any features of the definition which are immediately recognizable, but only by the consequences to which the definition leads.

There is another c-function c_w which at the first glance appears not less plausible than c*. The choice of this function may be suggested by the following consideration. Prior to experience, there seems to be no reason to regard one state-description as less probable than another. Accordingly, it might seem natural to assign equal m-values to all state-descriptions. Hence, if the number of the state-descriptions in L_N is n, we define for any state-description i

$$(6) \qquad \qquad \mathfrak{m}_W(i) = 1/n.$$

This definition (6) for $\mathfrak{m}_W$ is even simpler than the definition (5) for $\mathfrak{m}^*$. The measure ascribed to the ranges is here simply taken as proportional to the cardinal numbers of the ranges. On the basis of the $\mathfrak{m}_W$-values for the state-descriptions defined by (6), the values for the sentences are determined as before (see (2) in §3), and then $\mathfrak{c}_W$ is defined on the basis of $\mathfrak{m}_W$ (see (3) in §3).[13]

In spite of its apparent plausibility, the function $\mathfrak{c}_W$ can easily be seen to be entirely inadequate as a concept of degree of confirmation. As an example, consider the language L_{101} with 'P' as the only primitive predicate. Let the number of state-descriptions in this language be n (it is 2^{101}). Then for any state-description, $\mathfrak{m}_W = 1/n$. Let e be the conjunction $Pa_1 \cdot Pa_2 \cdot Pa_3 \cdots Pa_{100}$ and let h be 'Pa_{101}'. Then $e \cdot h$ is a state-description and hence $\mathfrak{m}_W(e \cdot h) = 1/n$. e holds only in the two state-descriptions $e \cdot h$ and $e \cdot \sim h$; hence $\mathfrak{m}_W(e) = 2/n$. Therefore $\mathfrak{c}_W(h, e) = \frac{1}{2}$. If e' is formed from e by replacing some or even all of the atomic sentences with their negations, we obtain likewise $\mathfrak{c}_W(h, e') = \frac{1}{2}$. Thus the $\mathfrak{c}_W$-value for the prediction that a_{101} is P is always the same, no matter whether among the hundred observed individuals the number of those which we have found to be P is 100 or 50 or 0 or any other number. Thus the choice of $\mathfrak{c}_W$ as the degree of confirmation would be tantamount to the principle never to let our past experiences influence our expectations for the future. This would obviously be in striking contradiction to the basic principle of all inductive reasoning.

§7. LANGUAGES WITH ONE-PLACE PREDICATES ONLY

The discussions in the rest of this paper concern only those language systems whose primitive predicates are one-place predicates and hence designate properties, not relations. It seems that all theories of probability constructed so far have restricted themselves, or at least all of their important theorems, to properties. Although the definition of c^* in the preceding section has been stated in a

[13] It seems that Wittgenstein meant this function $\mathfrak{c}_W$ in his definition of probability, which he indicates briefly without examining its consequences. In his *Tractatus Logico-Philosophicus*, he says: "A proposition is the expression of agreement and disagreement with the truth-possibilities of the elementary [i.e. atomic] propositions" (*4.4); "The world is completely described by the specification of all elementary propositions plus the specification, which of them are true and which false" (*4.26). The truth-possibilities specified in this way correspond to our state-descriptions. Those truth-possibilities which verify a given proposition (in our terminology, those state-descriptions in which a given sentence holds) are called the truth-grounds of that proposition (*5.101). "If T_r is the number of the truth-grounds of the proposition "r", T_{rs} the number of those truth-grounds of the proposition "s" which are at the same time truth-grounds of "r", then we call the ratio $T_{rs}:T_r$ the measure of the *probability* which the proposition "r" gives to the proposition "s" "(*5.15). It seems that the concept of probability thus defined coincides with the function $\mathfrak{c}_W$.

general way so as to apply also to languages with relations, the greater part of
our inductive logic will be restricted to properties. An extension of this part of
inductive logic to relations would require certain results in the deductive logic
of relations, results which this discipline, although widely developed in other
respects, has not yet reached (e.g. an answer to the apparently simple question
as to the number of structures in a given finite language system).

Let L_N^p be a language containing N individual constants 'a_1', $\cdots$ 'a_N', and p
one-place primitive predicates 'P_1', $\cdots$ 'P_p'. Let us consider the following expressions (sentential matrices). We start with '$P_1x \cdot P_2x \cdots P_px$'; from this
expression we form others by negating some of the conjunctive components,
until we come to '$\sim P_1x \cdot \sim P_2x \cdots \sim P_px$', where all components are negated.
The number of these expressions is $k = 2^p$; we abbreviate them by 'Q_1x', $\cdots$
'Q_kx'. We call the k properties expressed by those k expressions in conjunctive
form and now designated by the k new Q-predicates the Q-properties with respect to the given language L_N^p. We see easily that these Q-properties are the
strongest properties expressible in this language (except for the L-empty, i.e.,
logically self-contradictory, property); and further, that they constitute an exhaustive and non-overlapping classification, that is to say, every individual has
one and only one of the Q-properties. Thus, if we state for each individual which
of the Q-properties it has, then we have described the individuals completely.
Every state-description can be brought into the form of such a statement, i.e.
a conjunction of N Q-sentences, one for each of the N individuals. Suppose
that in a given state-description i the number of individuals having the property
Q_1 is N_1, the number for Q_2 is N_2, $\cdots$ that for Q_k is N_k. Then we call the
numbers N_1, N_2, $\cdots$ N_k the Q-numbers of the state-description i; their sum is N.
Two state-descriptions are isomorphic if and only if they have the same Q-num-
bers. Thus here a structure-description is a statistical description giving the
Q-numbers N_1, N_2, etc., without specifying which individuals have the properties Q_1, Q_2, etc.

Here—in contradistinction to languages with relations—it is easy to find an
explicit function for the number m of structure-descriptions and, for any given
state-description i with the Q-numbers N_1, $\cdots$ N_k, an explicit function for the
number n_i of state-descriptions isomorphic to i, and hence also a function for
$m^*(i)$.[14]

Let j be a non-general sentence (i.e. one without variables) in L_N^p. Since

[14] The results are as follows.

(1)
$$m = \frac{(N + k - 1)!}{N!(k - 1)!}$$

(2)
$$n_i = \frac{N!}{N_1! N_2! \cdots N_k!}$$

Therefore (according to (5) in §6):

(3)
$$m^*(i) = \frac{N_1! N_2! \cdots N_k!(k - 1)!}{(N + k - 1)!}$$

there are effective procedures (that is, sets of fixed rules furnishing results in a finite number of steps) for constructing all state-descriptions in which j holds and for computing $\mathfrak{m}^*$ for any given state-description, these procedures together yield an effective procedure for computing $\mathfrak{m}^*(j)$ (according to (2) in §3). However, the number of state-descriptions becomes very large even for small language systems (it is k^N, hence, e.g., in L_7^3 it is more than two million.) Therefore, while the procedure indicated for the computation of $\mathfrak{m}^*(j)$ is effective, nevertheless in most ordinary cases it is impracticable; that is to say, the number of steps to be taken, although finite, is so large that nobody will have the time to carry them out to the end. I have developed another procedure for the computation of $\mathfrak{m}^*(j)$ which is not only effective but also practicable if the number of individual constants occurring in j is not too large.

The value of $\mathfrak{m}^*$ for a sentence j in the infinite language has been defined (see (6) in §3) as the limit of its values for the same sentence j in the finite languages. The question arises whether and under what conditions this limit exists. Here we have to distinguish two cases. (i) Suppose that j contains no variable. Here the situation is simple; it can be shown that in this case $\mathfrak{m}^*(j)$ is the same in all finite languages in which j occurs; hence it has the same value also in the infinite language. (ii) Let j be general, i.e., contain variables. Here the situation is quite different. For a given finite language with N individuals, j can of course easily be transformed into an L-equivalent sentence j'_N without variables, because in this language a universal sentence is L-equivalent to a conjunction of N components. The values of $\mathfrak{m}^*(j'_N)$ are in general different for each N; and although the simplified procedure mentioned above is available for the computation of these values, this procedure becomes impracticable even for moderate N. Thus for general sentences the problem of the existence and the practical computability of the limit becomes serious. It can be shown that for every general sentence the limit exists; hence $\mathfrak{m}^*$ has a value for all sentences in the infinite language. Moreover, an effective procedure for the computation of $\mathfrak{m}^*(j)$ for any sentence j in the infinite language has been constructed. This is based on a procedure for transforming any given general sentence j into a non-general sentence j' such that j and j', although not necessarily L-equivalent, have the same $\mathfrak{m}^*$-value in the infinite language and j' does not contain more individual constants than j; this procedure is not only effective but also practicable for sentences of customary length. Thus, the computation of $\mathfrak{m}^*(j)$ for a general sentence j is in fact much simpler for the infinite language than for a finite language with a large N.

With the help of the procedure mentioned, the following theorem is obtained:

If j is a purely general sentence (i.e. one without individual constants) in the infinite language, then $\mathfrak{m}^*(j)$ is either 0 or 1.

§8. INDUCTIVE INFERENCES

One of the chief tasks of inductive logic is to furnish general theorems concerning inductive inferences. We keep the traditional term "inference"; however, we do not mean by it merely a transition from one sentence to another (viz.

from the evidence or premiss e to the hypothesis or conclusion h) but the determination of the degree of confirmation $c(h, e)$. In deductive logic it is sufficient to state that h follows with necessity from e; in inductive logic, on the other hand, it would not be sufficient to state that h follows—not with necessity but to some degree or other—from e. It must be specified to what degree h follows from e; in other words, the value of $c(h, e)$ must be given. We shall now indicate some results with respect to the most important kinds of inductive inference. These inferences are of special importance when the evidence or the hypothesis or both give statistical information, e.g. concerning the absolute or relative frequencies of given properties.

If a property can be expressed by primitive predicates together with the ordinary connectives of negation, disjunction, and conjunction (without the use of individual constants, quantifiers, or the identity sign), it is called an *elementary property*. We shall use 'M', 'M''', 'M_1', 'M_2', etc. for elementary properties. If a property is empty by logical necessity (e.g. the property designated by '$P \cdot \sim P$') we call it L-empty; if it is universal by logical necessity (e.g. '$P \lor \sim P$'), we call it L-universal. If it is neither L-empty nor L-universal (e.g. 'P_1', '$P_1 \cdot \sim P_2$'), we call it a *factual property*; in this case it may still happen to be universal or empty, but if so, then contingently, not necessarily. It can be shown that every elementary property which is not L-empty is uniquely analysable into a disjunction (i.e. or-connection) of Q-properties. If M is a disjunction of n Q-properties ($n \geq 1$), we say that the (logical) *width* of M is n; to an L-empty property we ascribe the width 0. If the width of M is w (≥ 0), we call w/k its *relative width* (k is the number of Q-properties).

The concepts of width and relative width are very important for inductive logic. Their neglect seems to me one of the decisive defects in the classical theory of probability which formulates its theorems "for any property" without qualification. For instance, Laplace takes the probability a priori that a given thing has a given property, no matter of what kind, to be $\frac{1}{2}$. However, it seems clear that this probability cannot be the same for a very strong property (e.g. '$P_1 \cdot P_2 \cdot P_3$') and for a very weak property (e.g. '$P_1 \lor P_2 \lor P_3$'). According to our definition, the first of the two properties just mentioned has the relative width $\frac{1}{8}$, and the second $\frac{7}{8}$. In this and in many other cases the probability or degree of confirmation must depend upon the widths of the properties involved. This will be seen in some of the theorems to be mentioned later.

§9. THE DIRECT INFERENCE

Inductive inferences often concern a situation where we investigate a whole population (of persons, things, atoms, or whatever else) and one or several samples picked out of the population. An inductive inference from the whole population to a sample is called a direct inductive inference. For the sake of simplicity, we shall discuss here and in most of the subsequent sections only the case of one property M, hence a classification of all individuals into M and $\sim M$. The theorems for classifications with more properties are analogous but more complicated. In the present case, the evidence e says that in a whole population of n individuals there are n_1 with the property M and $n_2 = n - n_1$ with $\sim M$; hence the relative frequency of M is $r = n_1/n$. The hypothesis h says that a sample of s individuals taken from the whole population will contain s_1 individuals with the property M and $s_2 = s - s_1$ with $\sim M$. Our theory yields in this case the same values as the classical theory.[15]

If we vary s_1, then c^* has its maximum in the case where the relative frequency s_1/s in the sample is equal or close to that in the whole population.

If the sample consists of only one individual c, and h says that c is M, then $c^*(h, e) = r$.

As an approximation in the case that n is very large in relation to s, Newton's theorem holds.[16] If furthermore the sample is sufficiently large, we obtain as an approximation Bernoulli's theorem in its various forms.

It is worthwhile to note two characteristics which distinguish the direct inductive inference from the other inductive inferences and make it, in a sense, more closely related to deductive inferences:

(i) The results just mentioned hold not only for c^* but likewise for all symmetrical c-functions; in other words, the results are independent of the particular m-metric chosen provided only that it takes all individuals on a par.

(ii) The results are independent of the width of M. This is the reason for the agreement between our theory and the classical theory at this point.

§10. THE PREDICTIVE INFERENCE

We call the inference from one sample to another the predictive inference. In this case, the evidence e says that in a first sample of s individuals, there are s_1 with the property M, and $s_2 = s - s_1$ with $\sim M$. The hypothesis h says that in a second sample of s' other individuals, there will be s_1' with M, and $s_2' = s' - s_1'$ with $\sim M$. Let the width of M be w_1; hence the width of $\sim M$ is $w_2 = k - w_1$.

---

[15] The general theorem is as follows:

$$c^*(h, e) = \frac{\binom{n_1}{s_1}\binom{n_2}{s_1}}{\binom{n}{s}}.$$

[16]

$$c^*(h, e) = \binom{s}{s_1} r^{s_1}(1 - r)^{s_2}.$$

[17] The general theorem is as follows:

$$c^*(h, e) = \frac{\binom{s_1 + s_1' + w_1 - 1}{s_1'}\binom{s_2 + s_2' + w_2 - 1}{s_2'}}{\binom{s + s' + k - 1}{s'}}.$$

The most important special case is that where h refers to one individual c only and says that c is M. In this case,

$$(1) \qquad\qquad c^*(h, e) \;=\; \frac{s_1 + w_1}{s + k}.$$

Laplace's much debated rule of succession gives in this case simply the value $\frac{s_1 + 1}{s + 2}$ for any property whatever; this, however, if applied to different properties, leads to contradictions. Other authors state the value s_1/s, that is, they take simply the observed relative frequency as the probability for the prediction that an unobserved individual has the property in question. This rule, however, leads to quite implausible results. If $s_1 = s$, e.g., if three individuals have been observed and all of them have been found to be M, the last-mentioned rule gives the probability for the next individual being M as 1, which seems hardly acceptable. According to (1), c^* is influenced by the following two factors (though not uniquely determined by them):

(i) w_1/k, the relative width of M;
(ii) s_1/s, the relative frequency of M in the observed sample.

The factor (i) is purely logical; it is determined by the semantical rules. (ii) is empirical; it is determined by observing and counting the individuals in the sample. The value of c^* always lies between those of (i) and (ii). Before any individual has been observed, c^* is equal to the logical factor (i). As we first begin to observe a sample, c^* is influenced more by this factor than by (ii). As the sample is increased by observing more and more individuals (but not including the one mentioned in h), the empirical factor (ii) gains more and more influence upon c^* which approaches closer and closer to (ii); and when the sample is sufficiently large, c^* is practically equal to the relative frequency (ii). These results seem quite plausible.[18]

The predictive inference is the most important inductive inference. The kinds of inference discussed in the subsequent sections may be construed as special cases of the predictive inference.

[18] Another theorem may be mentioned which deals with the case where, in distinction to the case just discussed, the evidence already gives some information about the individual c mentioned in h. Let M_1 be a factual elementary property with the width w_1 ($w_1 \geqq 2$); thus M_1 is a disjunction of w_1 Q-properties. Let M_2 be the disjunction of w_2 among those w_1 Q-properties ($1 \leqq w_2 < w_1$); hence M_2 L-implies M_1 and has the width w_2. e specifies first how the s individuals of an observed sample are distributed among certain properties, and, in particular, it says that s_1 of them have the property M_1 and s_2 of these s_1 individuals have also the property M_2; in addition, e says that c is M_1; and h says that c is also M_2. Then,

$$c^*(h, e) \;=\; \frac{s_2 + w_2}{s_1 + w_1}.$$

This is analogous to (1); but in the place of the whole sample we have here that part of it which shows the property M_1.

§11. THE INFERENCE BY ANALOGY

The inference by analogy applies to the following situation. The evidence known to us is the fact that individuals b and c agree in certain properties and, in addition, that b has a further property; thereupon we consider the hypothesis that c too has this property. Logicians have always felt that a peculiar difficulty is here involved. It seems plausible to assume that the probability of the hypothesis is the higher the more properties b and c are known to have in common; on the other hand, it is felt that these common properties should not simply be counted but weighed in some way. This becomes possible with the help of the concept of width. Let M_1 be the conjunction of all properties which b and c are known to have in common. The known similarity between b and c is the greater the stronger the property M_1, hence the smaller its width. Let M_2 be the conjunction of all properties which b is known to have. Let the width of M_1 be w_1, and that of M_2, w_2. According to the above description of the situation, we presuppose that M_2 L-implies M_1 but is not L-equivalent to M_1; hence $w_1 > w_2$. Now we take as evidence the conjunction $e \cdot j$; e says that b is M_2, and j says that c is M_1. The hypothesis h says that c has not only the properties ascribed to it in the evidence but also the one (or several) ascribed in the evidence to b only, in other words, that c has all known properties of b, or briefly that c is M_2. Then

$$(1) \qquad \mathfrak{c}^*(h, e \cdot j) = \frac{w_2 + 1}{w_1 + 1}.$$

j and h speak only about c; e introduces the other individual b which serves to connect the known properties of c expressed by j with its unknown properties expressed by h. The chief question is whether the degree of confirmation of h is increased by the analogy between c and b, in other words, by the addition of e to our knowledge. A theorem[19] is found which gives an affirmative answer to this question. However, the increase of $\mathfrak{c}^*$ is under ordinary conditions rather small; this is in agreement with the general conception according to which reasoning by analogy, although admissible, can usually yield only rather weak results.

Hosiasson[20] has raised the question mentioned above and discussed it in detail. She says that an affirmative answer, a proof for the increase of the degree of confirmation in the situation described, would justify the universally accepted reasoning by analogy. However, she finally admits that she does not find such a proof on the basis of her axioms. I think it is not astonishing that neither the classical theory nor modern theories of probability have been able to give a satisfactory account of and justification for the inference by analogy. For, as the theorems mentioned show, the degree of confirmation and its increase depend

---

[19]
$$\frac{\mathfrak{c}^*(h, e \cdot j)}{\mathfrak{c}^*(h, j)} = 1 + \frac{w_1 - w_2}{w_2(w_1 + 1)}.$$

This theorem shows that the ratio of the increase of $\mathfrak{c}^*$ is greater than 1, since $w_1 > w_2$.

[20] Janina Lindenbaum-Hosiasson, "Induction et analogie: Comparaison de leur fondement," *Mind*, vol. 50, 1941, pp. 351–365; see especially pp. 361–365.

here not on relative frequencies but entirely on the logical widths of the properties involved, thus on magnitudes neglected by both classical and modern theories.

The case discussed above is that of simple analogy. For the case of multiple analogy, based on the similarity of c not only with one other individual but with a number n of them, similar theorems hold. They show that c^* increases with increasing n and approaches 1 asymptotically. Thus, multiple analogy is shown to be much more effective than simple analogy, as seems plausible.

§12. THE INVERSE INFERENCE

The inference from a sample to the whole population is called the inverse inductive inference. This inference can be regarded as a special case of the predictive inference with the second sample covering the whole remainder of the population. This inference is of much greater importance for practical statistical work than the direct inference, because we usually have statistical information only for some samples and not for the whole population.

Let the evidence e say that in an observed sample of s individuals there are s_1 individuals with the property M and $s_2 = s - s_1$ with $\sim M$. The hypothesis h says that in the whole population of n individuals, of which the sample is a part, there are n_1 individuals with M and n_2 with $\sim M$ ($n_1 \geqq s_1$, $n_2 \geqq s_2$). Let the width of M be w_1, and that of $\sim M$ be $w_2 = k - w_1$. Here, in distinction to the direct inference, $c^*(h, e)$ is dependent not only upon the frequencies but also upon the widths of the two properties.[21]

§13. THE UNIVERSAL INFERENCE

The universal inductive inference is the inference from a report on an observed sample to a hypothesis of universal form. Sometimes the term 'induction' has been applied to this kind of inference alone, while we use it in a much wider sense for all non-deductive kinds of inference. The universal inference is not even the most important one; it seems to me now that the role of universal sentences in the inductive procedures of science has generally been overestimated. This will be explained in the next section.

Let us consider a simple law l, i.e. a factual universal sentence of the form "all M are M'" or, more exactly, "for every x, if x is M, then x is M'", where M and M' are elementary properties. As an example, take "all swans are white". Let us abbreviate '$M \cdot \sim M'$' ("non-white swan") by 'M_1' and let the width of

[21] The general theorem is as follows:

$$c^*(h, e) = \frac{\dbinom{n_1 + w_1 - 1}{s_1 + w_1 - 1}\dbinom{n_2 + w_2 - 1}{s_2 + w_2 - 1}}{\dbinom{n + k - 1}{n - s}}.$$

Other theorems, which cannot be stated here, concern the case where more than two properties are involved, or give approximations for the frequent case where the whole population is very large in relation to the sample.

M_1 be w_1. Then l can be formulated thus: "M_1 is empty", i.e. "there is no individual (in the domain of individuals of the language in question) with the property M_1" ("there are no non-white swans"). Since l is a factual sentence, M_1 is a factual property; hence $w_1 > 0$. To take an example, let w_1 be 3; hence M_1 is a disjunction of three Q-properties, say $Q \vee Q' \vee Q''$. Therefore, l can be transformed into: "Q is empty, and Q' is empty, and Q'' is empty". The weakest factual laws in a language are those which say that a certain Q-property is empty; we call them Q-laws. Thus we see that l can be transformed into a conjunction of w_1 Q-laws. Obviously l asserts more if w_1 is larger; therefore we say that the law l has the strength w_1.

Let the evidence e be a report about an observed sample of s individuals such that we see from e that none of these s individuals violates the law l; that is to say, e ascribes to each of the s individuals either simply the property $\sim M_1$ or some other property L-implying $\sim M_1$. Let l, as above, be a simple law which says that M_1 is empty, and w_1 be the width of M_1; hence the width of $\sim M_1$ is $w_2 = k - w_1$. For finite languages with N individuals, $c^*(l, e)$ is found to decrease with increasing N, as seems plausible.[22] If N is very large, c^* becomes very small; and for an infinite universe it becomes 0. The latter result may seem astonishing at first sight; it seems not in accordance with the fact that scientists often speak of "well-confirmed" laws. The problem involved here will be discussed later.

So far we have considered the case in which only positive instances of the law l have been observed. Inductive logic must, however, deal also with the case of negative instances. Therefore let us now examine another evidence e' which says that in the observed sample of s individuals there are s_1 which have the property M_1 (non-white swans) and hence violate the law l, and that $s_2 = s - s_1$ have $\sim M_1$ and hence satisfy the law l. Obviously, in this case there is no point in taking as hypothesis the law l in its original forms, because l is logically incom-

---

[22] The general theorem is as follows:

$$(1) \qquad c^*(l, e) = \frac{\dbinom{s + k - 1}{w_1}}{\dbinom{N + k - 1}{w_1}}.$$

In the special case of a language containing 'M_1' as the only primitive predicate, we have $w_1 = 1$ and $k = 2$, and hence $c^*(l, e) = \dfrac{s + 1}{N + 1}$. The latter value is given by some authors as holding generally (see Jeffreys, op.cit., p. 106 (16)). However, it seems plausible that the degree of confirmation must be smaller for a stronger law and hence depend upon w_1.

If s, and hence N, too, is very large in relation to k, the following holds as an approximation:

$$(2) \qquad c^*(l, e) = \left(\frac{s}{N}\right)^{w_1}.$$

For the infinite language L_∞ we obtain, according to definition (7) in §3:

$$(3) \qquad c^*(l, e) = 0.$$

patible with the present evidence e', and hence $c^*(l, e') = 0$. That all individuals satisfy l is excluded by e'; the question remains whether at least all unobserved individuals satisfy l. Therefore we take here as hypothesis the restricted law l' corresponding to the original unrestricted law l; l' says that all individuals not belonging to the sample of s individuals described in e' have the property $\sim M_1$. w_1 and w_2 are, as previously, the widths of M_1 and $\sim M_1$ respectively. It is found that $c^*(l', e')$ decreases with an increase of N and even more with an increase in the number s_1 of violating cases.[23] It can be shown that, under ordinary circumstances with large N, c^* increases moderately when a new individual is observed which satisfies the original law l. On the other hand, if the new individual violates l, c^* decreases very much, its value becoming a small fraction of its previous value. This seems in good agreement with the general conception.

For the infinite universe, c^* is again 0, as in the previous case. This result will be discussed in the next section.

§14. THE INSTANCE CONFIRMATION OF A LAW

Suppose we ask an engineer who is building a bridge why he has chosen the building materials he is using, the arrangement and dimensions of the supports, etc. He will refer to certain physical laws, among them some general laws of mechanics and some specific laws concerning the strength of the materials. On further inquiry as to his confidence in these laws he may apply to them phrases like "very reliable", "well founded", "amply confirmed by numerous experiences". What do these phrases mean? It is clear that they are intended to say something about probability₁ or degree of confirmation. Hence, what is meant could be formulated more explicitly in a statement of the form "$c(h, e)$ is high" or the like. Here the evidence e is obviously the relevant observational knowledge of the engineer or of all physicists together at the present time. But what is to serve as the hypothesis h? One might perhaps think at first that h is the law in question, hence a universal sentence l of the form: "For every space-time point x, if such and such conditions are fulfilled at x, then such and such is the case at x". I think, however, that the engineer is chiefly interested not in this sentence l, which speaks about an immense number, perhaps an infinite number, of instances dispersed through all time and space, but rather in one instance of l or a relatively small number of instances. When he says that the law is very reliable, he does not mean to say that he is willing to bet that among the billion of billions, or an infinite number, of instances to which the law applies there is not one counter-instance, but merely that this bridge will not be a counter-instance, or that among all bridges which he will construct during his lifetime, or among those which all engineers will construct during the next one

---

[23] The theorem is as follows:

$$c^*(l', e') = \frac{\dbinom{s + k - 1}{s_1 + w_1}}{\dbinom{N + k - 1}{s_1 + w_1}}.$$

thousand years, there will be no counter-instance. Thus h is not the law l itself but only a prediction concerning one instance or a relatively small number of instances. Therefore, what is vaguely called the reliability of a law is measured not by the degree of confirmation of the law itself but by that of one or several instances. This suggests the subsequent definitions. They refer, for the sake of simplicity, to just one instance; the case of several, say one hundred, instances can then easily be judged likewise. Let e be any non-L-false sentence without variables. Let l be a simple law of the form earlier described (§13). Then we understand by the *instance confirmation* of l on the evidence e, in symbols "c_i^* (l, e)", the degree of confirmation, on the evidence e, of the hypothesis that a new individual not mentioned in e fulfills the law l.[24]

The second concept, now to be defined, seems in many cases to represent still more accurately what is vaguely meant by the reliability of a law l. We suppose here that l has the frequently used conditional form mentioned earlier: "For every x, if x is M, then x is M'" (e.g. "all swans are white"). By the *qualified-instance confirmation* of the law that all swans are white we mean the degree of confirmation for the hypothesis h' that the next swan to be observed will likewise be white. The difference between the hypothesis h used previously for the instance confirmation and the hypothesis h' just described consists in the fact that the latter concerns an individual which is already qualified as fulfilling the condition M. That is the reason why we speak here of the qualified-instance confirmation, in symbols "c_{qi}^*".[25] The results obtained concerning instance confirmation and qualified-instance confirmation[26] show that the values of these two functions are independent of N and hence hold for all finite and infinite universes. It has been found that, if the number s_1 of observed counter-instances

---

[24] In technical terms, the definition is as follows:
$c_i^*(l, e) = _{Df} c^*(h, e)$, where h is an instance of l formed by the substitution of an individual constant not occurring in e.

[25] The technical definition will be given here. Let l be 'for every x, if x is M, then x is M''. Let l be non-L-false and without variables. Let 'c' be any individual constant not occurring in e; let j say that c is M, and h' that c is M'. Then the qualified-instance confirmation of l with respect to 'M'' and 'M''' on the evidence e is defined as follows: $c_{qi}^*('M', 'M''$, e) = _{Df} c^*(h', e \cdot j)$.

[26] Some of the theorems may here be given. Let the law l say, as above, that all M are M'. Let 'M_1' be defined, as earlier, by '$M \cdot \sim M'$' ("non-white swan") and 'M_2' by '$M \cdot M'$' ("white swan"). Let the widths of M_1 and M_2 be w_1 and w_2 respectively. Let e be a report about s observed individuals saying that s_1 of them are M_1 and s_2 are M_2, while the remaining ones are $\sim M$ and hence neither M_1 nor M_2. Then the following holds:

(1)
$$c_i^*(l, e) = 1 - \frac{s_1 + w_1}{s + k}.$$

(2)
$$c_{qi}^*('M', 'M''$, e) = 1 - \frac{s_1 + w_1}{s_1 + w_1 + s_2 + w_2}.$$

The values of c_i^* and c_{qi}^* for the case that the observed sample does not contain any individuals violating the law l can easily be obtained from the values stated in (1) and (2) by taking $s_1 = 0$.

is a fixed small number, then, with the increase of the sample s, both c_i^* and c_{qi}^* grow close to 1, in contradistinction to c^* for the law itself. This justifies the customary manner of speaking of "very reliable" or "well-founded" or "well confirmed" laws, provided we interpret these phrases as referring to a high value of either of our two concepts just introduced. Understood in this sense, the phrases are not in contradiction to our previous results that the degree of confirmation of a law is very small in a large domain of individuals and 0 in the infinite domain (§13).

These concepts will also be of help in situations of the following kind. Suppose a scientist has observed certain events, which are not sufficiently explained by the known physical laws. Therefore he looks for a new law as an explanation. Suppose he finds two incompatible laws l and l', each of which would explain the observed events satisfactorily. Which of them should he prefer? If the domain of individuals in question is finite, he may take the law with the higher degree of confirmation. In the infinite domain, however, this method of comparison fails, because the degree of confirmation is 0 for either law. Here the concept of instance confirmation (or that of qualified-instance confirmation) will help. If it has a higher value for one of the two laws, then this law will be preferable, if no reasons of another nature are against it.

It is clear that for any deliberate activity predictions are needed, and that these predictions must be "founded upon" or "(inductively) inferred from" past experiences, in some sense of those phrases. Let us examine the situation with the help of the following simplified schema. Suppose a man X wants to make a plan for his actions and, therefore, is interested in the prediction h that c is M'. Suppose further, X has observed (1) that many other things were M and that all of them were also M', let this be formulated in the sentence e; (2) that c is M, let this be j. Thus he knows e and j by observation. The problem is, how does he go from these premises to the desired conclusion h? It is clear that this cannot be done by deduction; an inductive procedure must be applied. What is this inductive procedure? It is usually explained in the following way. From the evidence e, X infers inductively the law l which says that all M are M'; this inference is supposed to be inductively valid because e contains many positive and no negative instances of the law l; then he infers h ("c is white") from l ("all swans are white") and j ("c is a swan") deductively. Now let us see what the procedure looks like from the point of view of our inductive logic. One might perhaps be tempted to transcribe the usual description of the procedure just given into technical terms as follows. X infers l from e inductively because $c^*(l, e)$ is high; since $l \cdot j$ L-implies h, $c^*(h, e \cdot j)$ is likewise high; thus h may be inferred inductively from $e \cdot j$. However, this way of reasoning would not be correct, because, under ordinary conditions, $c^*(l, e)$ is not high but very low, and even 0 if the domain of individuals is infinite. The difficulty disappears when we realize on the basis of our previous discussions that X does not need a high c^* for l in order to obtain the desired high c^* for h; all he needs is a high c_{qi}^* for l; and this he has by knowing e and j. To put it in another way, X need not take the roundabout way through the law l at all, as is usually believed; he can instead go from his observational knowledge $e \cdot j$ directly to the prediction h. That is to say, our inductive logic makes it possible to determine $c^*(h, e \cdot j)$ directly and to find that it has a high value, without making use of any law. Customary thinking in every-day life likewise often takes this short-cut, which is now justified by inductive logic. For instance, suppose somebody asks Mr. X what color he expects the next swan he will see to have. Then X may reason like this: he has seen many white swans and no non-white swans; therefore he presumes, admittedly not with certainty, that the next swan will likewise be white; and he is willing to bet on it. He does perhaps not even consider the question whether all swans in the universe without a single exception are white; and if he did, he would not be willing to bet on the affirmative answer.

We see that the use of laws is not indispensable for making predictions. Nevertheless it is expedient of course to state universal laws in books on physics, biology, psychology, etc. Although these laws stated by scientists do not have a high degree of confirmation, they have a high qualified-instance confirmation and thus serve us as efficient instruments for finding those highly confirmed singular predictions which we need for guiding our actions.

§15. THE VARIETY OF INSTANCES

A generally accepted and applied rule of scientific method says that for testing a given law we should choose a variety of specimens as great as possible. For instance, in order to test the law that all metals expand by heat, we should examine not only specimens of iron, but of many different metals. It seems clear that a greater variety of instances allows a more effective examination of the law. Suppose three physicists examine the law mentioned; each of them makes one hundred experiments by heating one hundred metal pieces and observing their expansion; the first physicist neglects the rule of variety and takes only pieces of iron; the second follows the rule to a small extent by examining iron and copper pieces; the third satisfies the rule more thoroughly by taking his one hundred specimens from six different metals. Then we should say that the third physicist has confirmed the law by a more thoroughgoing examination than the two other physicists; therefore he has better reasons to declare the law well-founded and to expect that future instances will likewise be found to be in accordance with the law; and in the same way the second physicist has more reasons than the first. Accordingly, if there is at all an adequate concept of degree of confirmation with numerical values, then its value for the law, or for the prediction that a certain number of future instances will fulfill the law, should be higher on the evidence of the report of the third physicist about the positive results of his experiments than for the second physicist, and higher for the second than for the first. Generally speaking, the degree of confirmation of a law on the evidence of a number of confirming experiments should depend not only on the total number of (positive) instances found but also on their variety, i.e. on the way they are distributed among various kinds.

Ernest Nagel[27] has discussed this problem in detail. He explains the difficulties involved in finding a quantitative concept of degree of confirmation that

[27] E. Nagel, *Principles of the Theory of Probability*. Int. Encycl. of Unified Science, vol. I, No. 6, 1939; see pp. 68–71.

would satisfy the requirement we have just discussed, and he therefore expresses his doubt whether such a concept can be found at all. He says (pp. 69f): "It follows, however, that the degree of confirmation for a theory seems to be a function not only of the absolute number of positive instances but also of the kinds of instances and of the relative number in each kind. It is not in general possible, therefore, to order degrees of confirmation in a linear order, because the evidence for theories may not be comparable in accordance with a simple linear schema; and a fortiori degrees of confirmation cannot, in general, be quantized." He illustrates his point by a numerical example. A theory T is examined by a number E of experiments all of which yield positive instances; the specimens tested are taken from two non-overlapping kinds K_1 and K_2. Nine possibilities $P_1, \cdots P_9$ are discussed with different numbers of instances in K_1 and in K_2. The total number E increases from 50 in P_1 to 200 in P_9. In P_1, 50 instances are taken from K_1 and none from K_2; in P_9, 198 from K_1 and 2 from K_2. It does indeed seem difficult to find a concept of degree of confirmation that takes into account in an adequate way not only the absolute number E of instances but also their distribution among the two kinds in the different cases. And I agree with Nagel that this requirement is important. However, I do not think it impossible to satisfy the requirement; in fact, it is satisfied by our concept c^*.

This is shown by a theorem in our system of inductive logic, which states the ratio in which the c^* of a law l is increased if s new positive instances of one or several different kinds are added by new observations to some former positive instances. The theorem, which is too complicated to be given here, shows that c^* is greater under the following conditions: (1) if the total number s of the new instances is greater, *ceteris paribus*; (2) if, with equal numbers s, the number of different kinds from which the instances are taken is greater; (3) if the instances are distributed more evenly among the kinds. Suppose a physicist has made experiments for testing the law l with specimens of various kinds and he wishes to make one more experiment with a new specimen. Then it follows from (2), that the new specimen is best taken from one of those kinds from which so far no specimen has been examined; if there are no such kinds, then we see from (3) that the new specimen should best be taken from one of those kinds which contain the minimum number of instances tested so far. This seems in good agreement with scientific practice. [The above formulations of (2) and (3) hold in the case where all the kinds considered have equal width; in the general and more exact formulation, the increase of c^* is shown to be dependent also upon the various widths of the kinds of instances.] The theorem shows further that c^* is much more influenced by (2) and (3) than by (1); that is to say, it is much more important to improve the variety of instances than to increase merely their number.

The situation is best illustrated by a numerical example. The computation of the increase of c^*, for the nine possible cases discussed by Nagel, under certain plausible assumptions concerning the form of the law l and the widths of the properties involved, leads to the following results. If we arrange the nine possibilities in the order of ascending values of c^*, we obtain this: P_1, P_3, P_7, P_9;

P_2, P_4, P_6, P_6, P_8. In this order we find first the four possibilities with a bad distribution among the two kinds, i.e. those where none or only very few (two) of the instances are taken from one of the two kinds, and these four possibilities occur in the order in which they are listed by Nagel; then the five possibilities with a good or fairly good distribution follow, again in the same order as Nagel's. Even for the smallest sample with a good distribution (viz., P_2, with 100 instances, 50 from each of the two kinds) c^* is considerably higher—under the assumptions made, more than four times as high—than for the largest sample with a bad distribution (viz. P_9, with 200 instances, divided into 198 and 2). This shows that a good distribution of the instances is much more important than a mere increase in the total number of instances. This is in accordance with Nagel's remark (p. 69): "A large increase in the number of positive instances of one kind may therefore count for less, in the judgment of skilled experimenters, than a small increase in the number of positive instances of another kind."

Thus we see that the concept c^* is in satisfactory accordance with the principle of the variety of instances.

§16. THE PROBLEM OF THE JUSTIFICATION OF INDUCTION

Suppose that a theory is offered as a more exact formulation—sometimes called a "rational reconstruction"—of a body of generally accepted but more or less vague beliefs. Then the demand for a justification of this theory may be understood in two different ways. (1) The first, more modest task is to validate the claim that the new theory is a satisfactory reconstruction of the beliefs in question. It must be shown that the statements of the theory are in sufficient agreement with those beliefs; this comparison is possible only on those points where the beliefs are sufficiently precise. The question whether the given beliefs are true or false is here not even raised. (2) The second task is to show the validity of the new theory and thereby of the given beliefs. This is a much deeper going and often much more difficult problem.

For example, Euclid's axiom system of geometry was a rational reconstruction of the beliefs concerning spatial relations which were generally held, based on experience and intuition, and applied in the practices of measuring, surveying, building, etc. Euclid's axiom system was accepted because it was in sufficient agreement with those beliefs and gave a more exact and consistent formulation for them. A critical investigation of the validity, the factual truth, of the axioms and the beliefs was only made more than two thousand years later by Gauss.

Our system of inductive logic, that is, the theory of c^* based on the definition of this concept, is intended as a rational reconstruction, restricted to a simple language form, of inductive thinking as customarily applied in everyday life and in science. Since the implicit rules of customary inductive thinking are rather vague, any rational reconstruction contains statements which are neither supported nor rejected by the ways of customary thinking. Therefore, a comparison is possible only on those points where the procedures of customary inductive thinking are precise enough. It seems to me, that on these points sufficient agreement is found to show that our theory is an adequate reconstruction;

171

this agreement is seen in many theorems, of which a few have been mentioned in this paper.

An entirely different question is the problem of the validity of our or any other proposed system of inductive logic, and thereby of the customary methods of inductive thinking. This is the genuinely philosophical problem of induction. The construction of a systematic inductive logic is an important step towards the solution of the problem, but still only a preliminary step. It is important because without an exact formulation of rules of induction, i.e. theorems on degree of confirmation, it is not clear what exactly is meant by "inductive procedures", and therefore the problem of the validity of these procedures cannot even be raised in precise terms. On the other hand, a construction of inductive logic, although it prepares the way towards a solution of the problem of induction, still does not by itself give a solution.

Older attempts at a justification of induction tried to transform it into a kind of deduction, by adding to the premises a general assumption of universal form, e.g. the principle of the uniformity of nature. I think there is fairly general agreement today among scientists and philosophers that neither this nor any other way of reducing induction to deduction with the help of a general principle is possible. It is generally acknowledged that induction is fundamentally different from deduction, and that any prediction of a future event reached inductively on the basis of observed events can never have the certainty of a deductive conclusion; and, conversely, the fact that a prediction reached by certain inductive procedures turns out to be false does not show that those inductive procedures were incorrect.

The situation just described has sometimes been characterized by saying that a theoretical justification of induction is not possible, and hence, that there is no problem of induction. However, it would be better to say merely that a justification in the old sense is not possible. Reichenbach[28] was the first to raise the problem of the justification of induction in a new sense and to take the first step towards a positive solution. Although I do not agree with certain other features of Reichenbach's theory of induction, I think it has the merit of having first emphasized these important points with respect to the problem of justification: (1) the decisive justification of an inductive procedure does not consist in its plausibility, i.e., its accordance with customary ways of inductive reasoning, but must refer to its success in some sense; (2) the fact that the truth of the predictions reached by induction cannot be guaranteed does not preclude a justification in a weaker sense; (3) it can be proved (as a purely logical result) that induction leads in the long run to success in a certain sense, provided the world is "predictable" at all, i.e. such that success in that respect is possible. Reichenbach shows that his rule of induction R leads to success in the following sense: R yields in the long run an approximate estimate of the relative frequency in the whole of any given property. Thus suppose that we observe the relative frequencies of a property M in an increasing series of samples, and that we determine on the basis of each sample with the help of the rule R the probability

[28] Hans Reichenbach, *Experience and Prediction*, 1938, §§38 ff., and earlier publications.

q that an unobserved thing has the property M, then the values q thus found approach in the long run the relative frequency of M in the whole. (This is, of course, merely a logical consequence of Reichenbach's definition or rule of induction, not a factual feature of the world.)

I think that the way in which Reichenbach examines and justifies his rule of induction is an important step in the right direction, but only a first step. What remains to be done is to find a procedure for the examination of any given rule of induction in a more thoroughgoing way. To be more specific, Reichenbach is right in the assertion that any procedure which does not possess the character-istic described above (viz. approximation to the relative frequency in the whole) is inferior to his rule of induction. However, his rule, which he calls "the" rule of induction, is far from being the only one possessing that characteristic. The same holds for an infinite number of other rules of induction, e.g., for Laplace's rule of succession (see above, §10; here restricted in a suitable way so as to avoid contradictions), and likewise for the corresponding rule of our theory of c* (as formulated in theorem (1), §10). Thus our inductive logic is justified to the same extent as Reichenbach's rule of induction, as far as the only criterion of justification so far developed goes. (In other respects, our inductive logic covers a much more extensive field than Reichenbach's rule; this can be seen by the theorems on various kinds of inductive inference mentioned in this paper.) However, Reichenbach's rule and the other two rules mentioned yield different numerical values for the probability under discussion, although these values converge for an increasing sample towards the same limit. Therefore we need a more general and stronger method for examining and comparing any two given rules of induction in order to find out which of them has more chance of success. I think we have to measure the success of any given rule of induction by the total balance with respect to a comprehensive system of wagers made according to the given rule. For this task, here formulated in vague terms, there is so far not even an exact formulation; and much further investigation will be needed before a solution can be found.

University of Chicago, Chicago, Ill.

VOL. LIV. No. 213.] [January, 1945.

MIND

A QUARTERLY REVIEW

OF

PSYCHOLOGY AND PHILOSOPHY

———— ⋙⋘ ————

I.—STUDIES IN THE LOGIC OF CON-
FIRMATION (I.).

To the memory of my wife, Eva Ahrends Hempel.

By CARL G. HEMPEL.

1. *Objective of the Study.*[1]—The defining characteristic of an empirical statement is its capability of being tested by a confrontation with experimental findings, *i.e.* with the results of suitable experiments or " focussed " observations. This feature distinguishes statements which have empirical content both from the statements of the formal sciences, logic and mathematics, which require no experiential test for their validation, and from

---

[1] The present analysis of confirmation was to a large extent suggested and stimulated by a co-operative study of certain more general problems which were raised by Dr. Paul Oppenheim, and which I have been investigating with him for several years. These problems concern the form and the function of scientific laws and the comparative methodology of the different branches of empirical science. The discussion with Mr. Oppenheim of these issues suggested to me the central problem of the present essay. The more comprehensive problems just referred to will be dealt with by Mr. Oppenheim in a publication which he is now preparing.

In my occupation with the logical aspects of confirmation, I have benefited greatly by discussions with several students of logic, including Professor R. Carnap, Professor A. Tarski, and particularly Dr. Nelson Goodman, to whom I am indebted for several valuable suggestions which will be indicated subsequently.

A detailed exposition of the more technical aspects of the analysis of confirmation presented in this article is included in my article " A Purely Syntactical Definition of Confirmation ", *The Journal of Symbolic Logic*, vol. 8 (1943).

1

the formulations of transempirical metaphysics, which do not admit of any.

The testability here referred to has to be understood in the comprehensive sense of " testability in principle " ; there are many empirical statements which, for practical reasons, cannot be actually tested at present. To call a statement of this kind testable in principle means that it is possible to state just what experiential findings, if they were actually obtained, would constitute favourable evidence for it, and what findings or " data ", as we shall say for brevity, would constitute unfavourable evidence ; in other words, a statement is called testable in principle, if it is possible to describe the kind of data which would confirm or disconfirm it.

The concepts of confirmation and of disconfirmation as here understood are clearly more comprehensive than those of conclusive verification and falsification.   Thus, *e.g.* no finite amount of experiential evidence can conclusively verify a hypothesis expressing a general law such as the law of gravitation, which covers an infinity of potential instances, many of which belong either to the as yet inaccessible future, or to the irretrievable past ; but a finite set of relevant data may well be " in accord with " the hypothesis and thus constitute confirming evidence for it. Similarly, an existential hypothesis, asserting, say, the existence of an as yet unknown chemical element with certain specified characteristics, cannot be conclusively proved false by a finite amount of evidence which fails to " bear out " the hypothesis ; but such unfavourable data may, under certain conditions, be considered as weakening the hypothesis in question, or as constituting disconfirming evidence for it.[1]

While, in the practice of scientific research, judgments as to the confirming or disconfirming character of experiential data obtained in the test of a hypothesis are often made without hesitation and with a wide consensus of opinion, it can hardly be said that these judgments are based on an explicit theory providing general criteria of confirmation and of disconfirmation.   In this respect, the situation is comparable to the manner in which deductive inferences are carried out in the practice of scientific research : This, too, is often done without reference to an explicitly stated system of rules of logical inference. But while criteria of valid deduction can be and have been supplied by

[1] This point as well as the possibility of conclusive verification and conclusive falsification will be discussed in some detail in section 10 of the present paper.

formal logic, no satisfactory theory providing general criteria of confirmation and disconfirmation appears to be available so far.

In the present essay, an attempt will be made to provide the elements of a theory of this kind. After a brief survey of the significance and the present status of the problem, I propose to present a detailed critical analysis of some common conceptions of confirmation and disconfirmation and then to construct explicit definitions for these concepts and to formulate some basic principles of what might be called the logic of confirmation.

2. *Significance and Present Status of the Problem.*—The establishment of a general theory of confirmation may well be regarded as one of the most urgent desiderata of the present methodology of empirical science.[1] Indeed, it seems that a precise analysis of the concept of confirmation is a necessary condition for an adequate solution of various fundamental problems concerning the logical structure of scientific procedure. Let us briefly survey the most outstanding of these problems.

(a) In the discussion of scientific method, the concept of relevant evidence plays an important part. And while certain " inductivist " accounts of scientific procedure seem to assume that relevant evidence, or relevant data, can be collected in the context of an inquiry prior to the formulation of any hypothesis, it should be clear upon brief reflection that relevance is a relative concept ; experiential data can be said to be relevant or irrelevant only with respect to a given hypothesis ; and it is the hypothesis which determines what kind of data or evidence are relevant for it. Indeed, an empirical finding is relevant for a hypothesis if and only if it constitutes either favourable or unfavourable evidence for it ; in other words, if it either confirms or disconfirms the hypothesis. Thus, a precise definition of relevance presupposes an analysis of confirmation and disconfirmation.

(b) A closely related concept is that of instance of a hypothesis. The so-called method of inductive inference is usually presented as proceeding from specific cases to a general hypothesis of which each of the special cases is an " instance " in the sense that it " conforms to " the general hypothesis in question, and thus constitutes confirming evidence for it.

Thus, any discussion of induction which refers to the establishment of general hypotheses on the strength of particular instances is fraught with all those logical difficulties—soon to be expounded

[1] Or of the " logic of science ", as understood by R. Carnap ; *cf. The Logical Syntax of Language* (New York and London, 1937), sect. 72, and the supplementary remarks in *Introduction to Semantics* (Cambridge, Mass., 1942), p. 250.

—which beset the concept of confirmation. A precise analysis ot this concept is, therefore, a necessary condition for a clear statement of the issues involved in the problem complex of induction and of the ideas suggested for their solution—no matter what their theoretical merits or demerits may be.

(c) Another issue customarily connected with the study of scientific method is the quest for " rules of induction ". Generally speaking, such rules would enable us to " infer ", from a given set of data, that hypothesis or generalization which accounts best for all the particular data in the given set. Recent logical analyses have made it increasingly clear that this way of conceiving the problem involves a misconception : While the process of invention by which scientific discoveries are made is as a rule *psychologically guided and stimulated* by antecedent knowledge of specific facts, its results are *not logically determined* by them ; the way in which scientific hypotheses or theories are discovered cannot be mirrored in a set of general rules of inductive inference.[1] One of the crucial considerations which lead to this conclusion is the following : Take a scientific theory such as the atomic theory of matter. The evidence on which it rests may be described in terms referring to directly observable phenomena, namely to certain " macroscopic " aspects of the various experimental and observational data which are relevant to the theory. On the other hand, the theory itself contains a large number of highly abstract, non-observational terms such as " atom ", " electron ", " nucleus ", " dissociation ", " valence " and others, none of which figures in the description of the observational data. An adequate rule of induction would therefore have to provide, for this and for every conceivable other case, mechanically applicable criteria determining unambiguously, and without any reliance on the inventiveness or additional scientific knowledge of its user, all those new abstract concepts which need to be created for the formulation of the theory that will account for the given evidence. Clearly, this requirement cannot be satisfied by any set of rules, however ingeniously devised ; there can be no general rules of induction in the above sense ; the demand for them rests on a confusion of logical and psychological issues. What determines the soundness of a hypothesis is not the way it

---

[1] See the lucid presentation of this point in Karl Popper's *Logik der Forschung* (Wien, 1935), esp. sect. 1, 2, 3, and 25, 26, 27 ; *cf.* also Albert Einstein's remarks in his lecture *On the Method of Theoretical Physics* (Oxford, 1933,) pp. 11 and 12. Also of interest in this context is the critical discussion of induction by H. Feigl in " The Logical Character of the Principle of Induction, " *Philosophy of Science*, vol. 1 (1934).

is arrived at (it may even have been suggested by a dream or a hallucination), but the way it stands up when tested, *i.e.* when confronted with relevant observational data. Accordingly, the quest for rules of induction in the original sense of canons of scientific discovery has to be replaced, in the logic of science, by the quest for general objective criteria determining (A) whether, and—if possible—even (B) to what degree, a hypothesis H may be said to be corroborated by a given body of evidence E. This approach differs essentially from the inductivist conception of the problem in that it presupposes not only E, but also H as given and then seeks to determine a certain logical relationship between them. The two parts of this latter problem can be restated in somewhat more precise terms as follows :

(A) To give precise definitions of the two non-quantitative relational concepts of confirmation and of disconfirmation ; *i.e.* to define the meaning of the phrases " E confirms H " and " E disconfirms H ". (When E neither confirms nor disconfirms H, we shall say that E is neutral, or irrelevant, with respect to H.)

(B) (1) To lay down criteria defining a metrical concept " degree of confirmation of H with respect to E ", whose values are real numbers ; or, failing this,

(2) To lay down criteria defining two relational concepts, " more highly confirmed than " and " equally well confirmed with ", which make possible a non-metrical comparison of hypotheses (each with a body of evidence assigned to it) with respect to the extent of their confirmation.

Interestingly, problem B has received much more attention in methodological research than problem A ; in particular, the various theories of the " probability of hypotheses " may be regarded as concerning this problem complex ; we have here adopted [1] the more neutral term " degree of confirmation " instead of " probability " because the latter is used in science in a definite technical sense involving reference to the relative frequency of the occurrence of a given event in a sequence, and it is at least an open question whether the degree of confirmation of a hypothesis can generally be defined, as a probability in this statistical sense.

The theories dealing with the probability of hypotheses fall into two main groups : the " logical " theories construe probability as a logical relation between sentences (or propositions ;

[1] Following R. Carnap's usage in Testability and Meaning, *Philosophy of Science*, vols. 3 (1936) and 4 (1937) ; esp. sect. 3 (in vol. 3).

it is not always clear which is meant)[1]; the "statistical" theories interpret the probability of a hypothesis in substance as the limit of the relative frequency of its confirming instances among all relevant cases.[2] Now it is a remarkable fact that none of the theories of the first type which have been developed so far provides an explicit general definition of the probability (or degree of confirmation) of a hypothesis H with respect to a body of evidence E; they all limit themselves essentially to the construction of an uninterpreted postulational system of logical probability. For this reason, these theories fail to provide a complete solution of problem B. The statistical approach, on the other hand, would, if successful, provide an explicit numerical definition of the degree of confirmation of a hypothesis; this definition would be formulated in terms of the numbers of confirming and disconfirming instances for H which constitute the body of evidence E. Thus, a necessary condition for an adequate interpretation of degrees of confirmation as statistical probabilities is the establishment of precise criteria of confirmation and disconfirmation, in other words, the solution of problem A.

However, despite their great ingenuity and suggestiveness, the attempts which have been made so far to formulate a precise statistical definition of the degree of confirmation of a hypothesis seem open to certain objections,[3] and several authors [4] have expressed doubts as to the possibility of defining the degree of confirmation of a hypothesis as a metrical magnitude, though

---

[1] This group includes the work of such writers as Janina Hosiasson-Lindenbaum (*cf.* for instance, her article "Induction et analogie: Comparàison de leur fondement", Mind, vol. L (1941); also see p. 21, n. 2), H. Jeffreys, J. M. Keynes, B. O. Koopman, J. Nicod (see p. 9, n. 2), St. Mazurkiewicz, F. Waismann. For a brief discussion of this conception of probability, see Ernest Nagel, *Principles of the Theory of Probability* (Internat. Encyclopedia of Unified Science, vol. i, no. 6, Chicago, 1939), esp. sects. 6 and 8.

[2] The chief proponent of this view is Hans Reichenbach; *cf.* especially Ueber Induktion und Wahrscheinlichkeit, *Erkenntnis*, vol. v (1935), and *Experience and Prediction* (Chicago, 1938), Ch. V.

[3] *Cf.* Karl Popper, *Logik der Forschung* (Wien, 1935), sect. 80; Ernest Nagel, *l.c.*, sect. 8, and "Probability and the Theory of Knowledge", *Philosophy of Science*, vol. 6 (1939); C. G. Hempel, "Le problème de la vérité", *Theoria* (Göteborg), vol. 3 (1937), sect. 5, and "On the Logical Form of Probability Statements", *Erkenntnis*, vol. 7 (1937-38), esp. sect. 5. *Cf.* also Morton White, "Probability and Confirmation", *The Journal of Philosophy*, vol. 36 (1939).

[4] See, for example, J. M. Keynes, *A Treatise on Probability*, London, 1929, esp. Ch. III; Ernest Nagel, *Principles of the Theory of Probability* (*cf.* n. 1 above), esp. p. 70; compare also the somewhat less definitely sceptical statement by Carnap, *l.c.* (see p. 5, n. 1), sect. 3, p. 427.

some of them consider it as possible, under certain conditions, to solve at least the less exacting problem B (2), *i.e.* to establish standards of non-metrical comparison between hypotheses with respect to the extent of their confirmation. An adequate comparison of this kind might have to take into account a variety of different factors [1] ; but again the numbers of the confirming and of the disconfirming instances which the given evidence includes will be among the most important of those factors.

Thus, of the two problems, A and B, the former appears to be the more basic one, first, because it does not presuppose the possibility of defining numerical degrees of confirmation or of comparing different hypotheses as to the extent of their confirmation ; and second because our considerations indicate that any attempt to solve problem B—unless it is to remain in the stage of an axiomatized system without interpretation—is likely to require a precise definition of the concepts of confirming and disconfirming instance of a hypothesis before it can proceed to define numerical degrees of confirmation, or to lay down non-metrical standards of comparison.

(*d*) It is now clear that an analysis of confirmation is of fundamental importance also for the study of the central problem of what is customarily called epistemology ; this problem may be characterized as the elaboration of " standards of rational belief " or of criteria of warranted assertibility. In the methodology of empirical science this problem is usually phrased as concerning the rules governing the test and the subsequent acceptance or rejection of empirical hypotheses on the basis of experimental or observational findings, while in its " epistemological " version the issue is often formulated as concerning the validation of beliefs by reference to perceptions, sense data, or the like. But no matter how the final empirical evidence is construed and in what terms it is accordingly expressed, the theoretical problem remains the same : to characterize, in precise and general terms, the conditions under which a body of evidence can be said to confirm, or to disconfirm, a hypothesis of empirical character ; and that is again our problem A.

(*e*) The same problem arises when one attempts to give a precise statement of the empiricist and operationalist criteria for the empirical meaningfulness of a sentence ; these criteria, as is well known, are formulated by reference to the theoretical

---

[1] See especially the survey of such factors given by Ernest Nagel in *Principles of the Theory of Probability* (*cf.* p. 6, n. 1), pp. 66-73.

testability of the sentence by means of experiential evidence [1] ;
and the concept of theoretical testability, as was pointed out
earlier, is closely related to the concepts of confirmation and disconfirmation.[2]

Considering the great importance of the concept of confirmation,
it is surprising that no systematic theory of the non-quantitative
relation of confirmation seems to have been developed so far.
Perhaps this fact reflects the tacit assumption that the concepts
of confirmation and of disconfirmation have a sufficiently clear
meaning to make explicit definitions unnecessary or at least
comparatively trivial.   And indeed, as will be shown below, there
are certain features which are rather generally associated with
the intuitive notion of confirming evidence, and which, at first,
seem well suited to serve as defining characteristics of confirmation.   Closer examination will reveal the definitions thus obtainable to be seriously deficient and will make it clear that an
adequate definition of confirmation involves considerable difficulties.

Now the very existence of such difficulties suggests the question
whether the problem we are considering does not rest on a false
assumption : Perhaps there are no objective criteria of confirmation ; perhaps the decision as to whether a given hypothesis is
acceptable in the light of a given body of evidence is no more
subject to rational, objective rules than is the process of inventing
a scientific hypothesis or theory ; perhaps, in the last analysis,
it is a " sense of evidence ", or a feeling of plausibility in view of
the relevant data, which ultimately decides whether a hypothesis
is scientifically acceptable.[3] This view is comparable to the
opinion that the validity of a mathematical proof or of a logical
argument has to be judged ultimately by reference to a feeling
of soundness or convincingness ; and both theses have to be
rejected on analogous grounds : They involve a confusion of
logical and psychological considerations.   Clearly, the occurrence

[1] Cf., for example, A. J. Ayer, *Language, Truth and Logic*, London and
New York, 1936, Ch. I ; R. Carnap, " Testability and Meaning " (*cf.*
p. 5, n. 1) sects. 1, 2, 3 ; H. Feigl, *Logical Empiricism* (in *Twentieth
Century Philosophy*, ed. by Dagobert D. Runes, New York, 1943) ;
P. W. Bridgman, *The Logic of Modern Physics*, New York, 1928.

[2] It should be noted, however, that in his essay "Testability and Meaning"
(*cf.* p. 5, n. 1) R. Carnap has constructed definitions of testability and
confirmability which avoid reference to the concept of confirming and of
disconfirming evidence ; in fact, no proposal for the definition of these
latter concepts is made in that study.

[3] A view of this kind has been expressed, for example, by M. Mandelbaum
in " Causal Analyses in History ", *Journal of the History of Ideas*, vol. 3
(1942) ; *cf.* esp. pp. 46-47.

or non-occurrence of a feeling of conviction upon the presentation of grounds for an assertion is a subjective matter which varies from person to person, and with the same person in the course of time ; it is often deceptive, and can certainly serve neither as a necessary nor as a sufficient condition for the soundness of the given assertion.[1] A rational reconstruction of the standards of scientific validation cannot, therefore, involve reference to a sense of evidence ; it has to be based on objective criteria. In fact, it seems reasonable to require that the criteria of empirical confirmation, besides being objective in character, should contain no reference to the specific subject-matter of the hypothesis or of the evidence in question ; it ought to be possible, one feels, to set up purely formal criteria of confirmation in a manner similar to that in which deductive logic provides purely formal criteria for the validity of deductive inferences.

With this goal in mind, we now turn to a study of the non-quantitative concept of confirmation. We shall begin by examining some current conceptions of confirmation and exhibiting their logical and methodological inadequacies ; in the course of this analysis, we shall develop a set of conditions for the adequacy of any proposed definition of confirmation ; and finally, we shall construct a definition of confirmation which satisfies those general standards of adequacy.

3. *Nicod's Criterion of Confirmation and its Shortcomings.*—We consider first a conception of confirmation which underlies many recent studies of induction and of scientific method. A very explicit statement of this conception has been given by Jean Nicod in the following passage : " Consider the formula or the law : *A entails B.* How can a particular proposition, or more briefly, a fact, affect its probability ? If this fact consists of the presence of B in a case of A, it is favourable to the law ' *A entails B* ' ; on the contrary, if it consists of the absence of B in a case of A, it is unfavourable to this law. It is conceivable that we have here the only two direct modes in which a fact can influence the probability of a law. . . . Thus, the entire influence of particular truths or facts on the probability of universal propositions or laws would operate by means of these two elementary relations which we shall call *confirmation* and *invalidation.*" [2] Note that the applicability of this criterion is restricted to hypotheses of

---

[1] See Karl Popper's pertinent statement, *l.c.*, sect. 8.

[2] Jean Nicod, *Foundations of Geometry and Induction* (transl. by P. P. Wiener), London, 1930 ; p. 219 ; *cf.* also R. M. Eaton's discussion of " Confirmation and Infirmation ", which is based on Nicod's views ; it is included in Ch. III of his *General Logic*, New York, 1931.

the form "*A entails B* ". Any hypothesis *H* of this kind may
be expressed in the notation of symbolic logic [1] by means of a
universal conditional sentence, such as, in the simplest case,

$$(x)(P(x) \supset Q(x)),$$

i.e. " For any object *x* : if *x* is a *P*, then *x* is a *Q*," or also
" Occurrence of the quality *P* entails occurrence of the quality
Q." According to the above criterion this hypothesis is confirmed by an object *a*, if *a* is *P* and *Q* ; and the hypothesis is
disconfirmed by *a* if *a* is *P*, but not *Q*. In other words, an object
confirms a universal conditional hypothesis if and only if it
satisfies both the antecedent (here : ' *P(x)* ') and the consequent
(here : ' *Q(x)* ') of the conditional ; it disconfirms the hypothesis
if and only if it satisfies the antecedent, but not the consequent
of the conditional ; and (we add this to Nicod's statement) it is
neutral, or irrelevant, with respect to the hypothesis if it does
not satisfy the antecedent.

This criterion can readily be extended so as to be applicable
also to universal conditionals containing more than one quantifier,
such as " Twins always resemble each other ", or, in symbolic
notation, ' $(x)(y)(\text{Twins}(x, y) \supset \text{Rsbl}(x, y))$ '. In these cases, a
confirming instance consists of an ordered couple, or triple, etc.,
of objects satisfying the antecedent and the consequent of the
conditional. (In the case of the last illustration, any two persons
who are twins and resemble each other would confirm the hypothesis ; twins who do not resemble each other would disconfirm
it ; and any two persons not twins—no matter whether they
resemble each other or not—would constitute irrelevant evidence.)

We shall refer to this criterion as Nicod's criterion.[2] It states
explicitly what is perhaps the most common tacit interpretation
of the concept of confirmation. While seemingly quite adequate,
it suffers from serious shortcomings, as will now be shown.

(*a*) First, the applicability of this criterion is restricted to
hypotheses of universal conditional form ; it provides no standards
of confirmation for existential hypotheses (such as " There exists
organic life on other stars ", or " Poliomyelitis is caused by some
virus ") or for hypotheses whose explicit formulation calls for
the use of both universal and existential quantifiers (such as

---

[1] In this paper, only the most elementary devices of this notation are
used ; the symbolism is essentially that of *Principia Mathematica*, except
that parentheses are used instead of dots, and that existential quantification is symbolized by ' (E) ' instead of by the inverted ' E '.

[2] This term is chosen for convenience, and in view of the above explicit
formulation given by Nicod ; it is not, of course, intended to imply that
this conception of confirmation originated with Nicod.

" Every human being dies some finite number of years after his birth ", or the psychological hypothesis, " You can fool all of the people some of the time and some of the people all of the time, but you cannot fool all of the people all of the time ", which may be symbolized by ' $(x)(Et)Fl(x, t) . (Ex)(t)Fl(x, t) . \sim (x)(t)Fl(x, t)$ ', (where ' $Fl(x, t)$ ' stands for " You can fool (person) x at time t "). We note, therefore, the desideratum of establishing a criterion of confirmation which is applicable to hypotheses of any form.[1]

(b) We now turn to a second shortcoming of Nicod's criterion. Consider the two sentences

$$S_1 : \text{‘ } (x)(\text{Raven}(x) \supset \text{Black}(x)) \text{ ’ ;}$$
$$S_2 : \text{‘ } (x)(\sim \text{Black}(x) \supset \sim \text{Raven}(x)) \text{ ’}$$

(i.e. " All ravens are black " and " Whatever is not black is not a raven "), and let a, b, c, d be four objects such that a is a raven and black, b a raven but not black, c not a raven but black, and d neither a raven nor black. Then, according to Nicod's criterion, a would confirm S_1, but be neutral with respect to S_2; b would disconfirm both S_1 and S_2; c would be neutral with respect to both S_1 and S_2, and d would confirm S_2, but be neutral with respect to S_1.

But S_1 and S_2 are logically equivalent; they have the same content, they are different formulations of the same hypothesis. And yet, by Nicod's criterion, either of the objects a and d would be confirming for one of the two sentences, but neutral with respect to the other. This means that Nicod's criterion makes confirmation depend not only on the content of the hypothesis, but also on its formulation.[2]

One remarkable consequence of this situation is that every hypothesis to which the criterion is applicable—i.e. every universal conditional—can be stated in a form for which there cannot possibly exist any confirming instances. Thus, e.g. the sentence

$$(x)[(\text{Raven}(x) . \sim \text{Black}(x)) \supset (\text{Raven}(x) . \sim \text{Raven}(x)]$$

is readily recognized as equivalent to both S_1 and S_2 above; yet no object whatever can confirm this sentence, i.e. satisfy both

[1] For a rigorous formulation of the problem, it is necessary first to lay down assumptions as to the means of expression and the logical structure of the language in which the hypotheses are supposed to be formulated ; the desideratum then calls for a definition of confirmation applicable to any hypothesis which can be expressed in the given language. Generally speaking, the problem becomes increasingly difficult with increasing richness and complexity of the assumed " language of science ".

[2] This difficulty was pointed out, in substance, in my article " Le problème de la vérité ", *Theoria* (Göteborg), vol. 3 (1937), esp. p. 222.

its antecedent and its consequent; for the consequent is contradictory. An analogous transformation is, of course, applicable to any other sentence of universal conditional form.

4. *The Equivalence Condition.*—The results just obtained call attention to a condition which an adequately defined concept of confirmation should satisfy, and in the light of which Nicod's criterion has to be rejected as inadequate : *Equivalence condition :* Whatever confirms (disconfirms) one of two equivalent sentences, also confirms (disconfirms) the other.

Fulfilment of this condition makes the confirmation of a hypothesis independent of the way in which it is formulated; and no doubt it will be conceded that this is a necessary condition for the adequacy of any proposed criterion of confirmation. Otherwise, the question as to whether certain data confirm a given hypothesis would have to be answered by saying : "That depends on which of the different equivalent formulations of the hypothesis is considered "—which appears absurd. Furthermore—and this is a more important point than an appeal to a feeling of absurdity —an adequate definition of confirmation will have to do justice to the way in which empirical hypotheses function in theoretical scientific contexts such as explanations and predictions; but when hypotheses are used for purposes of explanation or prediction,[1] they serve as premises in a deductive argument whose conclusion is a description of the event to be explained or predicted. The deduction is governed by the principles of formal logic, and according to the latter, a deduction which is valid will remain so if some or all of the premises are replaced by different, but equivalent statements ; and indeed, a scientist will feel free, in any theoretical reasoning involving certain hypotheses, to use the latter in whichever of their equivalent formulations is most convenient for the development of his conclusions. But if we adopted a concept of confirmation which did not satisfy the equivalence condition, then it would be possible, and indeed necessary, to argue in certain cases that it was sound scientific procedure to base a prediction on a given hypothesis if formulated in a sentence S_1, because a good deal of confirming evidence had

---

[1] For a more detailed account of the logical structure of scientific explanation and prediction, *cf.* C. G. Hempel, "The Function of General Laws in History", *The Journal of Philosophy*, vol. 39 (1942), esp. sects. 2, 3, 4. The characterization, given in that paper as well as in the above text, of explanations and predictions as arguments of a deductive logical structure, embodies an over-simplification : as will be shown in sect. 7 of the present essay, explanations and predictions often involve "quasi-inductive" steps besides deductive ones. This point, however, does not affect the validity of the above argument.

been found for S_1; but that it was altogether inadmissible to base the prediction (say, for convenience of deduction) on an equivalent formulation S_2, because no confirming evidence for S_2 was available. Thus, the equivalence condition has to be regarded as a necessary condition for the adequacy of any definition of confirmation.

5. *The " Paradoxes " of Confirmation.*—Perhaps we seem to have been labouring the obvious in stressing the necessity of satisfying the equivalence condition. This impression is likely to vanish upon consideration of certain consequences which derive from a combination of the equivalence condition with a most natural and plausible assumption concerning a sufficient condition of confirmation.

The essence of the criticism we have levelled so far against Nicod's criterion is that it certainly cannot serve as a necessary condition of confirmation ; thus, in the illustration given in the beginning of section 3, the object a confirms S_1 and should therefore also be considered as confirming S_2, while according to Nicod's criterion it is not. Satisfaction of the latter is therefore not a necessary condition for confirming evidence.

On the other hand, Nicod's criterion might still be considered as stating a particularly obvious and important sufficient condition of confirmation. And indeed, if we restrict ourselves to universal conditional hypotheses in one variable [1]—such as S_1

---

[1] This restriction is essential : In its general form, which applies to universal conditionals in any number of variables, Nicod's criterion cannot even be construed as expressing a sufficient condition of confirmation. This is shown by the following rather surprising example : Consider the hypothesis S_1 : $(x)(y)[\sim (R(x, y)) \supset (R(x, y) . \sim R(y, x))]$.

Let a, b be two objects such that $R(a, b)$ and $\sim R(b, a)$. Then clearly, the couple (a, b) satisfies both the antecedent and the consequent of the universal conditional S_1; hence, if Nicod's criterion in its general form is accepted as stating a sufficient condition of confirmation, (a, b) constitutes confirming evidence for S_1. However, S_1 can be shown to be equivalent to

$$S_2 : (x)(y)R(x, y)$$

Now, by hypothesis, we have $\sim R(b, a)$; and this flatly contradicts S_2 and thus S_1. Thus, the couple (a, b), although satisfying both the antecedent and the consequent of the universal conditional S_1 actually constitutes disconfirming evidence of the strongest kind (conclusively disconfirming evidence, as we shall say later) for that sentence. This illustration reveals a striking and—as far as I am aware—hitherto unnoticed weakness of that conception of confirmation which underlies Nicod's criterion. In order to realize the bearing of our illustration upon Nicod's original formulation, let A and B be $\sim (R(x, y) . R(y, x))$ and $R(x, y) . \sim R(y, x)$ respectively. Then S_1 asserts that A entails B, and the couple (a, b) is a case of the presence of B in the presence of A ; this should, according to Nicod, be favourable to S_1.

and S_2 in the above illustration—then it seems perfectly reasonable to qualify an object as confirming such a hypothesis if it satisfies both its antecedent and its consequent. The plausibility of this view will be further corroborated in the course of our subsequent analyses.

Thus, we shall agree that if a is both a raven and black, then a certainly confirms S_1 : ' (x) (Raven(x) $\supset$ Black(x)) ', and if d is neither black nor a raven, d certainly confirms S_2 :

$$' (x) (\sim \text{Black}(x) \supset \sim \text{Raven}(x)).'$$

Let us now combine this simple stipulation with the equivalence condition : Since S_1 and S_2 are equivalent, d is confirming also for S_1 ; and thus, we have to recognize as confirming for S_1 any object which is neither black nor a raven. Consequently, any red pencil, any green leaf, and yellow cow, etc., becomes confirming evidence for the hypothesis that all ravens are black. This surprising consequence of two very adequate assumptions (the equivalence condition and the above sufficient condition of confirmation) can be further expanded : The following sentence can readily be shown to be equivalent to S_1 : S_3 : ' (x) [(Raven(x) v $\sim$ Raven(x)) $\supset$ ($\sim$ Raven(x) v Black(x)))] ', *i.e.* " Anything which is or is not a raven is either no raven or black ". According to the above sufficient condition, S_3 is certainly confirmed by any object, say e, such that (1) e is or is not a raven and, in addition, (2) e is not a raven or also black. Since (1) is analytic, these conditions reduce to (2). By virtue of the equivalence condition, we have therefore to consider as confirming for S_1 any object which is either no raven or also black (in other words : any object which is no raven at all, or a black raven).

Of the four objects characterized in section 3, a, c and d would therefore constitute confirming evidence for S_1, while b would be disconfirming for S_1. This implies that any non-raven represents confirming evidence for the hypothesis that all ravens are black.

We shall refer to these implications of the equivalence criterion and of the above sufficient condition of confirmation as the *paradoxes of confirmation*.

How are these paradoxes to be dealt with ? Renouncing the equivalence condition would not represent an acceptable solution, as is shown by the considerations presented in section 4. Nor does it seem possible to dispense with the stipulation that an object satisfying two conditions, C_1 and C_2, should be considered as confirming a general hypothesis to the effect that any object which satisfies C_1, also satisfies C_2.

But the deduction of the above paradoxical results rests on

one other assumption which is usually taken for granted, namely, that the meaning of general empirical hypotheses, such as that all ravens are black, or that all sodium salts burn yellow, can be adequately expressed by means of sentences of universal conditional form, such as ' (x) (Raven(x) $\supset$ Black(x))' and ' (x) (Sod. Salt(x) $\supset$ Burn Yellow (x))', etc. Perhaps this customary mode of presentation has to be modified; and perhaps such a modification would automatically remove the paradoxes of confirmation? If this is not so, there seems to be only one alternative left, namely to show that the impression of the paradoxical character of those consequences is due to misunderstanding and can be dispelled, so that no theoretical difficulty remains. We shall now consider these two possibilities in turn: The sub-sections 5.11 and 5.12 are devoted to a discussion of two different proposals for a modified representation of general hypotheses; in subsection 5.2, we shall discuss the second alternative, *i.e.* the possibility of tracing the impression of paradoxicality back to a misunderstanding.

5.11. It has often been pointed out that while Aristotelian logic, in agreement with prevalent every day usage, confers " existential import " upon sentences of the form " All P's are Q's ", a universal conditional sentence, in the sense of modern logic, has no existential import; thus, the sentence

$$\text{' } (x) \text{ (Mermaid}(x) \supset \text{ Green}(x)) \text{ '}$$

does not imply the existence of mermaids; it merely asserts that any object either is not a mermaid at all, or a green mermaid; and it is true simply because of the fact that there are no mermaids. General laws and hypotheses in science, however—so it might be argued—are meant to have existential import; and one might attempt to express the latter by supplementing the customary universal conditional by an existential clause. Thus, the hypothesis that all ravens are black would be expressed by means of the sentence S_1 : ' (x) (Raven(x) $\supset$ Black(x)) . (Ex)Raven(x); and the hypothesis that no non-black things are ravens by S_2 : ' $(x)(\sim$ Black(x) $\supset$ $\sim$ Raven(x)) . $(Ex) \sim$ Black(x). Clearly, these sentences are not equivalent, and of the four objects a, b, c, d characterized in section 3, part (b), only a might reasonably be said to confirm S_1, and only d to confirm S_2. Yet this method of avoiding the paradoxes of confirmation is open to serious objections:

(a) First of all, the representation of every general hypothesis by a conjunction of a universal conditional and an existential sentence would invalidate many logical inferences which are generally accepted as permissible in a theoretical argument. Thus, for example, the assertions that all sodium salts burn yellow, and that whatever does not burn yellow is no sodium salt are logically equivalent according to customary understanding and usage ; and their representation by universal conditionals preserves this equivalence ; but if existential clauses are added, the two assertions are no longer equivalent, as is illustrated above by the analogous case of S_1 and S_2.

(b) Second, the customary formulation of general hypotheses in empirical science clearly does not contain an existential clause, nor does it, as a rule, even indirectly determine such a clause unambiguously. Thus, consider the hypothesis that if a person after receiving an injection of a certain test substance has a positive skin reaction, he has diphtheria. Should we construe the existential clause here as referring to persons, to persons receiving the injection, or to persons who, upon receiving the injection, show a positive skin reaction ? A more or less arbitrary decision has to be made ; each of the possible decisions gives a different interpretation to the hypothesis, and none of them seems to be really implied by the latter.

(c) Finally, many universal hypotheses cannot be said to imply an existential clause at all. Thus, it may happen that from a certain astrophysical theory a universal hypothesis is deduced concerning the character of the phenomena which would take place under certain specified extreme conditions. A hypothesis of this kind need not (and, as a rule, does not) imply that such extreme conditions ever were or will be realized ; it has no existential import. Or consider a biological hypothesis to the effect that whenever man and ape are crossed, the offspring will have such and such characteristics. This is a general hypothesis ; it might be contemplated as a mere conjecture, or as a consequence of a broader genetic theory, other implications of which may already have been tested with positive results ; but unquestionably the hypothesis does not imply an existential clause asserting that the contemplated kind of cross-breeding referred to will, at some time, actually take place.

While, therefore, the adjunction of an existential clause to the customary symbolization of a general hypothesis cannot be considered as an adequate *general* method of coping with the paradoxes of confirmation, there is a purpose which the use of an existential clause may serve very well, as was pointed out to me by Dr. Paul Oppenheim [1] : if somebody feels that objects of the

---

[1] This observation is related to Mr. Oppenheim's methodological studies referred to in p. 1, n. 1.

types c and d mentioned above are irrelevant rather than confirming for the hypothesis in question, and that qualifying them as confirming evidence does violence to the meaning of the hypothesis, then this may indicate that he is consciously or unconsciously construing the latter as having existential import; and this kind of understanding of general hypotheses is in fact very common. In this case, the " paradox " may be removed by pointing out that an adequate symbolization of the intended meaning requires the adjunction of an existential clause. The formulation thus obtained is more restrictive than the universal conditional alone ; and while we have as yet set up no criteria of confirmation applicable to hypotheses of this more complex form, it is clear that according to every acceptable definition of confirmation objects of the types c and d will fail to qualify as confirming cases. In this manner, the use of an existential clause may prove helpful in distinguishing and rendering explicit different possible interpretations of a given general hypothesis which is stated in non-symbolic terms.

5.12. Perhaps the impression of the paradoxical character of the cases discussed in the beginning of section 5 may be said to grow out of the feeling that the hypothesis that all ravens are black is about ravens, and not about non-black things, nor about all things. The use of an existential clause was one attempt at expressing this presumed peculiarity of the hypothesis. The attempt has failed, and if we wish to reflect the point in question, we shall have to look for a stronger device. The idea suggests itself of representing a general hypothesis by the customary universal conditional, supplemented by the indication of the specific " field of application " of the hypothesis ; thus, we might represent the hypothesis that all ravens are black by the sentence ' (x) (Raven$(x) \supset$ Black$(x))$ ' (or any one of its equivalents), plus the indication " Class of ravens " characterizing the field of application ; and we might then require that every confirming instance should belong to the field of application. This procedure would exclude the objects c and d from those constituting confirming evidence and would thus avoid those undesirable consequences of the existential-clause device which were pointed out in 5.11 (c). But apart from this advantage, the second method is open to objections similar to those which apply to the first : (a) The way in which general hypotheses are used in science never involves the statement of a field of application ; and the choice of the latter in a symbolic formulation of a given hypothesis thus introduces again a considerable measure of arbitrariness. In particular, for a scientific hypothesis to the effect that

2

all P's are Q's, the field of application cannot simply be said to be the class of all P's ; for a hypothesis such as that all sodium salts burn yellow finds important applications in tests with negative results ; *i.e.* it may be applied to a substance of which it is not known whether it contains sodium salts, nor whether it burns yellow ; and if the flame does not turn yellow, the hypothesis serves to establish the absence of sodium salts. The same is true of all other hypotheses used for tests of this type. (*b*) Again, the consistent use of a domain of application in the formulation of general hypotheses would involve considerable logical complications, and yet would have no counterpart in the theoretical procedure of science, where hypotheses are subjected to various kinds of logical transformation and inference without any consideration that might be regarded as referring to changes in the fields of application. This method of meeting the paradoxes would therefore amount to dodging the problem by means of an *ad hoc* device which cannot be justified by reference to actual scientific procedure.

5.2. We have examined two alternatives to the customary method of representing general hypotheses by means of universal conditionals ; neither of them proved an adequate means of precluding the paradoxes of confirmation. We shall now try to show that what is wrong does not lie in the customary way of construing and representing general hypotheses, but rather in our reliance on a misleading intuition in the matter : The impression of a paradoxical situation is not objectively founded ; it is a psychological illusion.

(*a*) One source of misunderstanding is the view, referred to before, that a hypothesis of the simple form " Every P is a Q " such as " All sodium salts burn yellow ", asserts something about a certain limited class of objects only, namely, the class of all P's. This idea involves a confusion of logical and practical considerations : Our interest in the hypothesis may be focussed upon its applicability to that particular class of objects, but the hypothesis nevertheless asserts something about, and indeed imposes restrictions upon, *all* objects (within the logical type of the variable occurring in the hypothesis, which in the case of our last illustration might be the class of all physical objects). Indeed, a hypothesis of the form " Every P is a Q " forbids the occurrence of any objects having the property P but lacking the property Q ; *i.e.* it restricts all objects whatsoever to the class of those which either lack the property P or also have the property Q. Now, every object either belongs to this class or falls outside it, and thus, every object—and not only the P's—either conforms to the hypothesis or violates it ; there is no object which is not implicitly " referred to " by a hypothesis of this type. In particular, every object which either is no sodium salt or burns yellow conforms to, and thus " bears out " the hypothesis that all sodium salts burn yellow ; every other object violates that hypothesis.

The weakness of the idea under consideration is evidenced also by the observation that the class of objects about which a hypothesis is supposed to assert something is in no way clearly determined, and that it changes with the context, as was shown in 5.12 (a).

(b) A second important source of the appearance of paradoxicality in certain cases of confirmation is exhibited by the following consideration.

Suppose that in support of the assertion " All sodium salts burn yellow " somebody were to adduce an experiment in which a piece of pure ice was held into a colourless flame and did not turn the flame yellow. This result would confirm the assertion, " Whatever does not burn yellow is no sodium salt ", and consequently, by virtue of the equivalence condition, it would confirm the original formulation. Why does this impress us as paradoxical ? The reason becomes clear when we compare the previous situation with the case of an experiment where an object whose chemical constitution is as yet unknown to us is held into a flame and fails to turn it yellow, and where subsequent analysis reveals it to contain no sodium salt. This outcome, we should no doubt agree, is what was to be expected on the basis of the hypothesis that all sodium salts burn yellow—no matter in which of its various equivalent formulations it may be expressed ; thus, the data here obtained constitute confirming evidence for the hypothesis. Now the only difference between the two situations here considered is that in the first case we are told beforehand the test substance is ice, and we happen to " know anyhow " that ice contains no sodium salt ; this has the consequence that the outcome of the flame-colour test becomes entirely irrelevant for the confirmation of the hypothesis and thus can yield no new evidence for us. Indeed, if the flame should not turn yellow, the hypothesis requires that the substance contain no sodium salt— and we know beforehand that ice does not—and if the flame should turn yellow, the hypothesis would impose no further restrictions on the substance ; hence, either of the possible outcomes of the experiment would be in accord with the hypothesis.

The analysis of this example illustrates a general point : In

the seemingly paradoxical cases of confirmation, we are often
not actually judging the relation of the given evidence, E alone
to the hypothesis H (we fail to observe the "methodological
fiction", characteristic of every case of confirmation, that we
have no relevant evidence for H other than that included in E);
instead, we tacitly introduce a comparison of H with a body of
evidence which consists of E in conjunction with an additional
amount of information which we happen to have at our disposal;
in our illustration, this information includes the knowledge (1) that
the substance used in the experiment is ice, and (2) that ice contains no sodium salt. If we assume this additional information
as given, then, of course, the outcome of the experiment can add
no strength to the hypothesis under consideration. But if we
are careful to avoid this tacit reference to additional knowledge
(which entirely changes the character of the problem), and if we
formulate the question as to the confirming character of the
evidence in a manner adequate to the concept of confirmation
as used in this paper, we have to ask: Given some object a
(it happens to be a piece of ice, but this fact is not included
in the evidence), and given the fact that a does not turn
the flame yellow and is no sodium salt—does a then constitute
confirming evidence for the hypothesis? And now—no matter
whether a is ice or some other substance—it is clear that the
answer has to be in the affirmative; and the paradoxes
vanish.

So far, in section (b), we have considered mainly that type of
paradoxical case which is illustrated by the assertion that any
non-black non-raven constitutes confirming evidence for the hypothesis, "All ravens are black." However, the general idea just
outlined applies as well to the even more extreme cases exemplified
by the assertion that any non-raven as well as any black object
confirms the hypothesis in question. Let us illustrate this by
reference to the latter case. If the given evidence E—i.e. in the
sense of the required methodological fiction, all our data relevant
for the hypothesis—consists only of one object which, in addition,
is black, then E may reasonably be said to support even the
hypothesis that all objects are black, and a fortiori E supports the
weaker assertion that all ravens are black. In this case, again,
our factual knowledge that not all objects are black tends to
create an impression of paradoxicality which is not justified on
logical grounds. Other "paradoxical" cases of confirmation
may be dealt with analogously, and it thus turns out that the
"paradoxes of confirmation", as formulated above, are due to
a misguided intuition in the matter rather than to a logical flaw

in the two stipulations from which the " paradoxes " were derived.[1],[2]

[1] The basic idea of sect. (b) in the above analysis of the " paradoxes of confirmation " is due to Dr. Nelson Goodman, to whom I wish to reiterate my thanks for the help he rendered me, through many discussions, in clarifying my ideas on this point.

[2] The considerations presented in section (b) above are also influenced by, though not identical in content with, the very illuminating discussion of the " paradoxes " by the Polish methodologist and logician Janina Hosiasson-Lindenbaum ; cf. her article " On Confirmation ", The Journal of Symbolic Logic, vol. 5 (1940), especially sect. 4. Dr. Hosiasson's attention had been called to the paradoxes by the article referred to in p. 11. n. 2, and by discussions with the author. To my knowledge, hers has so far been the only publication which presents an explicit attempt to solve the problem. Her solution is based on a theory of degrees of confirmation, which is developed in the form of an uninterpreted axiomatic system (cf. also p. 6, n. 1, and part (b) in sect. 1 of the present article), and most of her arguments presuppose that theoretical framework. I have profited, however, by some of Miss Hosiasson's more general observations which proved relevant for the analysis of the paradoxes of the non-gradated relation of confirmation which forms the object of the present study.

One point in those of Miss Hosiasson's comments which rest on her theory of degrees of confirmation is of particular interest, and I should like to discuss it briefly. Stated in reference to the raven-hypothesis, it consists in the suggestion that the finding of one non-black object which is no raven, while constituting confirming evidence for the hypothesis, would increase the degree of confirmation of the hypothesis by a smaller amount than the finding of one raven which is black. This is said to be so because the class of all ravens is much less numerous than that of all non-black objects, so that—to put the idea in suggestive though somewhat misleading terms—the finding of one black raven confirms a larger portion of the total content of the hypothesis than the finding of one non-black non-raven. In fact, from the basic assumptions of her theory, Miss Hosiasson is able to derive a theorem according to which the above statement about the relative increase in degree of confirmation will hold provided that actually the number of all ravens is small compared with the number of all non-black objects. But is this last numerical assumption actually warranted in the present case and analogously in all other " paradoxical " cases ? The answer depends in part upon the logical structure of the language of science. If a " co-ordinate language " is used, in which, say, finite space-time regions figure as individuals, then the raven-hypothesis assumes some such form as " Every space-time region which contains a raven, contains something black " ; and even if the total number of ravens ever to exist is finite, the class of space-time regions containing a raven has the power of the continuum, and so does the class of space-time regions containing something non-black ; thus, for a co-ordinate language of the type under consideration, the above numerical assumption is not warranted. Now the use of a co-ordinate language may appear quite artificial in this particular illustration ; but it will seem very appropriate in many other contexts, such as, e.g., that of physical field theories. On the other hand, Miss Hosiasson's numerical assumption may well be justified on the basis of a " thing language ", in which physical objects of finite size function

6. *Confirmation Construed as a Relation between Sentences.*—Our analysis of Nicod's criterion has so far led to two main results: The rejection of that criterion in view of several deficiencies, and the emergence of the equivalence condition as a necessary condition of adequacy for any proposed definition of confirmation. Another aspect of Nicod's criterion requires consideration now. In our formulation of the criterion, confirmation was construed as a dyadic relation between an object or an ordered set of objects, representing the evidence, and a sentence, representing the hypothesis. This means that confirmation was conceived of as a semantical relation [1] obtaining between certain extra-linguistic objects [2] on one hand and certain sentences on the other. It is possible, however, to construe confirmation in an alternative fashion as a relation between two sentences, one describing the given evidence, the other expressing the hypothesis. Thus, *e.g.* instead of saying that an object *a* which is both a raven and black (or the "fact" of *a* being both a raven and black) confirms the hypothesis, "All ravens are black", we may say that the evidence sentence, "*a* is a raven, and *a* is black", confirms the hypothesis-sentence (briefly, the hypothesis), "All ravens are black". We shall adopt this conception of confirmation as a relation between sentences here for the following reasons: First, the evidence adduced in support or criticism of a scientific hypothesis is always expressed in sentences, which frequently have the character of observation reports; and second, it will prove very fruitful to pursue the parallel, alluded to in section 2 above, between the concepts of confirmation and of logical consequence. And just as in the theory of the consequence relation, *i.e.* in deductive logic, the premisses of which a given conclusion is a consequence are construed as sentences rather than as "facts", so we propose to construe the data which confirm a given hypothesis as given in the form of sentences.

The preceding reference to observation reports suggests a certain restriction which might be imposed on evidence sentences. Indeed, the evidence adduced in support of a scientific hypothesis

---

as individuals. Of course, even on this basis, it remains an empirical question, for every hypothesis of the form "All *P*'s are *Q*'s", whether actually the class of non-*Q*'s is much more numerous than the class of *P*'s; and in many cases this question will be very difficult to decide.

[1] For a detailed account of this concept, see C. W. Morris, *Foundations of the Theory of Signs* (Internat. Encyclopedia of Unified Science, vol. i, no. 2, Chicago, 1938), and R. Carnap, *Introduction to Semantics* (Cambridge, Mass., 1942), esp. sects. 4 and 37.

[2] Instead of making the first term of the relation an object or a sequence of objects, we might construe it as a "state of affairs" (or perhaps as a "fact", or a "proposition", as Nicod puts it), such as that state of affairs which consists in *a* being a black raven, etc.

or theory consists, in the last analysis, in data accessible to what is loosely called " direct observation ", and such data are expressible in the form of " observation reports ". In view of this consideration, we shall restrict the evidence sentences which form the domain of the relation of confirmation, to sentences of the character of observation reports. In order to give a precise meaning to the concept of observation report, we shall assume that we are given a well-determined " language of science ", in terms of which all sentences under consideration, hypotheses as well as evidence sentences, are formulated. We shall further assume that this language contains, among other terms, a clearly delimited " observational vocabulary " which consists of terms designating more or less directly observable attributes of things or events, such as, say, " black ", " taller than ", " burning with a yellow light ", etc., but no theoretical constructs such as " aliphatic compound ", " circularly polarized light ", " heavy hydrogen ", etc.

We shall now understand by a hypothesis any sentence which can be expressed in the assumed language of science, no matter whether it is a generalized sentence, containing quantifiers, or a particular sentence referring only to a finite number of particular objects. An observation report will be construed as a finite class (or a conjunction of a finite number) of observation sentences ; and an observation sentence as a sentence which either asserts or denies that a given object has a certain observable property (such as " a is a raven ", " d is not black "), or that a given sequence of objects stand in a certain observable relation (such as " a is between b and c ").

Now the concept of observability itself obviously is relative to the techniques of observation used. What is unobservable to the unaided senses may well be observable by means of suitable devices such as telescopes, microscopes, polariscopes, lie-detectors, Gallup-polls, etc. If by direct observation we mean such observational procedures as do not make use of auxiliary devices, then such property terms as " black ", " hard ", " liquid ", " cool ", and such relation terms as " above ", " between ", " spatially coincident ", etc., might be said to refer to directly observable attributes ; if observability is construed in a broader sense, so as to allow for the use of certain specified instruments or other devices, the concept of observable attribute becomes more comprehensive. If, in our study of confirmation, we wanted to analyze the manner in which the hypotheses and theories of empirical science are ultimately supported by " evidence of the senses ", then we should have to require that observation reports refer exclusively to directly observable attributes. This view was taken, for simplicity and concreteness, in the preceding parts of this section. Actually, however, the general logical character- istics of that relation which obtains between a hypothesis and a group of empirical statements which " support " it, can be studied in isolation from this restriction to direct observability. All we will assume here is that in the context of the scientific test of a given hypothesis or theory, certain specified techniques of observation have been agreed upon ; these determine an ob- servational vocabulary, namely a set of terms designating proper- ties and relations observable by means of the accepted techniques. For our purposes it is entirely sufficient that these terms, con- stituting the " observational vocabulary ", be given. An ob- servation sentence is then defined simply as a sentence affirming or denying that a given object, or sequence of objects, possesses one of those observable attributes.[1]

Let it be noted that we do not require an observation sentence to be true, nor to be accepted on the basis of actual observations ; rather, an observation sentence expresses something that is de- cidable by means of the accepted techniques of observation ; in other words : An observation sentence describes a possible out- come of the accepted observational techniques ; it asserts some- thing that might conceivably be established by means of those

---

[1] The concept of observation sentence has, in the context of our study, a status and a logical function closely akin to that of the concepts of pro- tocol statement or basis sentence, etc., as used in many recent studies of empiricism. However, the conception of observation sentence which is being proposed in the present study is more liberal in that it renders the discussion of the logical problems of testing and confirmation independent of various highly controversial epistemological issues ; thus, *e.g.* we do not stipulate that observation reports must be about psychic acts, or about sense perceptions (*i.e.* that they have to be expressed in terms of a vocab- ulary of phenomenology, or of introspective psychology). According to the conception of observation sentence adopted in the present study, the " objects " referred to in an observation sentence may be construed in any one of the senses just referred to, or in various other ways ; for example, they might be space-time regions, or again physical objects such as stones, trees, etc. (most of the illustrations given throughout this article represent observation sentences belonging to this kind of " thing-language ") ; all that we require is that the few very general conditions stated above be satisfied.

These conditions impose on observation sentences and on observation reports certain restrictions with respect to their form ; in particular, neither kind of sentence may contain any quantifiers. This stipulation recommends itself for the purposes of the logical analysis here to be undertaken ; but we do not wish to claim that this formal restriction is indispensable. On the contrary, it is quite possible and perhaps desirable also to allow for observation sentences containing quantifiers : our simplifying assumption is introduced mainly in order to avoid considerable logical complications in the definition of confirmation.

techniques. Possibly, the term "observation-type sentence" would be more suggestive; but for convenience we give preference to the shorter term. An analogous comment applies, of course, to our definition of an observation report as a class or a conjunction of observation sentences. The need for this broad conception of observation sentences and observation reports is readily recognized: Confirmation as here conceived is a logical relationship between sentences, just as logical consequence is. Now whether a sentence S_2 is a consequence of a sentence S_1 does not depend on whether S_1 is true (or known to be true), or not; and analogously, the criteria of whether a given statement, expressed in terms of the observational vocabulary, confirms a certain hypothesis cannot depend on whether the statements in the report are true, or based on actual experience, or the like. Our definition of confirmation must enable us to indicate what kind of evidence *would* confirm a given hypothesis *if* it were available; and clearly the sentence characterizing such evidence can be required only to express something that might be observed, but not necessarily something that has actually been established by obsrvation.

It may be helpful to carry the analogy between confirmation and consequence one step further. The truth or falsity of S_1 is irrelevant for the question of whether S_2 is a consequence of S_1 (whether S_2 can be validly inferred from S_1); but in a logical inference which justifies a sentence S_2 by showing that it is a logical consequence of a conjunction of premisses, S_1, we can be certain of the truth of S_2 only if we know S_1 to be true. Analogously, the question of whether an observation report stands in the relation of confirmation to a given hypothesis does not depend on whether the report states actual or fictitious observational findings; but for a decision as to the soundness or acceptability of a hypothesis which is confirmed by a certain report, it is of course necessary to know whether the report is based on actual experience or not. Just as a conclusion of a logical inference, in order to be reliably true must be (a1) validly inferred from (a2) a set of true premisses, so a hypothesis, in order to be scientifically acceptable, must be (b1) formally confirmed by (b2) reliable reports on observational findings.

The central problem of this essay is to establish general criteria for the formal relation of confirmation as referred to in (b1); the analysis of the concept of a reliable observation report, which belongs largely to the field of pragmatics,[1] falls outside the scope of the present study. One point, however, deserves mention here: A statement of the form of an observation report (for

[1] An account of the concept of pragmatics may be found in the publications listed in p. 22, n. 1.

example, about the position of the pointer of a certain thermo-graph at 3 a.m.) may be accepted or rejected in science either on the basis of direct observation, or because it is indirectly confirmed or disconfirmed by other accepted observation sentences (in the example, these might be sentences describing the curve traced by the pointer during the night), and because of this possibility of indirect confirmation, our study has a bearing also on the question of the acceptance of hypotheses which have themselves the form of observation reports.

The conception of confirmation as a relation between sentences analogous to that of logical consequence suggests yet another specification for the attempted definition of confirmation : While logical consequence has to be conceived of as a basically semantical relation between sentences, it has been possible, for certain languages, to establish criteria of logical consequence in purely syntactical terms.[1] Analogously, confirmation may be conceived of as a semantical relation between an observation report and a hypothesis ; but the parallel with the consequence relation suggests that it should be possible, for certain languages, to establish purely syntactical criteria of confirmation. The subsequent considerations will indeed eventuate in a definition of confirmation based on the concept of logical consequence and other purely syntactical concepts.

The interpretation of confirmation as a logical relation between sentences involves no essential change in the central problem of the present study. In particular, all the points made in the preceding sections can readily be rephrased in accordance with this interpretation. Thus, for example, the assertion that an object a which is a swan and white confirms the hypothesis ' (x) (Swan(x) ⊃ White(x)) ' can be expressed by saying that the observation report ' Swan(a) . White(a) ' confirms that hypothesis. Similarly, the equivalence condition can be reformulated as follows : If an observation report confirms a certain sentence, then it also confirms every sentence which is logically equivalent with the latter. Nicod's criterion as well as our grounds for rejecting it can be re-formulated along the same lines. We presented Nicod's concept of confirmation as referring to a relation between non-linguistic objects on one hand and sentences on the other because this approach seemed to approximate most closely Nicod's own formulations, and because it enabled us to avoid certain technicalities which are actually unnecessary in that context.

(To be concluded)

[1] *Cf.* especially the two publications by R. Carnap listed in p. 3, n. 1.

VOL. LIV. No. 214.]   [April, 1945.

MIND

A QUARTERLY REVIEW

OF

PSYCHOLOGY AND PHILOSOPHY

———⸙———

I.—STUDIES IN THE LOGIC OF CON-FIRMATION (II.).

By Carl G. Hempel.

7. *The Prediction-criterion of Confirmation and its Short-comings.*—We are now in a position to analyze a second conception of confirmation which is reflected in many methodological discussions and which can claim a great deal of plausibility. Its basic idea is very simple : General hypotheses in science as well as in everyday usage are intended to enable us to anticipate future events ; hence, it seems reasonable to count any prediction which is borne out by subsequent observation as confirming evidence for the hypothesis on which it is based, and any prediction that fails as disconfirming evidence. To illustrate : Let H_1 be the hypothesis that all metals, when heated, expand ; symbolically : ' (x) ((Metal (x) . Heated (x)) $\supset$ Exp(x)) '. If we are given an observation report to the effect that a certain object a is a metal and is heated, then by means of H_1 we can derive the prediction that a expands. Suppose that this is borne out by observation and described in an additional observation statement. We should then have the total observation report. {Metal(a), Heated(a), Exp.(a)}.[1] This report would be qualified as confirming evidence for H_1 because its last sentence bears out what could be predicted, or derived, from the first two by means of

---

[1] An (observation) report, it will be recalled, may be represented by a conjunction or by a class of observation sentences ; in the latter case, we characterize it by writing the sentences between braces ; the quotation marks which normally would be used are, for convenience, assumed to be absorbed by the braces.

7

H_1 ; more explicitly : because the last sentence can be derived
from the first two in conjunction with H_1.—Now let H_2 be the
hypothesis that all swans are white ; symbolically : ' (x) (Swan
$(x) \supset$ White$(x))$ ' ; and consider the observation report {Swan(a),
$\sim$ White(a)}. This report would constitute disconfirming evidence for H_2 because the second of its sentences contradicts (and
thus fails to bear out) the prediction ' White(a) ' which can be
deduced from the first sentence in conjunction with H_2 ; or,
symmetrically, because the first sentence contradicts the consequence ' $\sim$ Swan(a) ' which can be derived from the second in
conjunction with H_2. Obviously, either of these formulations
implies that H_2 is incompatible with the given observation report.

These illustrations suggest the following general definition of
confirmation as successful prediction :

Prediction-criterion of Confirmation : Let H be a hypothesis, B
an observation report, *i.e.* a class of observation sentences. Then

(a) B is said to confirm H if B can be divided into two mutually
exclusive subclasses B_1 and B_2 such that B_2 is not empty, and every
sentence of B_2 can be logically deduced from B_1 in conjunction
with H, but not from B_1 alone.

(b) B is said to disconfirm H if H logically contradicts B.[1]

(c) B is said to be neutral with respect to H if it neither confirms nor disconfirms H.[2]

But while this criterion is quite sound as a statement of sufficient conditions of confirmation for hypotheses of the type illustrated above, it is considerably too narrow to serve as a general
definition of confirmation. Generally speaking, this criterion
would serve its purpose if all scientific hypotheses could be construed as asserting regular connections of observable features in
the subject-matter under investigation ; *i.e.* if they all were of

---

[1] It might seem more natural to stipulate that B disconfirms H if it
can be divided into two mutually exclusive classes B_1 and B_2 such that
the denial of at least one sentence in B_2 can be deduced from B_1 in conjunction with H ; but this condition can be shown to be equivalent to (b)
above.

[2] The following quotations from A. J. Ayer's book *Language, Truth and
Logic* (London, 1936) formulate in a particularly clear fashion the conception of confirmation as successful prediction (although the two are not
explicitly identified by definition): " . . . the function of an empirical
hypothesis is to enable us to anticipate experience. Accordingly, if an
observation to which a given proposition is relevant conforms to our
expectations, . . . that proposition is confirmed " (*loc. cit.* pp. 142-143).
" . . . it is the mark of a genuine factual proposition . . . that some
experiential propositions can be deduced from it in conjunction with
certain premises without being deducible from those other premises alone ".
(*loc. cit.* p. 26).

the form " Whenever the observable characteristic P is present in an object or a situation, then the observable characteristic Q will also be present." But actually, most scientific hypotheses and laws are not of this simple type ; as a rule, they express regular connections of characteristics which are not observable in the sense of direct observability, nor even in a much more liberal sense. Consider, for example, the following hypothesis : " Whenever plane-polarized light of wave length λ traverses a layer of quartz of thickness d, then its plane of polarization is rotated through an angle α which is proportional to $\frac{d}{\lambda}$."—Let us assume that the observational vocabulary, by means of which our observation reports have to be formulated, contains exclusively terms referring to directly observable attributes. Then, since the question of whether a given ray of light is plane-polarized and has the wave length λ cannot be decided by means of direct observation, no observation report of the kind here admitted could include information of this type. This in itself would not be crucial if at least we could assume that the fact that a given ray of light is plane-polarized, etc., could be logically inferred from some possible observation report ; for then, from a suitable report of this kind, in conjunction with the given hypothesis, one would be able to predict a rotation of the plane of polarization ; and from this prediction, which itself is not yet expressed in exclusively observational terms, one might expect to derive further predictions in the form of genuine observation sentences. But actually, a hypothesis to the effect that a given ray of light is plane-polarized has to be considered as a general hypothesis which entails an unlimited number of observation sentences ; thus it cannot be logically inferred from, but at best be confirmed by, a suitable set of observational findings. The logically essential point can best be exhibited by reference to a very simple abstract case : Let us assume that R_1 and R_2 are two relations of a kind accessible to direct observation, and that the field of scientific investigation contains infinitely many objects. Consider now the hypothesis

(H) $(x)((y)R_1(x, y) \supset (\textsf{E}z)R_2(x, z)),$

i.e. : Whenever an object x stands in R_1 to every object y, then it stands in R_2 to at least one object z.—This simple hypothesis has the following property : However many observation sentences may be given, H does not enable us to derive any new observation sentences from them. Indeed—to state the reason in suggestive though not formally rigorous terms—in order to make a prediction concerning some specific object a, we should first have to know that a stands in R_1 to every object ; and this necessary information clearly cannot be contained in any finite number, however large, of observation sentences, because a finite set of observation sentences can tell us at best for a finite number of objects that a stands in R_1 to them. Thus an observation report, which always involves only a finite number of observation sentences, can never provide a sufficiently broad basis for a prediction by means of H.[1]—Besides, even if we did know that a stood in R_1 to every object, the prediction derivable by means of H would not be an observation sentence ; it would assert that a stands in R_2 to *some* object, without specifying which, and where to find it. Thus, H would be an empirical hypothesis, containing, besides purely logical terms, only expressions belonging to the observational vocabulary, and yet the predictions which it renders possible neither start from nor lead to observation reports.

It is, therefore, a considerable over-simplification to say that scientific hypotheses and theories enable us to derive predictions of future experiences from descriptions of past ones. Unquestionably, scientific hypotheses do have a predictive function ; but the way in which they perform this function, the manner in which they establish logical connections between observation reports, is logically more complex than a deductive inference. Thus, in the last illustration, the predictive use of H may assume the following form : On the basis of a number of individual tests, which show that a does stand in R_1 to three objects b, c, and d, we may accept the hypothesis that a stands in R_1 to all objects ; or, in terms of our formal mode of speech : In view of the observation report $\{R_1(a, b), R_1(a, c), R_1(a, d)\}$, the hypothesis that $(y)R_1(a, y)$ is accepted as confirmed by, though not logically inferable from, that report.[2] This process might be referred to as quasi-induction.[3] From the hypothesis thus established we

[1] To illustrate : a might be an iron object which possibly is a magnet ; R_1 might be the relation of attracting ; the objects under investigation might be iron objects. Then a finite number of observation reports to the effect that a did attract a particular piece of iron is insufficient to *infer* that a will attract every piece of iron.

[2] Thus, in the illustration given in the preceding footnote, the hypothesis that the object a will attract every piece of iron might be accepted as sufficiently well substantiated by, though by no means derivable from, an observation report to the effect that in tests a did attract the iron objects b, c, and d.

[3] The prefix " quasi " is to contradistinguish the procedure in question from so-called induction, which is usually supposed to be a method of discovering, or arriving at, general regularities on the basis of a finite

can then proceed to derive, by means of H, the prediction that a stands in R_2 to at least one object. This again, as was pointed out above, is not an observation sentence ; and indeed no observation sentence can be derived from it ; but it can, in turn, be confirmed by a suitable observation sentence, such as ' $R_2(a, b)$ '. —In other cases, the prediction of actual observation sentences may be possible ; thus if the given hypothesis asserts that $(x)((y)R_1(x, y) \supset (z)R_2(x, z))$, then after quasi-inductively accepting, as above, that $(y)R_1(a, y)$, we can derive, by means of the given hypothesis, the sentence that a stands in R_2 to every object, and thence, we can deduce special predictions such as ' $R_2(a, b)$ ', etc., which do have the form of observation sentences.

Thus, the chain of reasoning which leads from given observational findings to the " prediction " of new ones actually involves, besides deductive inferences, certain quasi-inductive steps each of which consists in the acceptance of an intermediate statement on the basis of confirming, but usually not logically conclusive, evidence. In most scientific predictions, this general pattern occurs in multiple re-iteration ; an analysis of the predictive use of the hypothesis mentioned above, concerning plane-polarized light, could serve as an illustration. In the present context, however, this general account of the structure of scientific prediction is sufficient : it shows that a general definition of confirmation by reference to successful prediction becomes circular ; indeed, in order to make the original formulation of the prediction-criterion of confirmation sufficiently comprehensive, we should have to replace the phrase " can be logically deduced " by " can be obtained by a series of steps of deduction and quasi-induction " ; and the definition of " quasi-induction " in the above sense presupposes the concept of confirmation.

Let us note, as a by-product of the preceding consideration, the fact that an adequate analysis of scientific prediction (and analogously, of scientific explanation, and of the testing of empirical hypotheses) requires an analysis of the concept of confirmation. The reason for this fact may be restated in general terms as follows : Scientific laws and theories, as a rule, connect terms which lie on the level of abstract theoretical constructs rather than on that of direct observation ; and from observation sentences, no merely deductive logical inference leads

number of instances. In quasi-induction, the hypothesis is not " discovered " but has to be *given* in addition to the observation report : the process consists in the acceptance of the hypothesis if it is deemed sufficiently confirmed by the observation report. *Cf.* also the discussion in section 1c, above.

to statements about those theoretical constructs which are the starting point for scientific predictions ; statements about logical constructs, such as " This piece of iron is magnetic " or " Here, a plane-polarized ray of light traverses a quartz crystal " can be confirmed, but not entailed, by observation reports, and thus, even though based on general scientific laws, the " prediction " of new observational findings on the basis of given ones is a process involving confirmation in addition to logical deduction.[1]

8. *Conditions of Adequacy for any Definition of Confirmation.*— The two most customary conceptions of confirmation, which were rendered explicit in Nicod's criterion and in the prediction criterion, have thus been found unsuitable for a general definition of confirmation. Besides this negative result, the preceding analysis has also exhibited certain logical characteristics of scientific prediction, explanation, and testing, and it has led to the establishment of certain standards which an adequate definition of confirmation has to satisfy. These standards include the equivalence condition and the requirement that the definition of confirmation be applicable to hypotheses of any degree of logical complexity, rather than to the simplest type of universal conditional only. An adequate definition of confirmation, however, has to satisfy several further logical requirements, to which we now turn.

First of all, it will be agreed that any sentence which is entailed by—*i.e.* a logical consequence of—a given observation report has to be considered as confirmed by that report : Entailment is a special case of confirmation. Thus, *e.g.*, we want to say that the observation report " *a* is black " confirms the sentence (hypothesis) " *a* is black or grey " ; and—to refer to one of the illustrations given in the preceding section—the observation sentence ' $R_2(a, b)$ ' should certainly be confirming evidence for the sentence ' $(Ez)R_2(a, z)$ '. We are therefore led to the stipulation that any adequate definition of confirmation must insure the fulfilment of the

---

[1] In the above sketch of the structure of scientific prediction, we have disregarded the fact that in practically every case where a prediction is said to be obtained by means of a certain hypothesis or theory, a considerable mass of auxiliary theories is used in addition ; thus, *e.g.* the prediction of observable effects of the deflection of light in the gravitational field of the sun on the basis of the general theory of relativity, requires such auxiliary theories as mechanics and optics. But an explicit consideration of this fact would not affect our result that scientific predictions, even when based on hypotheses or theories of universal form, still are not purely deductive in character, but involve quasi-inductive steps as well.

(8.1) *Entailment condition:* Any sentence which is entailed by an observation report is confirmed by it.[1]

This condition is suggested by the preceding consideration, but of course not proved by it. To make it a standard of adequacy for the definition of confirmation means to lay down the stipulation that a proposed definition of confirmation will be rejected as logically inadequate if it is not constructed in such a way that (8.1) is unconditionally satisfied. An analogous remark applies to the subsequently proposed further standards of adequacy.—

Second, an observation report which confirms certain hypotheses would invariably be qualified as confirming any consequence of those hypotheses. Indeed : any such consequence is but an assertion of all or part of the combined content of the original hypotheses and has therefore to be regarded as confirmed by any evidence which confirms the original hypotheses. This suggests the following condition of adequacy :

(8.2) *Consequence Condition:* If an observation report confirms every one of a class K of sentences, then it also confirms any sentence which is a logical consequence of K.

If (8.2) is satisfied, then the same is true of the following two more special conditions :

(8.21) *Special Consequence Condition:* If an observation report confirms a hypothesis H, then it also confirms every consequence of H.

(8.22) *Equivalence Condition:* If an observation report confirms a hypothesis H, then it also confirms every hypothesis which is logically equivalent with H.

(This follows from (8.21) in view of the fact that equivalent hypotheses are mutual consequences of each other.) Thus, the satisfaction of the consequence condition entails that of our earlier equivalence condition, and the latter loses its status of an independent requirement.

In view of the apparent obviousness of these conditions, it is interesting to note that the definition of confirmation in terms of successful prediction, while satisfying the equivalence condition, would violate the consequence condition. Consider, for example, the formulation of the prediction-criterion given in the earlier

---

[1] As a consequence of this stipulation, a contradictory observation report, such as {Black(a), ∼ Black(a)} confirms every sentence, because it has every sentence as a consequence. Of course, it is possible to exclude the possibility of contradictory observation reports altogether by a slight restriction of the definition of " observation report ". There is, however, no important reason to do so.

part of the preceding section. Clearly, if the observational findings B_2 can be predicted on the basis of the findings B_1 by means of the hypothesis H, the same prediction is obtainable by means of any equivalent hypothesis, but not generally by means of a weaker one.

On the other hand, any prediction obtainable by means of H can obviously also be established by means of any hypothesis which is stronger than H, i.e. which logically entails H. Thus, while the consequence condition stipulates in effect that whatever confirms a given hypothesis also confirms any weaker hypothesis, the relation of confirmation defined in terms of successful prediction would satisfy the condition that whatever confirms a given hypothesis, also confirms every stronger one.

But is this " converse consequence condition ", as it might be called, not reasonable enough, and should it not even be included among our standards of adequacy for the definition of confirmation ? The second of these two suggestions can be readily disposed of : The adoption of the new condition, in addition to (8.1) and (8.2), would have the consequence that any observation report B would confirm any hypothesis H whatsoever. Thus, e.g., if B is the report " a is a raven " and H is Hooke's law, then, according to (8.1), B confirms the sentence " a is a raven ", hence B would, according to the converse consequence condition, confirm the stronger sentence " a is a raven, and Hooke's law holds " ; and finally, by virtue of (8.2), B would confirm H, which is a consequence of the last sentence. Obviously, the same type of argument can be applied in all other cases.

But is it not true, after all, that very often observational data which confirm a hypothesis H are considered also as confirming a stronger hypothesis ? Is it not true, for example, that those experimental findings which confirm Galileo's law, or Kepler's laws, are considered also as confirming Newton's law of gravitation ? [1] This is indeed the case, but this does not justify the acceptance of the converse entailment condition as a general rule of the logic of confirmation ; for in the cases just mentioned, the weaker hypothesis is connected with the stronger one by a logical bond of a particular kind : it is essentially a substitution instance of the stronger one ; thus, e.g., while the law of gravitation refers to the force obtaining between any two bodies, Galileo's law is a specialization referring to the case where one of

---

[1] Strictly speaking, Galileo's law and Kepler's laws can be deduced from the law of gravitation only if certain additional hypotheses—including the laws of motion—are presupposed ; but this does not affect the point under discussion.

the bodies is the earth, the other an object near its surface. In the preceding case, however, where Hooke's law was shown to be confirmed by the observation report that a is a raven, this situation does not prevail ; and here, the rule that whatever confirms a given hypothesis also confirms any stronger one becomes an entirely absurd principle. Thus, the converse consequence condition does not provide a sound general condition of adequacy.[1]

A third condition remains to be stated : [2]

(8.3) *Consistency Condition :* Every logically consistent observation report is logically compatible with the class of all the hypotheses which it confirms.

The two most important implications of this requirement are the following :

(8.31) Unless an observation report is self-contradictory,[3] it does not confirm any hypothesis with which it is not logically compatible.

(8.32) Unless an observation report is self-contradictory, it does not confirm any hypotheses which contradict each other.

The first of these corollaries will readily be accepted ; the second, however,—and consequently (8.3) itself—will perhaps be

---

[1] William Barrett, in a paper entitled " Discussion on Dewey's Logic " (*The Philosophical Review*, vol. 50, 1941, pp. 305 ff., esp. p. 312) raises some questions closely related to what we have called above the consequence condition and the converse consequence condition. In fact, he invokes the latter (without stating it explicitly) in an argument which is designed to show that " not every observation which confirms a sentence need also confirm all its consequences ", in other words, that the special consequence condition (8.21) need not always be satisfied. He supports his point by reference to " the simplest case : the sentence ' C ' is an abbreviation of ' A.B ', and the observation O confirms ' A ', *and so* ' C ', but is irrelevant to ' B ', which is a consequence of ' C '." (Italics mine.)

For reasons contained in the above discussion of the consequence condition and the converse consequence condition, the application of the latter in the case under consideration seems to us unjustifiable, so that the illustration does not prove the author's point ; and indeed, there seems to be every reason to preserve the unrestricted validity of the consequence condition. As a matter of fact, Mr. Barrett himself argues that " the degree of confirmation for the consequence of a sentence cannot be less than that of the sentence itself " ; this is indeed quite sound ; but it is hard to see how the recognition of this principle can be reconciled with a renunciation of the special consequence condition, since the latter may be considered simply as the correlate, for the non-gradated relation of confirmation, of the former principle which is adapted to the concept of degree of confirmation.

[2] For a fourth condition, see n. 1, p. 110.

[3] A contradictory observation report confirms every hypothesis (*cf.* n. 1, p. 103) and is, of course, incompatible with every one of the hypotheses it confirms.

felt to embody a too severe restriction. It might be pointed out, for example, that a finite set of measurements concerning the variation of one physical magnitude, x, with another, y, may conform to, and thus be said to confirm, several different hypotheses as to the particular mathematical function in terms of which the relationship of x and y can be expressed ; but such hypotheses are incompatible because to at least one value of x, they will assign different values of y.

No doubt it is possible to liberalize the formal standards of adequacy in line with these considerations. This would amount to dropping (8.3) and (8.32) and retaining only (8.31). One of the effects of this measure would be that when a logically consistent observation report B confirms each of two hypotheses, it does not necessarily confirm their conjunction ; for the hypotheses might be mutually incompatible, hence their conjunction self-contradictory ; consequently, by (8.31), B could not confirm it.—This consequence is intuitively rather awkward, and one might therefore feel inclined to suggest that while (8.3) should be dropped and (8.31) retained, (8.32) should be replaced by the requirement (8.33) : If an observation sentence confirms each of two hypotheses, then it also confirms their conjunction. But it can readily be shown that by virtue of (8.2) this set of conditions entails the fulfilment of (8.32).

If, therefore, the condition (8.3) appears to be too rigorous, the most obvious alternative would seem to lie in replacing (8.3) and its corollaries by the much weaker condition (8.31) alone ; and it is an important problem whether an intuitively adequate definition of confirmation can be constructed which satisfies (8.1), (8.2) and (8.31), but not (8.3).—One of the great advantages of a definition which satisfies (8.3) is that it sets a limit, so to speak, to the strength of the hypotheses which can be confirmed by given evidence.[1]

The remainder of the present study, therefore, will be concerned exclusively with the problem of establishing a definition of confirmation which satisfies the more severe formal conditions represented by (8.1), (8.2), and (8.3) together.

The fulfilment of these requirements, which may be regarded as general laws of the logic of confirmation, is of course only a necessary, not a sufficient, condition for the adequacy of any proposed definition of confirmation. Thus, e.g., if " B confirms

[1] This was pointed out to me by Dr. Nelson Goodman. The definition later to be outlined in this essay, which satisfies conditions (8.1), (8.2) and (8.3), lends itself, however, to certain generalizations which satisfy only the more liberal conditions of adequacy just considered.

H " were defined as meaning " B logically entails *H* ", then the above three conditions would clearly be satisfied ; but the definition would not be adequate because confirmation has to be a more comprehensive relation than entailment (the latter might be referred to as the special case of *conclusive* confirmation). Thus, a definition of confirmation, to be acceptable, also has to be materially adequate : it has to provide a reasonably close approximation to that conception of confirmation which is implicit in scientific procedure and methodological discussion. That conception is vague and to some extent quite unclear, as I have tried to show in earlier parts of this paper ; therefore, it would be too much to expect full agreement as to the material adequacy of a proposed definition of confirmation ; on the other hand, there will be rather general agreement on certain points ; thus, *e.g.*, the identification of confirmation with entailment, or the Nicod criterion of confirmation as analyzed above, or any definition of confirmation by reference to a " sense of evidence ", will probably now be admitted not to be adequate approximations to that concept of confirmation which is relevant for the logic of science.

On the other hand, the soundness of the logical analysis (which, in a clear sense, always involves a logical reconstruction) of a theoretical concept cannot be gauged simply by our feelings of satisfaction at a certain proposed analysis ; and if there are, say, two alternative proposals for defining a term on the basis of a logical analysis, and if both appear to come fairly close to the intended meaning, then the choice has to be made largely by reference to such features as the logical properties of the two reconstructions, and the comprehensiveness and simplicity of the theories to which they lead.

9. *The Satisfaction Criterion of Confirmation.*—As has been mentioned before, a precise definition of confirmation requires reference to some definite " language of science ", in which all observation reports and all hypotheses under consideration are assumed to be formulated, and whose logical structure is supposed to be precisely determined. The more complex this language, and the richer its logical means of expression, the more difficult it will be, as a rule, to establish an adequate definition of confirmation for it. However, the problem has been solved at least for certain cases : With respect to languages of a comparatively simple logical structure, it has been possible to construct an explicit definition of confirmation which satisfies all of the above logical requirements, and which appears to be intuitively rather adequate. An exposition of the technical details of this

definition has been published elsewhere ; [1] in the present study, which is concerned with the general logical and methodological aspects of the problem of confirmation rather than with technical details, it will be attempted to characterize the definition of confirmation thus obtained as clearly as possible with a minimum of technicalities.

Consider the simple case of the hypothesis H : ' $(x)(\text{Raven}(x) \supset \text{Black}(x))$ ', where ' Raven ' and ' Black ' are supposed to be terms of our observational vocabulary. Let B be an observation report to the effect that $\text{Raven}(a) . \text{Black}(a) . \sim \text{Raven}(c) . \text{Black}(c) . \sim \text{Raven}(d) . \sim \text{Black}(d)$. Then B may be said to confirm H in the following sense : There are three objects altogether mentioned in B, namely a, c, and d ; and as far as these are concerned, B informs us that all those which are ravens (*i.e.* just the object a) are also black.[2] In other words, from the information contained in B we can infer that the hypothesis H does hold true within the finite class of those objects which are mentioned in B.

Let us apply the same consideration to a hypothesis of a logically more complex structure. Let H be the hypothesis " Everybody likes somebody " ; in symbols : ' $(x)(Ey)\text{Likes}(x, y)$ ',

---

[1] In my article referred to in n. 1, p. 1. The logical structure of the languages to which the definition in question is applicable is that of the lower functional calculus with individual constants, and with predicate constants of any degree. All sentences of the language are assumed to be formed exclusively by means of predicate constants, individual constants, individual variables, universal and existential quantifiers for individual variables, and the connective symbols of denial, conjunction, alternation, and implication. The use of predicate variables or of the identity sign is not permitted.

As to the predicate constants, they are all assumed to belong to the observational vocabulary, *i.e.* to denote a property or a relation observable by means of the accepted techniques. (" Abstract " predicate terms are supposed to be defined in terms of those of the observational vocabulary and then actually to be replaced by their *definientia*, so that they never occur explicitly.)

As a consequence of these stipulations, an observation report can be characterized simply as a conjunction of sentences of the kind illustrated by ' $P(a)$ ', ' $\sim P(b)$ ', ' $R(c, d)$ ', ' $\sim R(e, f)$', etc., where ' P ', ' R ', etc., belong to the observational vocabulary, and ' a ', ' b ', ' c ', ' d ', ' e ', ' f ', etc., are individual names, denoting specific objects. It is also possible to define an observation report more liberally as any sentence containing no quantifiers, which means that besides conjunctions also alternations and implication sentences formed out of the above kind of components are included among the observation reports.

[2] I am indebted to Dr. Nelson Goodman for having suggested this idea ; it initiated all those considerations which finally led to the definition to be outlined below.

i.e. for every (person) x, there exists at least one (not necessarily different person) y such that x likes y. (Here again, ' Likes ' is supposed to be a relation-term which occurs in our observational vocabulary.) Suppose now that we are given an observation report B in which the names of two persons, say ' e ' and ' f ', occur. Under what conditions shall we say that B confirms H ? The previous illustration suggests the answer : If from B we can infer that H is satisfied within the finite class $\{e, f\}$; *i.e.* that within $\{e, f\}$ everybody likes somebody. This in turn means that e likes e or f, and f likes e or f. Thus, B would be said to confirm H if B entailed the statement " e likes e or f, and f likes e or f ". This latter statement will be called the development of H for the finite class $\{e, f\}$.—

The concept of *development of a hypothesis, H, for a finite class of individuals*, C, can be defined in a general fashion ; the development of H for C states what H would assert if there existed exclusively those objects which are elements of C.—Thus, *e.g.*, the development of the hypothesis $H_1 = $ ' $(x)(P(x) \lor Q(x))$ ' (*i.e.* " Every object has the property P or the property Q ") for the class $\{a, b\}$ is ' $(P(a) \lor Q(a)) . (P(b) \lor Q(b))$ ' (*i.e.* " a has the property P or the property Q, and b has the property P or the property Q ") ; the development of the existential hypothesis H_2 that at least one object has the property P, *i.e.* ' $(Ex)P(x)$ ', for $\{a, b\}$ is ' $P(a) \lor P(b)$ ' ; the development of a hypothesis which contains no quantifiers, such as H_3 : ' $P(c) \lor Q(c)$ ' is defined as that hypothesis itself, no matter what the reference class of individuals is.

A more detailed formal analysis based on considerations of this type leads to the introduction of a general relation of confirmation in two steps ; the first consists in defining a special relation of direct confirmation along the lines just indicated ; the second step then defines the general relation of confirmation by reference to direct confirmation.

Omitting minor details, we may summarize the two definitions as follows :

(9.1 Df.) An observation report B directly confirms a hypothesis H if B entails the development of H for the class of those objects which are mentioned in B.

(9.2 Df.) An observation report B confirms a hypothesis H if H is entailed by a class of sentences each of which is directly confirmed by B.

The criterion expressed in these definitions might be called the satisfaction criterion of confirmation because its basic idea consists in construing a hypothesis as confirmed by a given

observation report if the hypothesis is satisfied in the finite class
of those individuals which are mentioned in the report.—Let us
now apply the two definitions to our last examples : The observation report B_1 : ' $P(a) . Q(b)$ ' directly confirms (and therefore
also confirms) the hypothesis H_1, because it entails the development of H_1 for the class $\{a, b\}$, which was given above.—The
hypothesis H_3 is not directly confirmed by B, because its development—*i.e.* H_3 itself—obviously is not entailed by B_1. However,
H_3 is entailed by H_1, which is directly confirmed by B_1 ; hence,
by virtue of (9.2), B_1 confirms H_3.

Similarly, it can readily be seen that B_1 directly confirms H_2.

Finally, to refer to the first illustration given in this section :
The observation report ' Raven(a) . Black(a) . $\sim$ Raven(c) . $\sim$
Black(c) . $\sim$ Raven(d) . $\sim$ Black(d) ' confirms (even directly) the
hypothesis ' (x)(Raven(x) $\supset$ Black(x)) ', for it entails the development of the latter for the class $\{a, c, d\}$, which can be written as
follows : ' (Raven(a) $\supset$ Black(a)) . (Raven(c) $\supset$ Black(c)) . (Raven
(d) $\supset$ Black(d)) '.

It is now easy to define disconfirmation and neutrality :

(9.3 Df.) An observation report B disconfirms a hypothesis H if
it confirms the denial of H.

(9.4 Df.) An observation report B is neutral with respect to
a hypothesis H if B neither confirms nor disconfirms H.

By virtue of the criteria laid down in (9.2), (9.3), (9.4), every
consistent observation report, B, divides all possible hypotheses
into three mutually exclusive classes : those confirmed by B, those
disconfirmed by B, and those with respect to which B is neutral.

The definition of confirmation here proposed can be shown to
satisfy all the formal conditions of adequacy embodied in (8.1),
(8.2), and (8.3) and their consequences ; for the condition (8.2)
this is easy to see ; for the other conditions the proof is more
complicated.[1]

[1] For these proofs, see the article referred to in n. 1, p. 1. I should like
to take this opportunity to point out and to remedy a certain defect of the
definition of confirmation which was developed in that article, and which
has been outlined above : this defect was brought to my attention by a
discussion with Dr. Olaf Helmer.

It will be agreed that an acceptable definition of confirmation should
satisfy the following further condition which might well have been included among the logical standards of adequacy set up in section 8 above :
(8.4). If B_1 and B_2 are logically equivalent observation reports and B_1
confirms (disconfirms, is neutral with respect to) a hypothesis H, then B_2,
too, confirms (disconfirms, is neutral with respect to) H. This condition is
indeed satisfied if observation reports are construed, as they have been in
this article, as classes or conjunctions of observation sentences. As was
indicated at the end of n. 1, p. 108, however, this restriction of observation

Furthermore, the application of the above definition of confirmation is not restricted to hypotheses of universal conditional form (as Nicod's criterion is, for example), nor to universal hypotheses in general ; it applies, in fact, to any hypothesis which can be expressed by means of property and relation terms of the observational vocabulary of the given language, individual names, the customary connective symbols for ' not ', ' and ', ' or ', ' if-then ', and any number of universal and existential quantifiers.

Finally, as is suggested by the preceding illustrations as well as by the general considerations which underlie the establishment of the above definition, it seems that we have obtained a definition

reports to a conjunctive form is not essential ; in fact, it has been adopted here only for greater convenience of exposition, and all the preceding results, including especially the definitions and theorems of the present section, remain applicable without change if observation reports are given the more liberal interpretation characterized at the end of n. 1, p. 108. (In this case, if ' P ' and ' Q ' belong to the observational vocabulary, such sentences as ' $P(a) \lor Q(a)$ ', ' $P(a) \lor \sim Q(b)$ ', etc., would qualify as observation reports.)   This broader conception of observation reports was therefore adopted in the article referred to in n. 1, p. 1 ; but it has turned out that in this case, the definition of confirmation summarized above does not generally satisfy the requirement (8.4).   Thus, e.g., the observation reports, $B_1 = $ ' $P(a)$ ' and $B_2 = $ ' $P(a) \cdot (Q(b) \lor \sim Q(b))$ ' are logically equivalent, but while B_1 confirms (and even directly confirms) the hypothesis $H_1 = $ ' $(x)P(x)$ ', the second report does not do so, essentially because it does not entail ' $P(a) \cdot P(b)$ ', which is the development of H_1 for the class of those objects mentioned in B_2.   This deficiency can be remedied as follows : The fact that B_2 fails to confirm H_1 is obviously due to the circumstance that B_2 contains the individual constant ' b ', without asserting anything about b : The object b is mentioned only in an analytic component of B_2.   The atomic constituent ' $Q(b)$ ' will therefore be said to occur (twice) inessentially in B_2.   Generally, an atomic constituent A of a molecular sentence S will be said to occur inessentially in S if by virtue of the rules of the sentential calculus S is equivalent to a molecular sentence in which A does not occur at all.   Now an object will be said to be mentioned inessentially in an observation report if it is mentioned only in such components of that report as occur inessentially in it.   The sentential calculus clearly provides mechanical procedures for deciding whether a given observation report mentions any object inessentially, and for establishing equivalent formulations of the same report in which no object is mentioned inessentially.   Finally, let us say that an object is mentioned essentially in an observation report if it is mentioned, but not only mentioned inessentially, in that report.   Now we replace 9.1 by the following definition :

(9.1a) An observation report B directly confirms a hypothesis H if B entails the development of H for the class of those objects which are mentioned essentially in B.

The concept of confirmation as defined by (9.1a) and (9.2) now satisfies (8.4) in addition to (8.1), (8.2), (8.3) even if observation reports are construed in the broader fashion characterized earlier in this footnote.

of confirmation which also is materially adequate in the sense of being a reasonable approximation to the intended meaning of confirmation.

A brief discussion of certain special cases of confirmation might serve to shed further light on this latter aspect of our analysis.

10. *The Relative and the Absolute Concepts of Verification and Falsification.*—If an observation report entails a hypothesis H, then, by virtue of (8.1), it confirms H. This is in good agreement with the customary conception of confirming evidence ; in fact, we have here an extreme case of confirmation, the case where B *conclusively confirms* H ; this case is realized if, and only if, B entails H. We shall then also say that B *verifies* H. Thus, verification is a special case of confirmation ; it is a logical relation between sentences ; more specifically, it is simply the relation of entailment with its domain restricted to observation sentences.

Analogously, we shall say that B *conclusively disconfirms* H, or B *falsifies* H, if and only if B is incompatible with H ; in this case, B entails the denial of H and therefore, by virtue of (8.1) and (9.3), confirms the denial of H and disconfirms H. Hence, falsification is a special case of disconfirmation ; it is the logical relation of incompatibility between sentences, with its domain restricted to observation sentences.

Clearly, the concepts of *verification* and *falsification* as here defined are *relative ;* a hypothesis can be said to be verified or falsified only with respect to some observation report ; and a hypothesis may be verified by one observation report and may not be verified by another. There are, however, hypotheses which cannot be verified and others which cannot be falsified by any observation report. This will be shown presently. We shall say that a given *hypothesis is verifiable (falsifiable)* if it is possible to construct an observation report which verifies (falsifies) the hypothesis. Whether a hypothesis is verifiable, or falsifiable, in this sense depends exclusively on its logical form. Briefly, the following cases may be distinguished :

(*a*) If a hypothesis does not contain the quantifier terms " all " and " some " or their symbolic equivalents, then it is both verifiable and falsifiable. Thus, *e.g.*, the hypothesis " Object *a* turns blue or green " is entailed and thus verified by the report " Object *a* turns blue " ; and the same hypothesis is incompatible with, and thus falsified by, the report " Object *a* turns neither blue nor green ".

(*b*) A purely existential hypothesis (*i.e.* one which can be symbolized by a formula consisting of one or more existential quantifiers followed by a sentential function containing no quantifiers) is verifiable, but not falsifiable, if—as is usually assumed—the universe of discourse contains an infinite number of objects.—Thus, e.g., the hypothesis " There are blue roses " is verified by the observation report " Object a is a blue rose ", but no finite observation report can ever contradict and thus falsify the hypothesis.

(c) Conversely, a purely universal hypothesis (symbolized by a formula consisting of one or more universal quantifiers followed by a sentential function containing no quantifiers) is falsifiable but not verifiable for an infinite universe of discourse. Thus, e.g., the hypothesis " $(x)(\text{Swan}(x) \supset \text{White}(x))$ " is completely falsified by the observation report $\{\text{Swan}(a), \sim \text{White}(a)\}$; but no finite observation report can entail and thus verify the hypothesis in question.

(d) Hypotheses which cannot be expressed by sentences of one of the three types mentioned so far, and which in this sense require both universal and existential quantifiers for their formulation, are as a rule neither verifiable nor falsifiable.[1]   Thus, e.g., the hypothesis " Every substance is soluble in some solvent "— symbolically ' $(x)(Ey)\text{Soluble}(x, y)$ '—is neither entailed by, nor incompatible with any observation report, no matter how many cases of solubility or non-solubility of particular substances in particular solvents the report may list.   An analogous remark applies to the hypothesis " You can fool some of the people all of the time ", whose symbolic formulation ' $(Ex)(t)\text{Fl}(x,t)$ ' contains one existential and one universal quantifier.   But of course, all of the hypotheses belonging to this fourth class are capable of being confirmed or disconfirmed by suitable observation reports ; this was illustrated early in section 9 by reference to the hypothesis ' $(x)(Ey)\text{Likes}(x, y)$ '.

This rather detailed account of verification and falsification has been presented not only in the hope of further elucidating the meaning of confirmation and disconfirmation as defined above, but also in order to provide a basis for a sharp differentiation of two meanings of verification (and similarly of falsification) which have not always been clearly separated in recent discussions of the character of empirical knowledge.   One of the two meanings of verification which we wish to distinguish here is the relative concept just explained ; for greater clarity we shall sometimes

---

[1] A more precise study of the conditions of non-verifiability and non-falsifiability would involve technicalities which are unnecessary for the purposes of the present study.   Not all hypotheses of the type described in (d) are neither verifiable nor falsifiable ; thus, e.g., the hypothesis ' $(x)(Ey)(P(x) \lor Q(y))$ ' is verified by the report ' $Q(a)$ ', and the hypothesis ' $(x)(Ey)(P(x) . Q(y))$ ' is falsified by ' $\sim P(a)$ '.

8

refer to it as *relative verification*. The other meaning is what may be called *absolute or definitive verification*. This latter concept of verification does not belong to formal logic, but rather to pragmatics [1] : it refers to the acceptance of hypotheses by " observers " or " scientists ", etc., on the basis of relevant evidence. Generally speaking, we may distinguish three phases in the scientific test of a given hypothesis (which do not necessarily occur in the order in which they are listed here). The first phase consists in the performance of suitable experiments or observations and the ensuing acceptance of observation sentences, or of observation reports, stating the results obtained ; the next phase consists in confronting the given hypothesis with the accepted observation reports, *i.e.* in ascertaining whether the latter constitute confirming, disconfirming or irrelevant evidence with respect to the hypothesis ; the final phase consists either in accepting or rejecting the hypothesis on the strength of the confirming or disconfirming evidence constituted by the accepted observation reports, or in suspending judgment, awaiting the establishment of further relevant evidence.

The present study has been concerned almost exclusively with the second phase ; as we have seen, this phase is of a purely logical character ; the standards of evaluation here invoked— namely the criteria of confirmation, disconfirmation and neutrality—can be completely formulated in terms of concepts belonging to the field of pure logic.

The first phase, on the other hand, is of a pragmatic character ; it involves no logical confrontation of sentences with other sentences. It consists in performing certain experiments or systematic observations and noting the results. The latter are expressed in sentences which have the form of observation reports, and their acceptance by the scientist is connected (by causal, not by logical relations) with experiences occurring in those tests. (Of course, a sentence which has the form of an observation report may in certain cases be accepted not on the basis of direct observation, but because it is confirmed by other observation reports which were previously established ; but this process is illustrative of the second phase, which was discussed before. Here we are considering the case where a sentence is accepted directly " on the basis of experiential findings " rather than because it is supported by previously established statements.)

The third phase, too, can be construed as pragmatic, namely as consisting in a decision on the part of the scientist or a group of

---

[1] In the sense in which the term is used by Carnap in the work referred to in n. 1, p. 22.

scientists to accept (or reject, or leave in suspense, as the case may be) a given hypothesis after ascertaining what amount of confirming or of disconfirming evidence for the hypothesis is contained in the totality of the accepted observation sentences. However, it may well be attempted to give a reconstruction of this phase in purely logical terms. This would require the establishment of general " rules of acceptance " ; roughly speaking, these rules would state how well a given hypothesis has to be confirmed by the accepted observation reports to be scientifically acceptable itself ; [1] i.e. the rules would formulate criteria for the acceptance or rejection of a hypothesis by reference to the kind and amount of confirming or disconfirming evidence for it embodied in the totality of accepted observation reports ; possibly, these criteria would also refer to such additional factors as the " simplicity " of the hypothesis in question, the manner in which it fits into the system of previously accepted theories, etc. It is at present an open question to what extent a satisfactory system of such rules can be formulated in purely logical terms. [2]

[1] A stimulating discussion of some aspects of what we have called rules of acceptance is contained in an article by Felix Kaufmann, ' The logical rules of scientific procedure', *Philosophy and Phenomenological Research*, June, 1942.

If an explicit definition of the degree of confirmation of a hypothesis were available, then it might be possible to formulate criteria of acceptance in terms of the degree to which the accepted observation reports confirm the hypothesis in question.

[2] The preceding division of the test of an empirical hypothesis into three phases of different character may prove useful for the clarification of the question whether or to what extent an empiricist conception of confirmation implies a " coherence theory of truth ". This issue has recently been raised by Bertrand Russell, who, in ch. x of his *Inquiry into Meaning and Truth*, has levelled a number of objections against the views of Otto Neurath on this subject (*cf.* the articles mentioned in the next footnote), and against statements made by myself in articles published in *Analysis* in 1935 and 1936. I should like to add here a few, necessarily brief, comments on this issue.

(1) While, in the articles in *Analysis*, I argued in effect that the only possible interpretation of the phrase " Sentence S is true " is " S is highly confirmed by accepted observation reports ", I should now reject this view. As the work of A. Tarski, R. Carnap, and others has shown, it is possible to define a semantical concept of truth which is not synonymous with that of strong confirmation, and which corresponds much more closely to what has customarily been referred to as truth, especially in logic, but also in other contexts. Thus, e.g., if S is any empirical sentence, then either S or its denial is true in the semantical sense, but clearly it is possible that neither S nor its denial is highly confirmed by available evidence. To assert that a hypothesis is true is equivalent to asserting the hypothesis

At any rate, the acceptance of a hypothesis on the basis of a sufficient body of confirming evidence will as a rule be tentative, and will hold only " until further notice ", *i.e.* with the proviso that if new and unfavourable evidence should turn up (in other words, if new observation reports should be accepted which disconfirm the hypothesis in question) the hypothesis will be abandoned again.

Are there any exceptions to this rule ? Are there any empirical hypotheses which are capable of being established definitively, hypotheses such that we can be sure that once accepted on the basis of experiential evidence, they will never have to be revoked ? Hypotheses of this kind will be called absolutely or definitively verifiable ; and the concept of absolute or definitive falsifiability will be construed analogously.

While the existence of hypotheses which are relatively verifiable or relatively falsifiable is a simple logical fact, which was illustrated in the beginning of this section, the question of the existence of absolutely verifiable, or absolutely falsifiable, hypotheses is a highly controversial issue which has received a great deal of attention in recent empiricist writings.[1] As the problem

itself ; therefore the truth of an empirical hypothesis can be ascertained only in the sense in which the hypothesis itself can be established : *i.e.* the hypothesis—and thereby *ipso facto* its truth—can be more or less well confirmed by empirical evidence ; there is no other access to the question of the truth of a hypothesis.

In the light of these considerations, it seems advisable to me to reserve the term ' truth ' for the semantical concept ; I should now phrase the statements in the *Analysis* articles as dealing with confirmation. (For a brief and very illuminating survey of the distinctive characteristics of truth and confirmation, see R. Carnap, " Wahrheit and Bewährung," *Actes I*ᵉʳ *Congrès Internat. de Philosophie Scientifique 1935*, vol. 4 ; Paris, 1936.)

(2) It is now clear also in what sense the test of a hypothesis is a matter of confronting sentences with sentences rather than with " facts ", or a matter of the " coherence " of the hypothesis and the accepted basic sentences : All the logical aspects of scientific testing, *i.e.* all the criteria governing the second and third of the three phases distinguished above, are indeed concerned only with certain relationships between the hypotheses under test and certain other sentences (namely the accepted observation reports) ; no reference to extra-linguistic " facts " is needed. On the other hand, the first phase, the acceptance of certain basic sentences in connection with certain experiments or observations, involves, of course, extra-linguistic procedures ; but this had been explicitly stated by the author in the articles referred to before. The claim that the views concerning truth and confirmation which are held by contemporary logical empiricism involve a coherence theory of truth is therefore mistaken.

[1] *Cf.* especially A. Ayer, *The Foundations of Empirical Knowledge* (New York, 1940) ; see also the same author's article, " Verification and Experience ", *Proceedings of the Aristotelian Society* for 1937 ; R. Carnap,

is only loosely connected with the subject of this essay, we shall restrict ourselves here to a few general observations.

Let it be assumed that the language of science has the general structure characterized and presupposed in the previous discussions, especially in section 9. Then it is reasonable to expect that only such hypotheses can possibly be absolutely verifiable as are relatively verifiable by suitable observation reports; hypotheses of universal form, for example, which are not even capable of relative verification, certainly cannot be expected to be absolutely verifiable : In however many instances such a hypothesis may have been borne out by experiential findings, it is always possible that new evidence will be obtained which disconfirms the hypothesis. Let us, therefore, restrict our search for absolutely verifiable hypotheses to the class of those hypotheses which are relatively verifiable.

Suppose now that H is a hypothesis of this latter type, and that it is relatively verified, *i.e.* logically entailed, by an observation report B, and that the latter is accepted in science as an account of the outcome of some experiment or observation. Can we then say that H is absolutely confirmed, that it will never be revoked ? Clearly, that depends on whether the report B has been accepted irrevocably, or whether it may conceivably suffer the fate of being disavowed later. Thus the question as to the existence of absolutely verifiable hypotheses leads back to the question of whether all, or at least some, observation reports become irrevocable parts of the system of science once they have been accepted in connection with certain observations or experiments. This question is not simply one of fact ; it cannot adequately be answered by a descriptive account of the research behaviour of scientists. Here, as in all other cases of logical analysis of science, the problem calls for a " rational reconstruction " of scientific procedure, *i.e.* for the construction of a consistent and comprehensive theoretical model of scientific inquiry, which is then to serve as a system of reference, or a standard, in the examination of any particular scientific research. The

" Ueber Protokollsätze ", *Erkenntnis*, vol. 3 (1932), and § 82 of the same author's *The Logical Syntax of Language* (see n. 1, p. 3). O. Neurath, " Protokollsätze ", *Erkenntnis*, vol. 3 (1932) ; " Radikaler Physikalismus und ' wirkliche Welt ' ", *Erkenntnis*, vol. 4 (1934) ; " Pseudorationalismus der Falsifikation ", *Erkenntnis*, vol. 5 (1935). K. Popper, *Logik der Forschung* (see n. 1, p. 4). H. Reichenbach, *Experience and Prediction* (Chicago, 1938), ch. iii. Bertrand Russell, *An Inquiry into Meaning and Truth* (New York, 1940), especially chs. x and xi. M. Schlick, " Ueber das Fundament der Erkenntnis ", *Erkenntnis*, vol. 4 (1934).

construction of the theoretical model has, of course, to be oriented by the characteristics of actual scientific procedure, but it is not determined by the latter in the sense in which a descriptive account of some scientific study would be. Indeed, it is generally agreed that scientists sometimes infringe the standards of sound scientific procedure ; besides, for the sake of theoretical comprehensiveness and systematization, the abstract model will have to contain certain idealized elements which cannot possibly be determined in detail by a study of how scientists actually work. This is true especially of observation reports : A study of the way in which laboratory reports, or descriptions of other types of observational findings, are formulated in the practice of scientific research is of interest for the choice of assumptions concerning the form and the status of observation sentences in the model of a " language of science " ; but clearly, such a study cannot completely determine what form observation sentences are to have in the theoretical model, nor whether they are to be considered as irrevocable once they are accepted.

Perhaps an analogy may further elucidate this view concerning the character of logical analysis : Suppose that we observe two persons whose language we do not understand playing a game on some kind of chess board ; and suppose that we want to " reconstruct " the rules of the game. A mere descriptive account of the playing-behaviour of the individuals will not suffice to do this ; indeed, we should not even necessarily reject a theoretical reconstruction of the game which did not always characterize accurately the actual moves of the players : we should allow for the possibility of occasional violations of the rules. Our reconstruction would rather be guided by the objective of obtaining a consistent and comprehensive system of rules which are as simple as possible, and to which the observed playing behaviour conforms at least to a large extent. In terms of the standard thus obtained, we may then describe and critically analyze any concrete performance of the game.

The parallel is obvious ; and it appears to be clear, too, that in both cases the decision about various features of the theoretical model will have the character of a convention, which is influenced by considerations of simplicity, consistency, and comprehensiveness, and not only by a study of the actual procedure of scientists at work.[1]

---

[1] A clear account of the sense in which the results of logical analysis represent conventions can be found in §§ 9-11 and 25-30 of K. Popper's *Logik der Forschung*. An illustration of the considerations influencing the

This remark applies in particular to the specific question under consideration, namely whether " there are " in science any irrevocably accepted observation reports (all of whose consequences would then be absolutely verified empirical hypotheses). The situation becomes clearer when we put the question into this form : Shall we allow, in our rational reconstruction of science, for the possibility that certain observation reports may be accepted as irrevocable, or shall the acceptance of all observation reports be subject to the " until further notice " clause ? In comparing the merits of the alternative stipulations, we should have to investigate the extent to which each of them is capable of elucidating the structure of scientific inquiry in terms of a simple, consistent theory. We do not propose to enter into a discussion of this question here except for mentioning that various considerations militate in favour of the convention that no observation report is to be accepted definitively and irrevocably.[1] If this alternative is chosen, then not even those hypotheses which are entailed by accepted observation reports are absolutely verified, nor are those hypotheses which are found incompatible with accepted observation reports thereby absolutely falsified : in fact, in this case, no hypothesis whatsoever would be absolutely verifiable or absolutely falsifiable. If, on the other hand, some—or even all—observation sentences are declared irrevocable once they have been accepted, then those hypotheses entailed by or incompatible with irrevocable observation sentences will be absolutely verified, or absolutely falsified, respectively.

It should now be clear that the concepts of absolute and of relative verifiability (and falsifiability) are of an entirely different character. Failure to distinguish them has caused considerable misunderstanding in recent discussions on the nature of scientific knowledge. Thus, e.g., K. Popper's proposal to admit as scientific hypotheses exclusively sentences which are (relatively) falsifiable by suitable observation reports has been criticized by means of arguments which, in effect, support the claim that scientific hypotheses should not be construed as being absolutely falsifiable—a point that Popper had not denied.—As can be seen from our earlier discussion of relative falsifiability, however, Popper's proposal to limit scientific hypotheses to the form of (relatively) falsifiable sentences involves a very severe restriction

determination of various features of the theoretical model is provided by the discussion in n. 1, p. 24.

[1] *Cf.* especially the publications by Carnap, Neurath, and Popper mentioned in n. 1. p. 116 ; also Reichenbach, *loc. cit.*, ch. ii, § 9.

of the possible forms of scientific hypotheses [1] ; in particular, it rules out all purely existential hypotheses as well as most hypotheses whose formulation requires both universal and existential quantification ; and it may be criticized on this account ; for in terms of this theoretical reconstruction of science it seems difficult or altogether impossible to give an adequate account of the status and function of the more complex scientific hypotheses and theories.—

With these remarks let us conclude our study of the logic of confirmation. What has been said above about the nature of the logical analysis of science in general, applies to the present analysis of confirmation in particular : It is a specific proposal for a systematic and comprehensive logical reconstruction of a concept which is basic for the methodology of empirical science as well as for the problem area customarily called " epistemology ". The need for a theoretical clarification of that concept was evidenced by the fact that no general theoretical account of confirmation has been available so far, and that certain widely accepted conceptions of confirmation involve difficulties so serious that it might be doubted whether a satisfactory theory of the concept is at all attainable.

It was found, however, that the problem can be solved : A general definition of confirmation, couched in purely logical terms, was developed for scientific languages of a specified and relatively simple logical character. The logical model thus obtained appeared to be satisfactory in the sense of the formal and material standards of adequacy that had been set up previously.

I have tried to state the essential features of the proposed analysis and reconstruction of confirmation as explicitly as possible in the hope of stimulating a critical discussion and of facilitating further inquiries into the various issues pertinent to this problem area. Among the open questions which seem to deserve careful consideration, I should like to mention the exploration of concepts of confirmation which fail to satisfy the general consistency condition ; the extension of the definition of confirmation to the case where even observation sentences containing quantifiers are permitted ; and finally the development of

---

[1] This was pointed out by R. Carnap ; *cf.* his review of Popper's book in *Erkenntnis*, vol. 5 (1935), and " Testability and Meaning " (see n. 1, p. 5) §§ 25, 26. For the discussion of Popper's falsifiability criterion, see for example H. Reichenbach, " Ueber Induktion und Wahrscheinlichkeit ", *Erkenntnis*, vol. 5 (1935) ; O. Neurath, " Pseudorationalismus der Falsifikation ", *Erkenntnis*, vol. 5 (1935).

a definition of confirmation for languages of a more complex logical structure than that incorporated in our model.[1]   Languages of this kind would provide a greater variety of means of expression and would thus come closer to the high logical complexity of the language of empirical science.

[1] The languages to which our definition is applicable have the structure of the lower functional calculus without identity sign (*cf.* n. 1, p. 108) ; it would be highly desirable so to broaden the general theory of confirmation as to make it applicable to the lower functional calculus with identity sign, or even to the higher functional calculus ; for it seems hardly possible to give a precise formulation of more complex scientific theories without the logical means of expression provided by the higher functional calculus.

A QUERY ON CONFIRMATION

Hempel, Carnap, Oppenheim, and Helmer [1] have recently made important contributions towards the precise definition of the concepts of confirmation and degree of confirmation. Yet they seem to me to leave untouched one basic problem that must be solved before we can say that the proposed definitions are intuitively adequate— even in an approximate sense and for very limited languages.

Induction might roughly be described as the projection of characteristics of the past into the future, or more generally of characteristics of one realm of objects into another. But exact expression of this vague principle is exceedingly difficult. Some of the contradictions that result from seemingly straightforward formulations of it were explained and overcome in Hempel's papers. Unfortunately, equally serious difficulties remain.

Suppose we had drawn a marble from a certain bowl on each of the ninety-nine days up to and including VE day, and each marble drawn was red. We would expect that the marble drawn on the following day would also be red. So far all is well. Our evidence may be expressed by the conjunction "$Ra_1 \cdot Ra_2 \cdot \ldots \cdot Ra_{99}$," which well confirms the prediction "Ra_{100}." But increase of credibility, projection, "confirmation" in any intuitive sense, does not occur in the case of every predicate under similar circumstances. Let "S" be the predicate "is drawn by VE day and is red, or is drawn later and is non-red." The evidence of the same drawings above assumed may be expressed by the conjunction "$Sa_1 \cdot Sa_2 \cdot \ldots \cdot Sa_{99}$." By the theories of confirmation in question this well confirms the prediction "Sa_{100}"; but actually we do not expect that the hundredth marble will be non-red. "Sa_{100}" gains no whit of credibility from the evidence offered.

It is clear that "S" and "R" can not both be projected here, for that would mean that we expect that a_{100} will and will not be red. It is equally clear which predicate is actually projected and which is not. But how can the difference between projectible and non-projectible predicates be generally and rigorously defined?

That one predicate used in this example refers explicitly to

[1] In the following papers: C. G. Hempel, "A Purely Syntactical Definition of Confirmation," *Journal of Symbolic Logic*, Vol. 8 (1943), pp. 122–143; "Studies in the Logic of Confirmation," *Mind*, n.s., Vol. 54 (1945), pp. 1–26; Hempel and Paul Oppenheim, "A Definition of 'Degree of Confirmation,'" *Philosophy of Science*, Vol. 12 (1945), pp. 98–115; Oppenheim and Olaf Helmer, "A Syntactical Definition of Probability and of Degree of Confirmation," *Journal of Symbolic Logic*, Vol. 10 (1945), pp. 25–60; Rudolf Carnap, "On Inductive Logic," *Philosophy of Science*, Vol. 12 (1945), pp. 72–97; and "The Two Concepts of Probability," *Philosophy and Phenomenological Research*, Vol. V (1945), pp. 513–532.

temporal order is inessential. The same difficulty can be illustrated without the supposition of any order. Using the same letters as before, we need only suppose that the subscripts are merely for identification, having no ordinal significance, and that "S" means "is red and is not a_{100}, or is not red and is a_{100}."

The theories of confirmation in question require the primitive predicates to be logically independent.[2] This is perhaps a dubious stipulation since it places a logical requirement upon the informal, extrasystematic explanation of the predicates. Such doubts aside, the requirement would make it impossible for the predicates "R" and "S" to belong to the same system. Hence the conflicting confirmations would not occur in any one system. But this is of little help, since the system containing the predicate "S" alone is quite as admissible as the one containing "R" alone; and in the former system, as we have seen, "Sa_{100}" will be formally confirmed by the very evidence which intuitively disconfirms it. Carnap's concept of the "width" of a predicate does not bear on this point, since all atomic predicates are of the same width.[3]

More complex examples illustrating various phases of the same general question can easily be invented. I give only one more, to show how the theory of degree of confirmation is affected.

Suppose [4] that a certain unfamiliar machine tosses up one ball a minute and that every third one and only every third one is red. We observe ninety-six tosses. How much confidence does this lead us to place in the prediction that the next three tosses will produce a non-red ball, another non-red ball, and then a red ball? Plainly a good deal. But what degree of formal confirmation does the prediction derive from the observations according to the theories under consideration? The answer seems to be that this varies widely with the way the given evidence is described.

(i) If we let "a_1," "a_2," and so on represent in temporal order the individual tosses, our evidence may be expressed by

"$- Ra_1 \cdot - Ra_2 \cdot Ra_3 \cdot - Ra_4 \cdot - Ra_5 \cdot Ra_6 \cdot \ldots \cdot - Ra_{94} \cdot - Ra_{95} \cdot Ra_{96}$."

This imparts to the prediction "$- Ra_{97} \cdot - Ra_{98} \cdot Ra_{99}$" the degree [5]

---

[2] Although this requirement is not explicitly stated in the articles cited, Dr. Hempel tells me that its necessity was recognized by all the authors concerned.

[3] See page 84 of the first article by Carnap listed in footnote 1.

[4] The example in its present form is due to Dr. Hempel. He constructed it as the result of a conversation with Dr. W. V. Quine and the present writer concerning the problems here explained.

[5] The degrees of confirmation given in this paper are computed according to the Hempel-Oppenheim theory. The values under Carnap's theory would differ somewhat, but not in a way that appreciably affects the general question under discussion.

of confirmation $\frac{2}{3} \cdot \frac{2}{3} \cdot \frac{1}{3}$, or $\frac{4}{27}$. This figure seems intuitively much too low.

(ii) If we let "b_1" stand for the discontinuous individual consisting of the first three tosses, "b_2" for the individual consisting of the next three tosses, and so on, and let "S" mean "consists of three temporally separated parts ('tosses') of which the earliest and second are non-red and the latest red," our evidence may be expressed by

$$``Sb_1 \cdot \ldots \cdot Sb_{32}.''$$

This gives to "Sb_{33}" the degree of confirmation 1. Yet "Sb_{33}" expresses the same thing as "$- Ra_{94} \cdot - Ra_{95} \cdot Ra_{96}$," and we have assumed the same observations. Hence we seem to get different degrees of confirmation for the same prediction on the basis of the same evidence.

Now it may be argued that in (i) we ignored the fact of temporal order in stating our evidence, and that it is thus not surprising that we get a lower degree of confirmation than when we take this fact into account, as in (ii). However, it would be fatal to accept the implied thesis that an intuitively satisfactory degree of confirmation will result only when all the observed facts are expressed as evidence. Suppose the first ninety-six tosses exhibited a wholly irregular distribution of colors; the hypothesis that this distribution would be exactly repeated in the next ninety-six tosses would have the degree of confirmation 1. What is worse, if we are to express *all* the observed data in our statement of evidence, we shall have to include such particularized information—e.g., the unique date of each toss—that repetition in the future will be impossible.

Undoubtedly we do make predictions by projecting the patterns of the past into the future, but in selecting the patterns we project from among all those that the past exhibits, we use practical criteria that so far seem to have escaped discovery and formulation. The problem is not peculiar to the work of the authors I have named; so far as I am aware, no one has as yet offered any satisfactory solution. What we have in the papers cited is an ingenious and valuable logico-mathematical apparatus that we may apply to the sphere of projectible or confirmable predicates whenever we discover what a projectible or confirmable predicate is.

NELSON GOODMAN

TUFTS COLLEGE

A NOTE ON THE PARADOXES OF CONFIRMATION.

In MIND for April 1945, Mr. C. H. Whiteley discusses certain peculiar features of the concept of confirmation, which were pointed out and analysed in my article, " Studies in the Logic of Confirmation " in the January and April issues of MIND for 1945, especially on pages 13-21. The peculiarities in question, which I called paradoxes of confirmation, pertain to the non-graduated relation of being-confirming-evidence-for a hypothesis, as contradistinguished from the concept of degree of confirmation of a hypothesis relatively to given empirical evidence. The paradoxes result from the following consideration : It seems reasonable to say that any proposed analysis or rational reconstruction of the non-graduated relation of confirmation, in order to be adequate, has to satisfy the following two requirements :

R1 : Whenever an object has two attributes C_1, C_2, it constitutes confirming evidence for the hypothesis that every object which has the attribute C_1 also has the attribute C_2.

R2 : An object which constitutes confirming evidence for one of two logically equivalent sentences, is confirming evidence also for the other (for the two sentences in this case express the same hypothesis).

Now R2 entails in particular that the following three sentences, being logically equivalent, have to be confirmed by the same class of objects :

S1 : Whatever is a raven is black,
S2 : Whatever is not black is not a raven,
S3 : Whatever is or is not a raven is either no raven or black.

But according to R1, S3 is confirmed by any object which has the attributes (C_1) of being or not being a raven, and (C_2) of being no raven or black. As C_1 is analytic, it follows that each of the three sentences must be confirmed by any object that is either no raven or black. Thus, e.g., to take Mr. Whiteley's illustrations, a red herring or a white elephant would constitute confirming evidence for S1. The general principle here illustrated together with its various consequences, which at first blush appear highly counter-intuitive, constitute the paradoxes of confirmation. In my article I tried to show that upon closer analysis the results thus arrived at prove to be reasonable, and that the impression of paradoxicality arises from a misguided intuition in the matter ; the details of this argument, however, are not needed for a discussion of Mr. Whiteley's note.

Mr. Whiteley states that the results in question are " not only paradoxical, but false ". If this is so, then obviously at least one of the premises R1, R2, from which they were derived, must be false, and Mr. Whiteley argues indeed that R1 does not generally hold. In elaborating this point, he argues that a distinction has to be made between the relation of being-evidence-for a hypothesis, which he claims does not satisfy R1, and the relation of being-an-instance-of an hypothesis; and while he does not provide a definition for this second relation, he seems to assume that at any rate it does satisfy R1 (when this condition is formulated for " instance of " instead of for " evidence for "), for he illustrates the instance relation by pointing out that any black raven is an instance of S1, any non-black non-raven an instance of S2, and anything that is not a non-black raven an instance of S3. This instance relation, Mr. Whiteley argues, is entirely different from the relation of being-evidence-for a hypo-thesis. Thus, e.g., one instance of S1 does not constitute confirming evidence for S1 because " the occurrence of one black raven is not improbable even if S1 is false, and therefore provides no evidence for it "; but a series of black ravens, he states, does constitute confirming evidence because " If S1 is false, i.e., if there are ravens which are not black, it is a priori unlikely that a collection of ravens taken at random will consist entirely of black ones ". In connection with his assertion, finally, that the confirming-evidence relation does not generally satisfy R1, Mr. Whiteley states that while " S1 is established by observing instances of it ", this is not true of S2, because " in S1 the conditions specified are positive, and in S2 they are negative "; and he proceeds to argue, in effect, that the instances of S2, whether taken singly or in collections, are irrelevant for the confirmation of S2.

Mr. Whiteley's suggestions do not seem to me to provide a basis for an adequate analysis of the paradoxes. My principal reasons for this opinion are as follows :

1. Mr. Whiteley does not give a definition or any other general characterisation of exactly what is to be understood by an instance of a hypothesis, although this concept is crucial for his argument. Let us assume however—and this seems to be a fair interpretation of his intentions—that he would call an object x an instance of a general hypothesis asserting that everything with the attribute C_1 also has the attribute C_2 if, and only if, x has both the attribute C_1 and the attribute C_2. This conception of the instance relation involves a serious difficulty : according to it, no instance of the sentence S1 above is an instance of S2, and some of the instances of S3 (namely those which are black non-ravens) are instances of neither S1 nor S2—despite the fact that all three sentences, being logically equivalent, express the same hypothesis. In fact, after having characterised the three different classes of objects which constitute instances of S1, S2, S3, respectively, Mr. Whiteley himself briefly

remarks that " in this respect the three sentences are not equivalent ", without, however, making any further comment on what is meant here by equivalence, or how the difficulty is to be met that sentences which make the same assertion differ in their instances.

2. These paradoxes of the instance relation also seem hardly reconcilable with Mr. Whiteley's assertion that " in order to know whether a particular fact is an instance of . . . a general hypothesis all that I need to know is the fact and the hypothesis " ; for, as the preceding illustration shows, one would also have to know by means of what particular sentence the hypothesis happens to be formulated ; and the class of instances of a given hypothesis would, in general, change with its formulation.

3. Besides, the general concept of instance, which seems intuitively so clear and unproblematic as hardly to require any explanation, actually involves very serious difficulties when applied to general hypotheses which contain relation terms. This can be seen from an argument on page 13, footnote, of my article, where it is shown that if for an ordered couple of objects, (a, b), both $R(a, b)$ and $\sim R(b, a)$ is the case, then the couple satisfies the two conditions $C_1(x, y) : \sim (R(x, y) . R(y, x))$ and $C_2(x, y) : (R(x, y) . \sim R(y, x))$; therefore, according to the intuitive notion of instance, any such couple (a, b) would constitute an instance of the following universal hypothesis : " Any couple which has the attribute C_1 also has the attribute C_2 " ; i.e., ' $(x)(y)(C_1(x, y) \supset C_2(x, y))$ '. Actually, however, this hypothesis can be shown to be logically incompatible with ' $\sim R(b, a)$ ', i.e., with one of the two statements which seem to make (a, b) an instance of the hypothesis.

In this context, I should like to correct a misprint in the footnote on page 13 of my article : S1 should read as follows :

$$(x)(y)[(\sim R(x, y) . R(y, x)) \supset (R(x, y) . \sim R(y, x))].$$

4. The argument which Mr. Whiteley uses to show that one black raven cannot constitute confirming evidence for S1, while a series of ravens can, makes use of a number of concepts such as " probable ", " improbable ", and " a priori unlikely ", which have been employed in philosophical discussions in a variety of different meanings. Mr. Whiteley does not indicate which of these meanings he has in mind, and I am not aware of any theory of probability which could serve as a basis for the establishment of the argument which Mr. Whiteley presents in this context. Indeed, I doubt that there is such a theory ; for clearly, the difference between one instance and a series of instances is one of degree only, and it seems that, therefore, in any adequate theory of probability in which several instances may confirm a hypothesis—i.e., in Mr. Whiteley's terms, " make it more or less probable "—a single instance must be able to do the same. (The degree of confirmation might, of course, be different, but that question is not involved here.)

6

5. The distinction between positive and negative conditions, which is presupposed but not explained in Mr. Whiteley's argument, appears to be quite problematic. In fact, the intended differentiation seems to be applicable to *predicate expressions*, which designate conditions, attributes, or properties, but *not to their designata :* It is obviously possible to distinguish between predicate expressions which begin with a denial sign and those which do not ; but the same condition or attribute may be referred to by positive as well as by negative expressions in this sense. Thus, *e.g.*, in our example, it would be possible to introduce two special terms denoting non-black things and non-ravens respectively ; and when formulated by means of these, the conditions specified in S2 would be " positive " and those in S1 " negative ". Any argument which presupposes a distinction of positive and negative conditions appears, therefore, to be untenable.

Carl G. Hempel.

DISCUSSION

ON INFIRMITIES OF CONFIRMATION-THEORY

Carnap's paper "On the Application of Inductive Logic"[1] sets forth certain assumptions on the basis of which he seeks to answer the question raised in my "Query on Confirmation."[2] Not much comment on these assumptions is necessary; the reader may decide for himself whether he finds them acceptable, as Carnap does, or quite unacceptable, as I do. The root assumption is that there are absolutely simple properties into which others may, and indeed for some purposes must, be analyzed. The nature of this simplicity is obscure to me, since the question whether or not a given property is analyzable seems to me quite as ambiguous as the question whether a given body is in motion. I regard "unanalyzability" as meaningful only with respect to a sphere of reference and a method of analysis, while Carnap seems to regard it as having an absolute meaning.

By way of partial justification for the restrictions Carnap places upon the interpretation of the predicates admissible in his system, he argues that these restrictions are also necessary for deductive logic. The analogy does not seem to me well-drawn. He says that in deductive logic, knowledge of such matters as the independence of predicates, etc., is necessary if we are to be able to determine whether or not a statement is analytic. But certainly we do not need such knowledge in order to carry out perfectly valid deductions; I can infer S_1 from $S_1 \cdot S_2$ quite safely without knowing anything about the independence of the predicates involved in these sentences. On the contrary, in the case of Carnap's system of inductive logic, I cannot safely make an inductive inference without such knowledge; I must have this knowledge before I can tell whether the computation of a degree of confirmation will be at all correct. The analogy Carnap seeks to draw would seem to me convincing only if he could show that the assumptions necessary to guarantee the correctness of inductive inference (by his methods) are likewise necessary to guarantee the validity of deductive inference.

Furthermore, even supposing all predicates to have been classified into purely qualitative, positional, and mixed, we are offered no evidence or argument in support of Carnap's conjecture that either the class of purely qualitative predicates is identical with the class of intuitively projectible predicates, or that such predicates as are intuitively projectible though not purely qualitative will also prove to be projectible by his definition. The

---

[1] *Philosophy and Phenomenological Research*, Vol. VIII, no. 1.

[2] *Journal of Philosophy*, Vol. XLIII (1946), pp. 383–385.

first alternative seems *prima facie* dubious since predicates like "solar," "arctic," and "Sung" appear to be intuitively projectible but not purely qualitative; the grounds for the second alternative are not evident.

In the last page or two of his article, Carnap seems almost to be claiming that no such question of intuitive adequacy any longer exists. He maintains that with his present restrictions on the interpretation of primitive predicates, his formal system of inductive logic provides a definition of projectibility; and he suggests therefore that anyone who has queries about projectibility can find out the answers by studying his system, just as one can learn about right triangles by studying Euclid. The catch is, though, that the question whether the formal system is intuitively adequate is quite pertinent both to Euclid's system and to Carnap's. In fact, the only difference in the two cases is that I am better satisfied that the triangles to which the Pythagorean theorem applies are just those I know as right triangles than I am that those properties to which Carnap's formula in terms of c^* applies are just those which are intuitively projectible.

Concerning the principle of total evidence, a rather complex discussion has grown out of a subsidiary point that I apparently did not explain very well. The point was just this: it might have been possible to claim that a criterion of projectibility is unnecessary because the evidence need never contain any non-projectible predicates. Examples show, however, that such predicates as those of order must be admitted into our evidence if we are to avoid counterintuitive results. If it is therefore required that order and all other properties be covered in our evidence-statement, then we shall have to have a criterion of projectibility in order to determine which are to be expected to attach to future cases; if no such distinction is made in our formal system, or in the rules for applying it, we shall reach the absurd conclusion that future cases will probably have all the properties common to past cases. (Incidentally, Carnap's example of an investigator who omits all cases unfavorable to the hypothesis being tested is not parallel to mine and does not bear on the same point; for in my example, no cases were omitted but only, quite consciously, some information about these cases.)

One further point concerning my examples having to do with order. I was in effect asking how the obviously relevant fact of order was to be taken into account by the theories in question, and pointing out that in trying to devise a method for doing this, we must face the fact that regular orders influence our expectations in a way that irregular ones do not. Thus if a method should give a high degree of confirmation for repetition of the pattern *red, red, not-red*, on the evidence that this pattern had repeatedly occurred, it would seem likely that this method would be in danger of giving a high degree of confirmation for the repetition of a wholly irregular pattern of, say, 96 tosses. In effect, if not technically, we would seem to be regarding each occurrence of a pattern as a confirming instance; hence in the irregular case we have the equivalent of one positive and no negative instances. I am not clear as to how Carnap's proposal would render regular order effective without rendering irregular order equally effective and thus leading to counterintuitive results. Carnap cannot say antecedently that regular order is relevant and irregular order isn't. This would mean just that the latter but not the former affects the c* computation, and would indeed, were it true, provide us with an interesting definition of degree of regularity. But I am afraid that if order is so to affect the computation as to give intuitive results when the order is regular, it will be difficult to avoid getting counterintuitive results when the order is irregular. It is worth noting, however, that this difficulty might be overcome by a definition of degree of confirmation for which the number of confirming cases is always important, since an irregular order is perhaps one that does not consist of repetitions of a smaller pattern.

But the consideration of the examples involving order are in any case quite secondary to my main point which is adequately illustrated by the other examples I gave. I regret that Carnap's method of dealing with it involves assumptions I cannot accept and that no other answer has been forthcoming. For until this problem is solved, we are seriously hampered in our efforts to solve certain other important problems.[3]

NELSON GOODMAN.

UNIVERSITY OF PENNSYLVANIA.

---

[3] For example, see my "Problem of Counterfactual Conditionals," *Journal of Philosophy*, Vol. XLIV (1947), especially pp. 127–128.

DISCUSSION
REPLY TO NELSON GOODMAN

In my earlier paper[1] I have indicated some requirements which I believe must be fulfilled in any application of a system of inductive logic to a given knowledge situation in order to lead to adequate results. In his discussion[2] Goodman regards these requirements as quite unacceptable; in particular he regards the simplicity of properties as meaningful only with respect to a sphere of reference. I must confess that I too have a rather uneasy feeling concerning the concepts of absolute simplicity and absolute completeness referred to in the requirements. I hope very much that it will be possible to find a way of avoiding these problematic concepts and replacing them by the kind of relative concepts with which we usually work. But at the present moment I do not see whether or how this can be done. Although those absolute concepts involve problems and difficulties, I do not think that they are meaningless. The question: "Are all properties of individuals in a given universe expressible in a certain language?" is formulated in what I, at an earlier time,[3] called the material mode of speech. After the appearance of the semantical method it became clear that questions of this kind can be formulated and dealt with in an exact way. We should certainly always look out for the dangers involved in the material mode, also in the present case; but it is not necessary to prohibit this mode completely.

I regard it as the task of deductive logic to supply not only positive but also negative answers to questions of logical truth and logical implication (e.g., "S_2 is *not* implied by S_1"). It is with respect to these negative results that the requirement of simplicity becomes relevant, as explained in my paper.

I feel, as Goodman does, that questions concerning the intuitive adequacy of any proposed system of inductive logic are of greatest importance, and I shall discuss in my book questions of this kind in detail with respect to other systems and to my own. Of course, this examination may center upon many different points. I have found that an examination of the subsequent two points, which are closely related, seems especially fruitful, because most methods proposed make it easily possible to calculate values for at least one of the two cases and we have often a fairly clear intuitive

---

[1] "On the Application of Inductive Logic," Vol. VIII, No. 1, pp. 133–148.
[2] "On Infirmities of Confirmation Theory," Vol. VIII, No. 1, pp. 149–151.
[3] *Logical Syntax of Language*, 1937, Ch. V.

feeling of adequacy or inadequacy concerning such results: (1) the degree of confirmation, interpreted as giving a fair betting quotient, for a hypothesis concerning a single unobserved individual,[4] (2) an estimate of the relative frequency of a property M in an unobserved sample or in the whole population on the basis of the relative frequency of M in an observed sample.[5] These results can then also be used for an examination of projectibility. No matter which points are chosen for an examination of adequacy with the help of examples and counter-examples, it seems advisable to use as *primitive* such properties as Red, Hot, Hard (or similar simple, directly observable properties). This has the advantage of eliminating the otherwise bothersome task of showing that the general requirements stated in my paper are fulfilled and, in particular, of showing the logical independence of the primitive properties. This procedure by no means excludes the examination of complex properties; the definition of the property M to be examined may have any complexity desired.

I agree with Goodman that the problem of projectibility, which he has pointed out, is interesting and important. As I see it, our difference with respect to this problem is only, or mainly, the following: Goodman seems to believe that the construction of an adequate system of inductive logic *pre-supposes* a solution of the problem and involves an explicit formulation of a criterion of projectibility. I think that this procedure, though possible, would be unnecessarily complicated. My definition of degree of confirmation (c^*) shows a different way of procedure. This definition is rather simple; it is based on the concepts of state-description and isomorphism. If this definition should be found to be inadequate, then, I believe, an adequate definition could be constructed by a similar procedure, based on the same concepts and not containing an explicit reference to projectibility. Even for this procedure however the problem of projectibility is not irrelevant; only the place of its appearance is changed. It remains pertinent for the examination of adequacy.

RUDOLF CARNAP.

University of Chicago.

---

[4] The so-called singular predictive inference; see my earlier paper §5.

[5] For an explanation of this estimate and its connection with the degree of confirmation see my paper "Probability as a Guide in Life," Journal of Philosophy, XLIV, 1947, pp. 141–148.

THE USE OF SIMPLICITY IN INDUCTION

I. THE PROBLEM

THE CONCEPT of simplicity plays a central role in inductive inferences. Given any inductive problem in which there are several "equally good" hypotheses, the scientist will choose the simplest one. In spite of the fact that this much is generally admitted, the concept of simplicity remains highly controversial in the philosophy of science.

By the "rule of simplicity" we mean a rule instructing the scientist to choose the simplest of several acceptable hypotheses. (This rule will be left vague for the time being.) Justifications for such a rule come from two diametrically opposed camps. At one extreme we find the belief that the rule involves an assumption about the "simplicity of nature"; the other extreme justifies it by saying that it is only a matter of convenience, a laborsaving device. And whatever the philosopher's attitude may be to the justification of such a rule, you will almost always find the firm opinion that no precise definition can be given for the concept of simplicity.

It is the purpose of this paper, first of all, to make a contribution to the explication (precise definition) of the concept of simplicity; secondly, to state a rule of simplicity in terms of this explicatum, and thirdly, to show the *methodological* advantages of this rule. It will be shown that such rules have a purpose deeper than mere convenience, and this will be shown without any metaphysical assumptions about nature.

The paper will first give a general discussion, and then apply the results to some well-known examples. It is convenient, however, to illustrate all the general points in terms of two simple examples. These will be included in the main body of the paper, but in order to preserve the continuity of the main line of argument, remarks referring exclusively to the illustrations will be marked off by brackets.

Although a certain amount of mathematical knowledge is pre-

391

supposed in the examples, the main argument requires no such background.

II. PRECISE FORMULATION

Let us begin with a precise description of the type of inductive problems here to be considered. The scientist is confronted with some definite problem in which he will perform a series of experiments, which will culminate in the acceptance of some theory. There are two distinct steps in the technique of theory-formation: first of all, the scientist must formulate alternate hypotheses (generally there will be an infinity of alternatives considered), and secondly, the selection of one definite hypothesis on the basis of the results of experiments. It is the latter step that will be considered here. The former step is just as important, and presumably much more difficult, but it will be ignored in this paper.

We start with a given set of alternative hypotheses, and a series of experiments designed to eliminate incorrect hypotheses. The problem can be stated as follows: given the result of the first n experiments, which hypothesis shall we select?

There is some question as to the number of alternate hypotheses to be considered, and the number of experiments we should plan. Both of these will be taken to be denumerably infinite. In the case of hypotheses, in all interesting examples there will be an infinity of possible explanations (certainly in any case involving measurement), and, on the other hand, there is serious doubt whether a scientist ever *really* considers a nondenumerable infinity of possibilities. As to the experiments, we consider a series (endless in principle) of planned experiments, though, of course, at any one time we will have the results of but a finite number of them.

[The first example to be considered is the classical inductive problem of the *urn*. There is a sealed urn, with an unknown number of white and black balls in it. Through an opening you take out one ball at a time, note its color, replace it, and shake the urn thoroughly. After n draws you must decide what fraction of the balls in the urn is white. Your experiments in this case consist of drawing balls, and the result of n experiments is to tell you that m of the n drawn balls were white. Your possible hypotheses state that some fraction r of the balls in the urn is white, where r is any

392

242

rational number between 0 and 1. Of course, in practice one can estimate an upper-bound for the number of balls in the urn, but for the sake of the illustration we assume that no such upper-bound is available, hence any proper fraction is possible (including 0 and 1 if we allow the cases where there are no white balls, or no black balls). On the basis of the report that m of n balls were white, we must select some r.

The second example is one involving measurements. We have two independently measurable quantities x and y, and we make a series of measurements (say for values of x between 0 and 1), and try to find how y varies as a function of x, e.g., one quantity may be temperature, the other pressure, for a gas of fixed volume. We will suppose that the scientist has selected *polynomials* with rational coefficients as his possible hypotheses. After n measurements, each one giving us an observed value of y for some preassigned value of x, we have to select some polynomial which is to describe the relation between the two quantities.]

It will be convenient to introduce a simple notation for a typical problem. Let us denote the hypotheses we must choose from by h_i, and the possible outcomes of n experiments by e_j^n. Let e^n be the actual outcome, and h^n the selected hypothesis. (We need the superscript for h, because after 100 experiments we may decide to choose a hypothesis different from the one chosen after 10 experiments.)

[In the urn example the various h_i differ only as to the number r, while e_j^n reports that j of n balls were white. If actually m of them turn out to be white, then e^n will be e_m^n and h^n is determined by the choice of an r.

In the polynomial example the h_i are polynomials with rational coefficients, such as x or $\frac{1}{2} - \frac{1}{4}x + \frac{1}{3}x^2$. e_j^n reports the result of measurement at n points. Thus e^s may tell us that at the points $x = \frac{1}{4}, \frac{1}{2}, \frac{3}{4}$ the values $y = \frac{1}{8}, \frac{1}{4}, \frac{3}{8}$ were observed. h^s must then select some particular polynomial; in this case it would undoubtedly be $\frac{1}{2}x$.]

There are certain assumptions underlying this procedure, which we must make explicit.

Assumption 1. One (and, of course, only one) of the hypotheses is true.

Of course this assumption may be false; if the evidence indicates that the assumption is likely to be false, then the scientist looks for a new (or wider) set of hypotheses. But then the assumption will again serve as a working hypothesis for the new set of hypotheses. [In the urn example we are certain that one of our hypotheses is true.] We will use h to denote the true hypothesis.

Secondly, we need some measure as to how well the hypothesis in question agrees with the result of the experiments. For this we introduce a measure of deviation. ($m(h_i, e_j^n)$ will be used to denote the measure.) The choice of this measure will be left, to a large degree, arbitrary to allow the scientist a maximum amount of freedom. But in many examples there is a well-accepted measure of deviation. [In the urn example this would be the absolute value of the difference between m/n and the predicted r. In the polynomial case we have some choice, but we may take the average of the absolute differences between predicted values and observed values.]

Assumption 2. The deviation between a given hypothesis and the observed results tends to 0 if and only if the hypothesis is the true one.

This assumption concerns the question whether the experiments are (at least in principle) adequate for the elimination of all false hypotheses.[1]

We are now in a position to formulate the ordinary rule of induction.

Rule 1. Select a hypothesis which is as well in agreement with the observed values as possible.

This is to be interpreted in the sense that for a given e^n we must select an h^n giving us the smallest possible deviation, i.e., making $m(h^n, e^n)$ a minimum.

[For the case of the urn we get "Reichenbach's Rule" exactly.[2] The rule tells us to select the hypothesis with $r = m/n$, which is Reichenbach's posit. (In this case the deviation is 0.)

In the polynomial example we can again make the deviation 0, by selecting a polynomial going through the observed values (as-

---

[1] We will not take into consideration the additional difficulties introduced by the Indeterminacy Principle.

[2] Hans Reichenbach, *The Theory of Probability* (University of California Press, 1949), section 87.

394

suming that the measurements, as is usual, give us rational values).
But here the choice is not uniquely determined—which will soon
lead to trouble. Reichenbach does not even consider such problems,
and we refrain from adding anything to the rule at this time.]

It is customary in such papers to base the adequacy of a rule
entirely on the question of "convergence." Although this will not
be our only criterion, we must consider convergence. For this we
will need some measure of the difference between hypotheses. (Denote this measure by $d(h_i, h_i')$.)

For the sake of logical economy we could *define* a suitable d in
terms of m. We will do this in two steps. First of all we introduce

$$d^n(h_i, h_i') = \min_j [m(h_i, e_j^n) + m(h_i', e_j^n)]$$

and then let

$$d(h_i, h_i') = \lim_{n \to \infty} d^n(h_i, h_i').$$

We must then make a special assumption about this measure,
though this assumption can be satisfied in all usual cases by a suitably chosen m-function.

Assumption 3. For any two hypotheses the limit in the definition
of d exists, and it is o only if the hypotheses are identical.

[In our two examples the method just described leads to the accepted measures of difference; the absolute difference in r's for the
urn hypotheses, and the area between the curves in the polynomial
example. In the latter case this is true only if the values of x are
taken from "all over" the unit interval, but this is necessary if assumption 2 is to be satisfied.]

We are now supposed to be able to prove that the selected hypotheses converge to the true hypothesis, but this is not always so.

If we introduce d by definition, as above, we can easily show
where the difficulty arises. It is easy to prove that $\lim_{n \to \infty} d^n(h^n, h)$
$= 0$,[3] but what we need for convergence is that $\lim_{n \to \infty} d(h^n, h) = 0$,
which does not follow.

[In the urn example we do get convergence, and this is the example usually discussed in the literature. However, in the polynomial example we get into trouble because the rule does not determine h^n uniquely. We can select any polynomial passing through

---

[3] By definition $d^n(h^n, h) \leqq m(h^n, e^n) + m(h, e^n)$, which in turn is at
most $2m(h, e^n)$ because of the way h^n was selected. But this tends to o by
assumption 2, hence $d(h^n, h)$ tends to o.

395

the n observed points. Suppose we systematically select a polynomial which oscillates greatly between the observed values, then we can easily get a series of h^n which does not converge.]

What we need in addition to rule 1 is some assurance that the values of the hypothesis at points not observed "fit in well" with the observed values. The usual way of assuring this is to do what any scientist would do in the application of rule 1, namely to use the following rule:

Rule 2. Select a hypothesis which is as well in agreement with the observed values as possible; if there is any choice left, choose the simplest possible hypothesis.

[In the polynomial case this would direct us to choose the polynomial of lowest degree passing through the n points, and then we get convergence.]

In most applications rule 2 will result in convergence, and this is the rule that is undoubtedly in the minds of most writers when they *prove* convergence. We could have introduced it immediately, in place of the first rule, but we wanted to emphasize the point that even in the traditional rule "simplicity" comes in from purely methodological considerations.

We must now deviate entirely from the traditional approach, and start a new line of attack.

III. FINDING THE TRUE HYPOTHESIS

We have so far discussed only the question of how well the chosen hypothesis approximates the true hypothesis. We will now consider what our chances are of selecting the true hypothesis itself.

We find the strange result that in the usual examples, if we use rule 2, our chance of selecting the true hypothesis gets worse and worse! And we even find that the requirements we must put on the results of experiments in order to have the true hypothesis selected are very strange indeed.

[Let there be ⅓ white balls in our urn. In order for rule 2 to select the true hypothesis, we need that $m = \frac{1}{3} n$. This has the very strange consequence that a necessary condition for the selection of h is that n be divisible by 3. And even for these values of n we find that while the probability of being within a preassigned percentage of the true value tends to 1, the probability of getting the true value itself tends to 0.

The situation in the polynomial case is even more counterintuitive. We can select the true hypothesis only if all n measurements are exactly correct, the probability for which even for $n = 1$ is 0. In short, we see that while rule 2 is designed to get an h^n near h, it is also so designed that it would take a miracle to give us h itself.]

When we take these facts into account, we must realize that rule 2 somehow fails to reproduce what scientists actually do. This could also have been noted from a particular example. [If 503 out of 1000 drawn balls were white, it is most unlikely that the scientist would select the value $r = 503/1000$ instead of $r = \frac{1}{2}$.[4]]

And there is a third difficulty. While we can prove convergence, we have said nothing at all about how many experiments are needed to arrive at a hypothesis close enough (for practical purposes) to the true hypothesis. Indeed in many fairly simple cases we note that the number of experiments needed is quite large, often too large from a practical standpoint. So we arrive at the following three criticisms of rule 2:

(1) The probability of selecting the true hypothesis tends to 0 as we increase the number of experiments.

(2) If we need high accuracy (h^n must be very close to h), we often have to make n too large for practical purposes.

(3) Rule 2 does not reproduce what scientists actually do.

Let us start with the last point. We note that in the examples quoted, scientists would have chosen a hypothesis different from the one selected by rule 2, even if this hypothesis was not as close to the observed values. Of course, they do not allow very large deviations, but they allow some deviation—recognizing that the results of the experiments need not be exactly the true values. They allow themselves some margin of deviation; they agree as to when a hypothesis is compatible with the evidence, and then they select the simplest one from all such compatible hypotheses.

Rule 3. Select the simplest hypothesis compatible with the observed values. (If there are several, select any one of them.)

---

[4] This criticism was used by Bertrand Russell in *Human Knowledge* (Simon and Schuster, 1948), p. 370.

397

The difficulty with this rule is that it contains two very vague terms, "simplest" and "compatible." Our immediate task is to explicate these terms.

Let us begin with the term "compatible," because in this case we have a reasonably clear usage in science. The scientist says that a hypothesis is compatible with the result of observations if these results do not definitely eliminate the hypothesis from the realm of possibilities. As has often been pointed out, this does not, in general, mean that the observations contradict the hypothesis, but only that they make the hypothesis very improbable. More precisely, for a given hypothesis we can calculate the probability of getting observational results within preassigned deviations, and we select some level of deviation such that the outcome is almost certain to fall within it; if it does not, then we eliminate the hypothesis. What this level should be is to a great degree arbitrary. For the sake of this paper we select the simple fixed level at which 99 per cent of the observations will fall within the level.[5]

Definition 1. The hypothesis h_i is *compatible* with the result e^n if, assuming the truth of h_i, there was at least a one per cent chance of getting a deviation as great as $m(h_i, e^n)$.

[In the urn example, given an h_i, i.e., given an r, it is a simple probabilistic calculation to find the compatible and the incompatible results e_j^n. Assuming that h_i is true, we can calculate the probabilities of each *possible* outcome of n experiments, and then we allow an interval around r just large enough so that the total probability of all outcomes in this interval is just 99 per cent. In the polynomial case we must invoke the theory of errors in measurement, which will tell you whether an observed deviation is within the one per cent level or not.]

So we see that under rule 3, for any e^n, there is a good deal of choice left—and we must select the simplest possible hypothesis. But which hypothesis is simplest? It is best to answer this question for examples first, and then try to give a general criterion.

---

[5] Other levels of compatibility could be used as well. As a matter of fact it might be much better to use a criterion that tends to 100 per cent slowly as n increases. This would give us an improvement on the theorem which we will prove later. But since this method leads to difficulties, we will take the simpler criterion for the sake of this paper.

398

[We have agreed to select ½ in preference to 503/1000, and we will readily agree that the selection of ⅓ is better than that of 331/1000. If we are not oversophisticated it should be pretty obvious that the simplicity in these cases lies in the fact that the simpler fractions can be expressed in terms of smaller numbers. Yet it would seem counterintuitive to select ⅓ over ⅔ because of greater simplicity. First of all, they occupy symmetric positions; secondly, ⅔ white balls is the same as ⅓ black balls; and finally, these two fractions would hardly ever be both compatible with the observations. So at least in the urn example we can safely state that simplicity consists in the smallness of the denominator of r. Of course it could be objected that the same fraction can be stated in an infinity of forms, but if we take advantage of this supposed difficulty, then we get a deeper insight into the simplicity ordering. Let us divide all fractions into classes, the kth class having all fractions with denominators $\leq 2^k$, $k = 0, 1, 2, \ldots$ Then there are few duplications within a given class, and each new class contains all the previous ones. ($\frac{1}{3} = 2/6 = 3/9$, etc.) Thus we see that allowing greater complexity (less simplicity) coincides with allowing ourselves more freedom of description and greater accuracy.] Hence we note that rule 3 can also be stated as a warning: do not use more precision in your theories than is necessary to explain the observations. For example, in the case of a simple measurement,[6] the rule tells you to use no more digits in your summing up of results than is necessary to be within the experimental error. Hence we get the rule of significant digits!

[In the polynomial case the theory of errors allows us some deviation from the observed values, and we must fit a polynomial of lowest possible degree within this deviation, e.g., if there is a straight line compatible with the observations, we must choose it. (Of course scientists very often choose straight lines even when there is none compatible with the evidence.) So we see that simplicity in this case consists first of all in choosing the lowest exponent possible, and only if there is more than one polynomial of lowest degree com-

---

[6] The possible hypotheses in this case are all finite decimal expansions. We order them according to the number of decimal places. Each new class contains the previous ones, since we can always put 0's at the end of the decimal expansions. Then, e.g., if we measure .1234 ± .005, we must select .12.

399

patible with the observations do we take the coefficients into account. We might do the latter by considering the largest denominator occurring in a coefficient.]

We have constructed rule 3 to bring our rules into better agreement with what scientists actually do. We are now in a position to ask: does this improved rule give us any better chance of selecting the true hypothesis? The answer turns out to be emphatically "yes."

There are two reasons why the hypothesis selected by rule 3 might not be the true one. The true hypothesis may not be compatible with the observations; or it might be compatible, but there could be a simpler hypothesis also compatible with the results. The former danger is very slight. According to definition 1 there is always a 99 per cent chance that e^n is compatible with the true hypothesis, h. This was the reason why we chose a percentage as high as 99; we wanted to be almost certain that we do not eliminate the true hypothesis. But is not there a serious danger that simpler hypotheses will also be acceptable (especially since it is so difficult to eliminate hypotheses)? This is indeed the case, but in each example this happens only for a finite number of n; from a certain value of n on it can never happen.

We must look into the reason for this. As we know from statistics, as n increases, the deviations allowed by the compatibility requirement decrease. Hence for high n we can find an interval around the observed values (an interval that can be made as small as required by increasing n) such that all compatible hypotheses lie within this interval. This, of course, is not enough; this only assures convergence under rule 3. But *the characteristic property of orderings according to simplicity* is that for any given hypothesis, even though there may be infinitely many hypotheses at least as simple, we can find an interval within which there is no other hypothesis as simple (or simpler). Hence, by making n large enough we can make the interval of compatibility sufficiently small so that if the true hypothesis lies within it, no other hypothesis as simple will be in it. Hence for all sufficiently high n we must select h if it is compatible with the observations. Hence for all sufficiently high n we are 99 per cent sure of selecting h. *This is the justification we offer for rule 3.*

[Let us suppose that ⅜ of the balls are white. There are only a

400

250

finite number of fractions at least as simple as ⅜, hence there is a closest one. It happens to be $2/5$. $2/5 - ⅜ = 1/40$. From the theory of probability we can calculate an n such that if we draw at least that many balls, then r must be within $1/80$ of m/n in order to be compatible. Then, if ⅜ is compatible, none of the others at least as simple are compatible with the observations. So from this point on if ⅜ is compatible with e^n, then it must be chosen. But (by the definition of compatibility) there is always a 99 per cent chance of that. Hence from this point on we are practically certain to select the true hypothesis.

The same result holds for the polynomials, but for different reasons. Given the true polynomial, it is not true that there is only a finite number of polynomials as simple or simpler; e.g., all polynomials of lower degree are simpler, and there are infinitely many of these in general. However, it is still true that there is a lower bound (greater than 0) for the differences between the true hypotheses and other polynomials at least as simple.[7] For sufficiently many measurements (the exact number determined by the theory of errors), we can narrow the interval of compatibility sufficiently that if the true hypothesis falls within it, none of the other polynomials which are at least as simple can be compatible with the observations. Hence, from this point on we are again 99 per cent sure of selecting the true hypothesis.]

Let us summarize these results in the form of a theorem.

Theorem: If the true hypothesis is one of the hypotheses under consideration, then—given enough experiments—we are 99 per sent sure of selecting it.[8]

The proof of this theorem was given in the main text above; all that was assumed was that the simplicity ordering has one fundamental property: for each hypothesis we can find an integer such that for n at least as great as this we are assured that, if the hypothesis is compatible with the observations, no other hypothesis as simple is compatible.

---

[7] The proof of this is not difficult in principle, but too long to be included in this paper.

[8] This result depends on the definition of compatibility (see footnote 5). E.g., had we selected the very common 95 per cent level, we would never be more than 95 per cent sure, but we would be that sure sooner.

401

We can now state precisely in what sense we propose to explicate the concept of simplicity. It will not be a unique definition, but for each inductive problem we will give a class of possible orders. This will be achieved by stating some necessary conditions for the ordering, the most important being that the order has the fundamental property just described.

But before stating the general definition, let us first see whether we have taken care of all three of the objections raised against the traditional rules. We started with the third objection, that the rules did not correspond to what scientists actually do. This was taken care of by inserting the "compatible with the observations" clause in place of "closest to the observed values," and by showing that the simplicity order corresponds to the order of going from less precision (less wealth of description) to greater precision. The former point is taken care of by the wording of rule 3, the latter by the first two conditions to be given.

Then we took up the first objection, that our chances of selecting the true hypothesis deteriorated. This is corrected by rule 3, as shown in the proven theorem. All we need for this is the fundamental property of the ordering, given in condition 3 below.

This still leaves one objection. Couldn't the "sufficiently large n" be too large for practical purposes? Actually, in many instances we can show that the n's are quite low. [In the urn example, if ½ of the balls are white, we are 99 per cent sure of selecting the true ratio for $n \geq 8$.] How high this n is depends, of course, on how simple the true hypothesis is.[9] The best we can do is require that the ordering should make the required n's as low as possible. This is not so easy, since one n may be lowered at the cost of raising others. This is a question that deserves a good deal more research, but at

---

[9] This is the point at which an assumption about the "Simplicity of Nature" can come in. But it has to be a relation between the true laws and our ability of discovering them. It would have to take the form: The laws of nature are sufficiently simple that, following the best available inductive methods, man can in a reasonable period of time find them (or find a good approximation of them). Thus it expresses a relation between the n assigned to the true law in our inductive method, and the number of experiments we can actually carry out during some era of science. This point will be developed elsewhere.

least the formulation here given to the problem should help in finding an answer. We will be content with the vague fourth condition below.

Definition 2. Given an inductive problem (as defined in this paper), we say that the hypotheses are *ordered according to simplicity* if we have the following four conditions satisfied:

Condition 1. The hypotheses are divided into sets $H_{(a, \ldots, z)}$, where $a, \ldots, z$ are natural numbers (called the *characteristic numbers*). These sets are ordered lexicographically according to the characteristic numbers. (The hypotheses in earlier sets being called simpler than the ones occurring only in later sets.)

Condition 2. The later sets include the earlier sets in the following sense: $H_{(a, \ldots, k, \ldots, z)}$ is always a subset of $H_{(a, \ldots, k+1, \ldots, z)}$.

Condition 3. For every set H there is an integer N_H such that, if a member of H is compatible with e^n for some $n \geq N_H$, then no member of H or of any earlier set can be compatible with this e^n.

Condition 4. N_H should in each case be as low possible. (Or better, each h_i should belong to a set H with an N_H as low as possible.)

This completes the explication. We arrived at a characterization of orders according to simplicity. Although these conditions do not determine the order uniquely in any given problem, very often we find that there is but one natural way of satisfying the conditions. It is hoped that with the improvement of the fourth condition, and with the addition of new conditions, the explication of simplicity will be considerably advanced in the future.

It is most important to note, however, that the order defined is relative to an inductive problem! It is not true that the number ¼ need always be simpler than 1/10. [In the urn example it is, but in the case of a simple measurement[10] the former is written as .25, the latter as .1, and hence the latter is simpler.] It should be noted, however, that the same numbers express entirely different hypotheses in different problems. This point may have been one factor that led many people to believe that no definition of simplicity can be given.

---

[10] See footnote 6.

IV. APPLICATIONS

So far we have only considered three extremely simple inductive problems: the urn and polynomial examples running through the text, and the simple measurement discussed in footnote 6. We are now ready to apply our method to some well-known historical problems.

Let us first take the problem of planetary motion in the time of Copernicus. Had Copernicus been familiar with the results of this paper, he could have argued as follows: the planets move in a plane closed curve around the sun. We must consider families of such curves, and order them according to their simplicity. We can in this follow the polynomial example: an ath order polynomial has $a+1$ parameters. The polynomials were ordered first according to the number of parameters (coefficients) and then according to the values of these parameters. We also had the inclusion theorem, a first order polynomial can be written as a second order one with a 0 coefficient. We must order our curves similarly, making sure that we count only the parameters determining the shape of the path, not those locating them in space. The simplest case is the one-parameter family of circles. So the Greeks were right in trying circular paths first. But by now we see that no circle is compatible with the observed positions. They next tried adding an epicycle to the circle. This family does include circles as special cases (the radius of the epicycle being 0), but it is a three-parameter family. And when they allowed any number of epicycles, they allowed hypotheses of arbitrary complexity, which is methodologically utterly unacceptable.[11] Instead we must search for a two-parameter family of curves having circles as a special case. This family is clearly the family of ellipses (determined by major and minor axes, circles being the special case where the two are equal). In this one step Copernicus could have anticipated Kepler's main result, *from purely methodological considerations*.[12]

---

[11] There is a theorem the author heard quoted (though he has seen no proof) that *any* closed convex curve can be approximated as closely as desired by a system of epicycles.

[12] Cf. Karl Popper, *Logik der Forschung* (Julius Springer, 1935), sections 39-46. The author wishes to acknowledge his indebtedness to this book, even though he is only in partial agreement with Dr. Popper.

404

So far our examples had one or two characteristic numbers: number of parameters and the complexity of the values of the parameters. In linear differential equations we still have the number of parameters determined directly, this time by the order of differentiation. When we come to nonlinear equations, we must take both the order of differentiation and the exponents into account. The resulting order will depend on the class of equations being considered, but in general this will require the introduction of a third characteristic number. In this case the hypotheses will be ordered first according to the order of differentiation, then according to the degree, and finally according to the coefficients.[13] All of these hypotheses are stated for one variable x so far. If there are several independent variables, then we must first order laws according to the number of variables, and then according to the above three characteristic numbers. (I.e., any hypothesis with two variables is simpler than one with three independent variables.) It is according to this last principle that we understand the methodological advantage of a single unified law that can take the place of many disconnected laws; e.g., Newton's laws contain many fewer independent variables than the many unrelated laws, which we had before Newton and which can be derived from Newton's laws, had together.

And finally let us apply our method to the analysis of Einstein's thinking in passing from the Special Theory to the General Theory of Relativity.[14] There were no new facts that failed to be explained by the Special Theory. Einstein was motivated by his conviction that the Special Theory was not the simplest theory that can explain all the observed facts. In accordance with the four characteristic numbers already discussed, we see that any reduction in the number of independent variables will simplify the law, even if its form

---

[13] This gives an order of ordinal type ω'. In general if we have k characteristic numbers, we get an ordering of type ω^k. No examples have so far been found which require order of higher type, but there is little difficulty in extending this method to type ω^ω and higher types.

[14] It was a conversation with Einstein that first started the author thinking about these problems. Einstein made no metaphysical assumption, and yet simplicity played a central role in his account of his own discovery. It seemed to the author that the revolutionary success of the General Theory cannot be due to a choice whose only justification is convenience.

becomes otherwise more complicated. By the requirement of general covariance Einstein succeeded in replacing the previously independent "gravitational mass" and "inertial mass" by a single concept. This is sufficient to justify a change, but why just *those* gravitational equations? It is possible to show that in order that the equations explain known facts, they had to contain differentiation of at least the second order, and that at most the highest order terms could be linear. This describes a large class of simplest possible hypotheses, which are however narrowed by the condition that the results must (at least in first approximation) agree with what the well-confirmed Special Theory says. It was then shown that there is only one such hypothesis in this simplest class. Since this was the simplest hypothesis compatible with the known facts, Einstein was methodologically perfectly right in saying that this is the law we *must* accept.[15]

V. FURTHER PROBLEMS

Let us return to the question of the relation of this method to Reichenbach's posits. We have pointed out that his rule of induction agrees with rule 1 (and 2) for the urn example. He points out that instead of his posit, m/n, we could equally well have posited $m/n + c_n$, where c_n, is an arbitrary function of n, tending to 0 with increasing n.[16] He believes that any choice of c_n is as good as any other, and hence he sets $c_n \equiv 0$, on grounds of "descriptive simplicity." Our posit, the simplest fraction compatible with m/n, can also be thought of as a choice of a function c_n (since the interval of compatibility tends to 0, the difference between our posit and Reichenbach's must also tend to 0), only c depends not only on n, but also on m. Hence we argue that Reichenbach is quite right in

---

[15] Einstein is still following the same methodology. He is now trying to reduce the number of independent variables further, by combining Maxwell's Theory with his own, in his Unified Field Theory. He also believes that with this new theory he will be able to achieve a further simplification, namely the elimination of singularities from the laws of Physics. (This would lead to a new characteristic number(s) relating to the number and type of singularities in the law, with a continuous law being the simplest one—having 0 singularities.) If a philosopher may venture to make a comment, the present paper indicates that Einstein is again using the right method, in spite of the opinion of the majority of his colleagues.

[16] *Op. cit.*, p. 447.

406

saying that the posit must be of the form $m/n + c_n$, but we argue that there is one choice of c_n superior to others (since it gives us not only convergence but an excellent chance of finding the true hypothesis), and that it is not the choice of $c_n \equiv 0$.

We must also show the relation between this paper and the work on degree of credibility (confirmation). Carnap in his recent monograph[17] treats Reichenbach's posits as estimates of the true ratio. This sounds reasonable, but when we consider the very special sense in which Carnap uses "estimate," it does not seem that he is correct in his interpretation. If he were correct, then some very reasonable posits would commit Reichenbach to a most unreasonable measure of credibility. Reichenbach, on the other hand, claims that he is not committed to any measure of credibility, and we see no reason for doubting this claim. A posit is a method of selecting one hypothesis from several possible ones, not a general method of estimation. We can find a connection if we adopt the very intuitive rule of selection: From a set of hypotheses select the one having the highest degree of credibility. Then a method of positing tells us only maximum credibility values, not all values. There are several credibility measures that would give us Reichenbach's posits, but not the counterintuitive measure that Carnap assigns to Reichenbach. If our interpretation is correct, then it is easy to show the connection between our rule and credibility. We adopted a rule of selection different from the one just stated. Ours would read: From a set of hypotheses select the simplest one having a high degree of credibility. Just how high the degree has to be depends on our definition of compatibility, and on how the particular credibility measure is constructed.

There are several problems left unsolved by this paper, and the author sincerely hopes that other people will be interested in working on these (often very difficult) questions. The most important of these is to find a precise form for the fourth simplicity-condition. The purpose of this condition is to assure fast convergence, that is to enable us to reach the 99 per cent level of certainty with as few experiments as possible. A great deal more work will have to be done in investigating the relation between the ordering of hy-

---

[17] Rudolf Carnap, *The Continuum of Inductive Methods* (University of Chicago Press, 1952), p. 44.

potheses in this method, and the resulting N_H's. This is the problem on which the recent statistical methods should prove most helpful.

Another possible method of getting faster convergence is a modification of rule 3. We might try to establish a level of simplicity and then select the hypothesis on this level which is closest to the observed values (this is the converse of our rule), or try to find a criterion combining an optimum of simplicity and compatibility.[18] The former method does not give us our theorem. But the combined method may very well give us our theorem, and faster convergence.

And finally, there is the problem that our conditions allow too many different orders. We should try the method out in many more examples, and try to find additional conditions. Ultimately we should be able to prove either that the order is uniquely determined by the conditions, or that the remaining orders are in some sense equally good.

These problems are of great methodological importance, and they are intrinsically interesting; they deserve careful study, far beyond these first considerations.

JOHN G. KEMENY

Princeton University

---

[18] I am indebted to Prof. Nelson Goodman for this suggestion.

408

THE AIM OF INDUCTIVE LOGIC

RUDOLF CARNAP

University of California, Los Angeles, California, U.S.A.

By inductive logic I understand a theory of logical probability providing rules for inductive thinking. I shall try to explain the nature and purpose of inductive logic by showing how it can be used in determining rational decisions.

I shall begin with the customary schema of decision theory involving the concepts of utility and probability. I shall try to show that we must understand "probability" in this context not in the objective sense, but in the subjective sense, i.e., as the degree of belief. This is a psychological concept in empirical decision theory, referring to actual beliefs of actual human beings. Later I shall go over to rational or normative decision theory by introducing some requirements of rationality. Up to that point I shall be in agreement with the representatives of the subjective conception of probability. Then I shall take a further step, namely, the transition from a quasi-psychological to a logical concept. This transition will lead to the theory which I call "inductive logic".

We begin with the customary model of decision making. A person X at a certain time T has to make a choice between possible acts $A_1, A_2, \cdots$. X knows that the possible states of nature are $S_1, S_2, \cdots$; but he does not know which of them is the actual state. For simplicity, we shall here assume that the number of possible acts and the number of possible states of nature

---

The author is indebted to the National Science Foundation for the support of research in inductive probability.

are finite. X knows the following: if he were to carry out the act A_m and if the state S_n were the actual state of nature, then the outcome would be $O_{m,n}$. This outcome $O_{m,n}$ is uniquely determined by A_m and S_n; and X knows how it is determined. We assume that there is a utility function U_X for the person X and that X knows his utility function so that he can use it in order to calculate subjective values.

Now we define the *subjective value* of a possible act A_m for X at time T:

(1) DEFINITION.

$$V_{X,T}(A_m) = \sum_n U_X(O_{m,n}) \times P(S_n),$$

where $P(S_n)$ is the probability of the state S_n, and the sum covers all possible states S_n.

In other words, we take as the subjective value of the act A_m for X the *expected utility* of the outcome of this act. (1) holds for the time T before any act is carried out. It refers to the contemplated act A_m; therefore it uses the utilities for the possible outcomes $O_{m,n}$ of act A_m in the various possible states S_n. [If the situation is such that the probability of S_n could possibly be influenced by the assumption that act A_m were carried out, we should take the conditional probability $P(S_n|A_m)$ instead of $P(S_n)$. Analogous remarks hold for our later forms of the definition of V].

We can now formulate the customary *decision principle* as follows:

(2) Choose an act so as to maximize the subjective value V.

This principle can be understood either as referring to *actual* decision making, or to *rational* decisions. In the first interpretation it would be a psychological law belonging to *empirical* decision theory as a branch of psychology; in the second interpretation, it would be a normative principle in the theory of *rational* decisions. I shall soon come back to this distinction. First we have to remove an ambiguity in the definition (1) of value, concerning the interpretation of the probability P. There are several conceptions of probability; thus the question arises which of them is adequate in the context of decision making.

The main conceptions of probability are often divided into two kinds, objectivistic and subjectivistic conceptions. In my view, these are not two incompatible doctrines concerning the same concept, but rather two theories concerning two different probability concepts, both of them legitimate and useful. The concept of *objective* (or statistical) *probability* is closely connected with relative frequencies in mass phenomena. It plays an important role in mathematical statistics, and it occurs in laws of various branches of empirical science, especially physics.

The second concept is *subjective* (or personal) *probability*. It is the probability assigned to a proposition or event H by a subject X, say a person or a group of persons, in other words, the degree of belief of X in H. Now it seems to me that we should clearly distinguish two versions of subjective probability, one representing the *actual* degree of belief and the other the *rational* degree of belief.

Which of these two concepts of probability, the objective or the subjective, ought to be used in the definition of subjective value and thereby in the decision principle? At the present time, the great majority of those who work in mathematical statistics still regard the statistical concept of probability as the only legitimate one. However, this concept refers to an objective feature of nature; a feature that holds whether or not the observer X knows about it. And in fact, the numerical values of statistical probability are in general not known to X. Therefore this concept is unsuitable for a decision principle. It seems that for this reason a number of those who work in the theory of decisions, be it actual decisions or rational decisions, incline toward the view that some version of the subjective concept of probability must be used here. I agree emphatically with this view.

The statistical concept of probability remains, of course, a legitimate and important concept both for mathematical statistics and for many branches of empirical science. And in the special case that X knows the statistical probabilities for the relevant states S_n but has no more specific knowledge about these states, the decision principle would use these values. There is general agreement on this point. And this is not in conflict with the view that the decision principle should refer to subjective probability, because in this special situation the subjective probability for X would be equal to the objective probability.

Once we recognize that decision theory needs the subjective concept of probability, it is clear that the theory of *actual* decisions involves the first version of this concept, i.e., the *actual* degree of belief, and the theory of *rational* decisions involves the second version, the *rational* degree of belief.

Let us first discuss the theory of *actual* decisions. The concept of probability in the sense of the *actual* degree of belief is a psychological concept; its laws are empirical laws of psychology, to be established by the investigation of the behavior of persons in situations of uncertainty, e.g., behavior with respect to bets or games of chance. I shall use for this psychological concept the technical term *"degree of credence"* or shortly *"credence"*. In symbols, I write '$Cr_{X,T}(H)$' for "the (degree of) credence of the proposition H for the person X at the time T". Different persons X and Y may have different credence functions $Cr_{X,T}$ and $Cr_{Y,T}$. And the same person X may have different credence functions Cr_{X,T_1} and Cr_{X,T_2} at different times T_1 and T_2; e.g., if X observes between T_1 and T_2 that H holds, then $Cr_{X,T_1}(H) \neq Cr_{X,T_2}(H)$. (Let the ultimate possible cases be represented by the points of a logical space, usually called the probability space. Then a proposition or event is understood, not as a sentence, but as the range of a sentence, i.e., the set of points representing those possible cases in which the sentence holds. To the conjunction of two sentences corresponds the intersection of the propositions.)

On the basis of credence, we can define *conditional credence*, "the credence of H with respect to the proposition E" (or "$\cdots$ given E"):

(3) Definition.

$$Cr'_{X,T}(H|E) = \frac{Cr_{X,T}(E \cap H)}{Cr_{X,T}(E)},$$

provided that $Cr_{X,T}(E) > 0$. $Cr'_{X,T}(H|E)$ is the credence which H would have for X at T if X ascertained that E holds.

Using the concept of credence, we now replace (1) by the following:

(4) Definition.

$$V_{X,T}(A_m) = \sum_n U_X(O_{m,n}) \times Cr_{X,T}(S_n).$$

As was pointed out by Ramsey, we can determine X's credence function by his betting behavior. A bet is a contract of the following form. X pays into the pool the amount u, his partner Y pays the amount v; they agree that the total stake $u+v$ goes to X if the hypothesis H turns out to be true, and to Y if it turns out to be false. If X accepts this contract, we say that he bets on H with the total stake $u+v$ and with the betting quotient $q = u/(u+v)$ (or, at odds of u to v). If we apply the decision principle with the definition (4) to the situation in which X may either accept or reject an offered bet on H with the betting quotient q, we find that X will accept the bet if q is not larger than his credence for H. Thus we may interpret $Cr_{X,T}(H)$ as the highest betting quotient at which X is willing to bet on H. (As is well known, this holds only under certain conditions and only approximately.)

Utility and credence are psychological concepts. The utility function of X represents the system of valuations and preferences of X; his credence function represents his system of beliefs (not only the content of each belief, but also its strength). Both concepts are theoretical concepts which characterize the state of mind of a person; more exactly, the non-observable micro-state of his central nervous system, not his consciousness, let alone his overt behavior. But since his behavior is influenced by his state, we can indirectly determine characteristics of his state from his behavior. Thus experimental methods have been developed for the determination of some values and some general characteristics of the utility function and the credence function ("subjective probability") of a person on the basis of his behavior with respect to bets and similar situations. Interesting investigations of this kind have been made by F. Mosteller and P. Nogee [13], and more recently by D. Davidson and P. Suppes [4], and others.

Now we take the step from empirical to *rational decision theory*. The latter is of greater interest to us, not so much for its own sake (its methodological status is in fact somewhat problematic), but because it is the connecting link between empirical decision theory and inductive logic. Rational decision theory is concerned not with actual credence, but with *rational* credence. (We should also distinguish here between actual utility and rational utility; but we will omit this.) The statements of a theory of this kind are not found by experiments but are established on the basis of requirements of rationality; the formal procedure usually consists in deducing theorems from axioms which are justified by general considerations of rationality, as we shall see. It seems fairly clear that the probability concepts used by the following authors are meant in the sense of rational credence (or rational credibility, which I shall explain presently): John Maynard Keynes (1921), Frank P. Ramsey (1928), Harold Jeffreys (1931), B. O. Koopman (1940), Georg Henrik von Wright (1941), I. G. Good (1950), and Leonard J. Savage (1954). I am inclined to include here also those authors who do not declare explicitly that their concept refers to rational rather than actual beliefs, but who accept general axioms and do not base their theories on psychological results. Bruno De Finetti (1931) satisfies these conditions; however, he says explicitly that his concept of "subjective probability" refers not to rational, but to actual beliefs. I find this puzzling.

The term "subjective probability" seems quite satisfactory for the actual degree of credence. It is frequently applied also to a probability concept interpreted as something like rational credence. But here the use of the word "subjective" might be misleading (comp. Keynes [9, p. 4] and Carnap [1, § 12A]). Savage has suggested the term "personal probability".

Rational credence is to be understood as the credence function of a completely rational person X; this is, of course, not any real person, but an imaginary, idealized person. We carry out the idealization step for step, by introducing *requirements of rationality* for the credence function. I shall now explain some of these requirements.

Suppose that X makes n simultaneous bets; let the ith bet $(i = 1, \cdots, n)$ be on the proposition H_i with the betting quotient q_i and the total stake s_i. Before we observe which of the propositions H_i are true and which are false, we can consider the *possible* cases. For any possible case, i.e., a logically possible distribution of truth-values among the H_i, we can calculate the gain or loss for each bet and hence the total balance of gains and losses from the n bets. If in *every* possible case X suffers a net loss, i.e., his total balance is negative, it is obviously unreasonable for X to make these n bets. Let X's credence function at a given time be Cr. By a (finite) betting system in accordance with Cr we mean a finite system of n bets on n arbitrary propositions H_i $(i = 1, \cdots, n)$ with n arbitrary (positive) stakes s_i, but with the betting quotients $q_i = Cr(H_i)$.

(5) DEFINITION. *A function Cr is coherent if and only if there is no betting system in accordance with Cr such that there is a net loss in every possible case.*

For X to make bets of a system of this kind would obviously be unreasonable. Therefore we lay down the *first requirement* as follows:

R1. *In order to be rational, Cr must be coherent.*

Now the following important result holds:

(6) A function Cr from propositions to real numbers is coherent if and only if Cr is a normalized probability measure.

(A real-valued function of propositions is said to be a probability measure if it is a non-negative, finitely additive set function; it is normalized if its value for the necessary proposition is 1. In other words, a normalized probability measure is a function which satisfies the basic axioms of the calculus of probability, e.g., the axioms I through V in Kolmogoroff's system [10, § 1].)

The first part of (6) ("... coherent if ...") was stated first by Ramsey [15] and was later independently stated and proved by De Finetti [5]. The much more complicated proof for the second part ("... only if ...") was found independently by John G. Kemeny [8, p. 269] and R. Sherman Lehman [12, p. 256].

Let Cr' be the conditional credence function defined on the basis of Cr by (3). As ordinary bets are based on Cr, conditional bets are based on Cr'. The concept of coherence can be generalized so as to be applicable also to conditional credence functions. (6) can then easily be extended by the result that a conditional credence function Cr' is coherent if and only if Cr' is a normalized conditional probability measure, in other words, if and only if Cr' satisfies the customary basic axioms of conditional probability, including the general multiplication axiom.

Following Shimony [17], we introduce now a concept of coherence in a stronger sense, for which I use the term "strict coherence":

(7) DEFINITION. *A function Cr is strictly coherent if and only if Cr is coherent and there is no (finite) system of bets in accordance with Cr on molecular propositions such that the result is a net loss in at least one possible case, but not a net gain in any possible case.*

It is clear that it would be unreasonable to make a system of bets of the kind just specified. Therefore we lay down the *second requirement*:

R2. *In order to be rational, a credence function must be strictly coherent.*

We define *regular credence function* (essentially in the sense of Carnap [1, § 55A]):

(8) DEFINITION. *A function Cr is regular if and only if Cr is a normalized probability measure and, for any molecular proposition H, Cr(H) = 0 only if H is impossible.*

By analogy with (6) we have now the following important theorem; its first part is due to Shimony, its second part again to Kemeny and Lehman:

(9) A function Cr is strictly coherent if and only if Cr is regular.

Most of the authors of systems for subjective or logical probability adopt only the basic axioms; thus they require nothing but coherence. A few go one step further by including an axiom for what I call regularity; thus they require in effect strict coherence, but nothing more. Axiom systems of both kinds are extremely weak; they yield no result of the form "$P(H|E) = r$", except in the trivial cases where r is 0 or 1. In my view, much more should be required.

The two previous requirements apply to any credence function that holds for X at any time T of his life. We now consider two of these functions, Cr_n for the time T_n and Cr_{n+1} for a time T_{n+1} shortly after T_n. Let the proposition E represent the observation data received by X between these two time points. The *third requirement* refers to the transition from Cr_n to Cr_{n+1}:

R3. (a) *The transformation of Cr_n into Cr_{n+1} depends only on the proposition E.*

(b) *More specifically, Cr_{n+1} is determined by Cr_n and E as follows: for any H, $Cr_{n+1}(H) = Cr_n(E \cap H)/Cr_n(E)$ (hence $= Cr'_n(H|E)$ by definition (3)).*

Part (a) is of course implied by (b). I have separated part (a) from (b) because X's function Cr might satisfy (a) without satisfying (b). Part (a) requires merely that X be rational to the extent that changes in his credence function are influenced only by his observational results, but not by any other factors, e.g., feelings like his hopes or fears concerning a possible future event H, feelings which in fact influence the beliefs of all actual human beings. Part (b) specifies exactly the transformation of Cr_n into Cr_{n+1}; the latter is the conditional credence Cr'_n with respect to E. The rule (b) can be used only if $Cr_n(E) \neq 0$; this condition is fulfilled for any possible observational result, provided that Cr_n satisfies the requirement of strict coherence.

Let the proposition E_{n+2} represent the data obtained between T_{n+1} and a later time point T_{n+2}. Let Cr_{n+2} be the credence function at T_{n+2} obtained by R3b from Cr_{n+1} with respect to E_{n+2}. It can easily be shown that the same function Cr_{n+2} results if R3b is applied to Cr_n with respect to the combined data $E_{n+1} \cap E_{n+2}$. In the same way we can determine any later credence function Cr_{n+m} from the given function Cr_n either in m steps, applying the rule R3b in each step with one datum of the sequence $E_{n+1}, E_{n+2}, \cdots, E_{n+m}$, or in one step with the intersection $\bigcap_{p=1}^{m} E_{n+p}$. If m is large so that the intersection contains thousands of single data, the objection might be raised that it is unrealistic to think of a procedure of this kind, because a man's memory is unable to retain and reproduce at will so many items. However, since our goal is not the psychology of actual human behavior in the field of inductive reasoning, but rather inductive logic as a system of rules, we do not aim at realism. We make the further idealization that X is not only perfectly rational but has also an infallible memory. Our assumptions deviate from reality very much if the observer and agent is a natural human being, but not so much if we think of X as a robot with organs of perception, data processing, decision making, and acting. Thinking about the design of a robot will help us in finding rules of rationality. Once found, these rules can be applied not only in the construction of a robot but also in advising human beings in their effort to make their decisions as rational as their limited abilities permit.

Consider now the whole sequence of data obtained by X up to the present time T_n: $E_1, E_2, \cdots, E_n$. Let K_{X, T_n} or, for short, K_n be the proposition representing the combination of all these data:

(10) DEFINITION.

$$K_n = \bigcap_{i=1}^{n} E_i.$$

Thus K_n represents, under the assumption of infallible memory, the total observational knowledge of X at the time T_n. Now consider the sequence of X's credence functions. In the case of a human being we would hesitate to ascribe to him a credence function at a very early time point, before his abilities of reason and deliberate action are sufficiently developed. But again we disregard this difficulty by thinking either of an idealized human baby or of a robot. We ascribe to him a credence function Cr_1 for the time point T_1; Cr_1 represents X's personal probabilities based upon the datum E_1 as his only experience. Going even one step further, let us ascribe to him an *initial credence function* Cr_0 for the time point T_0 before he obtains his first datum E_1. Any later function Cr_n for a time point T_n is uniquely determined by Cr_0 and K_n:

(11) For any H, $Cr_n(H) = Cr_0'(H|K_n)$, where Cr_0' is the conditional function based on Cr_0.

$Cr_n(H)$ is thus seen to be the *conditional initial credence of H given K_n.* How can we understand the function Cr_0? In terms of the robot, Cr_0 is the credence function that we originally build in and that he transforms step for step, with regard to the incoming data, into the later credence functions. In the case of a human being X, suppose that we find at the time T_n his credence function Cr_n. Then we can, under suitable conditions, reconstruct a sequence $E_1, \cdots, E_n$, the proposition K_n, and a function Cr_0 such that (a) $E_1, \cdots, E_n$ are possible observation data, (b) K_n is defined by (10), (c) Cr_0 satisfies all requirements of rationality for initial credence functions, and (d) the application of (11) to the assumed function Cr_0 and K_n would lead to the ascertained function Cr_n. We do not assert that X actually experienced the data $E_1, \cdots, E_n$, and that he actually had the initial credence function Cr_0, but merely that, under idealized conditions, his function Cr_n could have evolved from Cr_0 by the effect of the data $E_1, \cdots, E_n$.

For the conditional initial credence (Cr_0') we shall also use the term "*credibility*" and the symbol 'Cred'. As an alternative to defining 'Cred' on the basis of 'Cr_0', we could introduce it as a primitive term. In this case we may take the following universal statement as the main postulate for the theoretical primitive term 'Cred':

(12) Let *Cred* be any function from pairs of propositions to real numbers, satisfying all requirements which we have laid down or shall lay down for credibility functions. Let H and A be any propositions (A not empty). Let X be any observer and T any time point. If X's credibility function is *Cred* and his total observational knowledge at T is A, then his credence for H and T is $Cred(H|A)$.

Note that (12) is much more general than (11). There the function *Cred*
(or Cr_0') was applied only to those pairs H, A, in which A is a proposition of
the sequence $K_1, K_2, \cdots$, and thus represents the actual knowledge of X at
some time point. In (12), however, A may be any non-empty proposition.
Let A_1 be a certain proposition which does not occur in the sequence K_1,
$K_2, \cdots$, and H_1 some proposition. Then the statement

$$Cr_T(H_1) = Cred(H_1|A_1)$$

is to be understood as a counterfactual conditional as follows:

(13) If the total knowledge of X at T had been A_1, then his credence for
H_1 at T would have been equal to $Cred(H_1|A_1)$.

This is a true counterfactual based on the postulate (12), analogous to
ordinary counterfactuals based on physical laws.

Applying (12) to X's actual total observational knowledge $K_{X,T}$ at time T,
we have:

(14) For any H, $Cr_{X,T}(H) = Cred_X(H|K_{X,T})$.

Now we can use credibility instead of credence in the definition of the
subjective value of an act A_m, and thereby in the decision rule. Thus we
have instead of (4):

(15) DEFINITION.

$$V_{X,T}(A_m) = \sum_n U_X(O_{m,n}) \times Cred_X(S_n|K_{X,T}).$$

(If the situation is such that the assumption of A_m could possibly change the
credence of S_n, we have to replace '$K_{X,T}$' by '$K_{X,T} \cap A_m$'; see the remark
on (1).)

If *Cred* is taken as primitive, Cr_0 can be defined as follows:

(16) DEFINITION. *For any H, $Cr_0(H) = Cred(H/Z)$, where Z is the
necessary proposition (the tautology).*

This is the special case of (12) for the initial time T_0, when X's knowledge
K_0 is the tautology.

While $Cr_{X,T}$ characterizes the momentary state of X at time T with
respect to his beliefs, his function $Cred_X$ is a trait of his underlying permanent
intellectual character, namely his permanent disposition for forming beliefs
on the basis of his observations.

Since each of the two functions Cr_0 and *Cred* is definable on the basis of
the other one, there are two alternative equivalent procedures for specifying
a basic belief-forming disposition, namely either by Cr_0 or by *Cred*.

Most of those who have constructed systems of subjective or personal
probability (in the narrower sense, in contrast to logical probability), e.g.,
Ramsey, De Finetti, and Savage, have concentrated their attention on what
we might call "adult" credence functions, i.e., those of persons sufficiently
developed to communicate by language, to play games, make bets, etc.,

hence persons with an enormous amount of experience. In empirical decision theory it has great practical advantages to take adult persons as subjects of investigation, since it is relatively easy to determine their credence functions on the basis of their behavior with games, bets, and the like. When I propose to take as a basic concept, not adult credence but either initial credence or credibility, I must admit that these concepts are less realistic and remoter from overt behavior and may therefore appear as elusive and dubious. On the other hand, when we are interested in *rational* decision theory, these concepts have great methodological advantages. Only for these concepts, not for credence, can we find a sufficient number of requirements of rationality as a basis for the construction of a system of inductive logic.

If we look at the development of theories and concepts in various branches of science, we find frequently that it was possible to arrive at powerful laws of great generality only when the development of concepts, beginning with directly observable properties, had progressed step by step to more abstract concepts, connected only indirectly with observables. Thus physics proceeds from concepts describing visible motion of bodies to the concept of a momentary electric force, and then to the still more abstract concept of a permanent electric field. In the sphere of human action we have first concepts describing overt behavior, say of a boy who is offered the choice of an apple or an ice cream cone and takes the latter; then we introduce the concept of an underlying momentary inclination, in this case the momentary preference of ice cream over apple; and finally we form the abstract concept of an underlying permanent disposition, in our example the general utility function of the boy.

What I propose to do is simply to take the same step from momentary inclination to the permanent disposition for forming momentary inclinations also with the second concept occurring in the decision principle, namely, personal probability or degree of belief. This is the step from credence to credibility.

When we wish to judge the morality of a person, we do not simply look at some of his acts, we study rather his character, the system of his moral values, which is part of his utility function. Single acts without knowledge of motives give little basis for a judgment. Similarly, if we wish to judge the rationality of a person's beliefs, we should not simply look at his present beliefs. Beliefs without knowledge of the evidence out of which they arose tell us little. We must rather study the way in which the person forms his beliefs on the basis of evidence. In other words, we should study his credibility function, not simply his present credence function. For example, let X have the evidence E that from an urn containing white and black balls ten balls have been drawn, two of them white and eight black. Let Y have the evidence E' which is similar to E, but with seven balls white and three black. Let H be the prediction that the next ball drawn will be white. Suppose that for both X and Y the credence of H is $\frac{2}{3}$. Then we would judge this same credence $\frac{2}{3}$ to be unreasonable for X, but reasonable for Y. We would condemn a credibility function $Cred$ as non-rational if $Cred(H|E) = \frac{2}{3}$; while the result $Cred(H|E') = \frac{2}{3}$ would be no ground for condemnation.

Suppose X has the credibility function $Cred$, which leads him, on the basis of his knowledge K_n at time T_n to the credence function Cr_n, and thereby, with his utility function U, to the act A_m. If this act seems to us unreasonable in view of his evidence K_n and his utilities, we shall judge that $Cred$ is non-rational. But for such a judgment on $Cred$ it is not necessary that X is actually led to an unreasonable act. Suppose that for E and H as in the above example, K_n contains E and otherwise only evidence irrelevant for H. Then we have $Cr_n(H) = Cred(H|K_n) = Cred(H|E) = \frac{2}{3}$; and this result seems unreasonable on the given evidence. If X bets on H with betting quotient $\frac{2}{3}$, this bet is unreasonable, even if he wins it. But his credence $\frac{2}{3}$ is anyway unreasonable, no matter whether he acts on it or not. It is unreasonable because there are possible situations, no matter whether real or not, in which the result $Cred(H|E) = \frac{2}{3}$ would lead him to an unreasonable act. Furthermore, it is not necessary for our condemnation of the function $Cred$ that it actually leads to unreasonable Cr-values. Suppose that another man X' has the same function $Cred$, but is not led to the unreasonable Cr-value in the example, because he has an entirely different life history, and at no time is his total knowledge either E or a combination of E with data irrelevant for H. Then we would still condemn the function $Cred$ and the man X' characterized by this function. Our argument would be as follows: if the total knowledge of X' had at some time been E, or E together with irrelevant data, then his credence for H would have had the unreasonable value $\frac{2}{3}$. The same considerations hold, of course, for the initial credence function Cr_0 corresponding to the function $Cred$; for, on the basis of any possible knowledge proposition K, Cr_0 and $Cred$ would lead to the same credence function.

The following is an example of a requirement of rationality for Cr_0 (and hence for $Cred$) which has no analogue for credence functions. As we shall see later, this requirement leads to one of the most important axioms of inductive logic. (The term "individual" means "element of the universe of discourse", or "element of the population" in the terminology of statistics.)

R4. *Requirement of symmetry. Let a_i and a_j be two distinct individuals. Let H and H' be two propositions such that H' results from H by taking a_j for a_i and vice versa. Then Cr_0 must be such that $Cr_0(H) = Cr_0(H')$.* (In other words, Cr_0 must be invariant with respect to any finite permutation of individuals.)

This requirement seems indispensable. H and H' have exactly the same logical form; they differ merely by their reference to two distinct individuals. These individuals may happen to be quite different. But since their differences are not known to X at time T_0, they cannot have any influence on the Cr_0-values of H and H'. But suppose that at a later time T_n, X's knowledge K_n contains information E relevant to H and H', say information making H

more probable than H' (as an extreme case, E may imply that H is true and H' is false). Then X's credence function Cr_n at T_n will have different values for H and for H'. Thus it is clear that R4 applies only to Cr_0, but is not generally valid for other credence functions $Cr_n (n > 0)$.

Suppose that X is a robot constructed by us. Because H and H' are alike in all their logical properties, it would be entirely arbitrary and therefore inadmissible for us to assign to them different Cr_0-values.

A function Cr_0 is suitable for being built into a robot only if it fulfills the requirements of rationality; and most of these requirements (e.g., R4 and all those not yet mentioned) apply only to Cr_0 (and $Cred$) but not generally to other credence functions.

Now we are ready to take the step to *inductive logic*. This step consists in the transition from the concepts of the Cr_0-function and the $Cred$-function of an imaginary subject X to corresponding purely logical concepts. The former concepts are quasi-psychological; they are assigned to an imaginary subject X supposed to be equipped with perfect rationality and an unfailing memory; the logical concepts, in contrast, have nothing to do with observers and agents, whether natural or constructed, real or imaginary. For a logical function corresponding to Cr_0, I shall use the symbol '$\mathcal{M}$' and I call such functions (inductive) measure functions or $\mathcal{M}$-functions; for a logical function corresponding to $Cred$, I shall use the symbol '$\mathcal{C}$', and I call these functions (inductive) confirmation functions or $\mathcal{C}$-functions. I read '$\mathcal{C}(H|E)$' as "the degree of confirmation (or briefly "the confirmation") of H with respect to E" (or: "... given E"). An $\mathcal{M}$-function is a function from propositions to real numbers. A $\mathcal{C}$-function is a function from pairs of propositions to real numbers. Any $\mathcal{M}$-function $\mathcal{M}$ is supposed to be defined in a purely logical way, i.e., on the basis of concepts of logic (in the wide sense, including set-theory and hence the whole of pure mathematics). Therefore the value $\mathcal{M}(A)$ for any proposition A depends merely on the logical (set-theoretic) properties of A (which is a set in a probability space) but not on any contingent facts of nature (e.g., the truth of A or of other contingent propositions). Likewise any $\mathcal{C}$-function is supposed to be defined in purely logical terms.

Inductive logic studies those $\mathcal{M}$-functions which correspond to rational Cr_0-functions, and those $\mathcal{C}$-functions which correspond to rational $Cred$-functions. Suppose $\mathcal{M}$ is a logically defined $\mathcal{M}$-function. Let us imagine a subject X whose function Cr_0 corresponds to $\mathcal{M}$, i.e., for every proposition H, $Cr_0(H) = \mathcal{M}(H)$. If we find that Cr_0 violates one of the rationality requirements, say R4, then we would reject this function Cr_0, say for a robot we plan to build. Then we wish also to exclude the corresponding function $\mathcal{M}$ from those treated as admissible in the system of inductive logic we plan to construct. Therefore, we set up axioms of inductive logic about $\mathcal{M}$-functions so that these axioms correspond to the requirements of rationality which we find in the theory of rational decision making about Cr_0-functions.

For example, we shall lay down as the basic axioms of inductive logic those which say that $\mathcal{M}$ is a non-negative, finitely additive, and normalized measure function. These axioms correspond to the requirement R1 of coherence, by virtue of theorem (6). Further we shall have an axiom saying that $\mathcal{M}$ is regular. This axiom corresponds to the requirement R2 of strict coherence by theorem (9).

Then we shall have in inductive logic, in analogy to the requirement R4 of symmetry, the following:

(17) AXIOM OF SYMMETRY. $\mathcal{M}$ *is invariant with respect to any finite permutation of individuals.*

All axioms of inductive logic state relations among values of $\mathcal{M}$ or $\mathcal{C}$ as dependent only upon the logical properties and relations of the propositions involved (with respect to language-systems with specified logical and semantical rules). Inductive logic is the theory based upon these axioms. It may be regarded as a part of logic in view of the fact that the concepts occurring are logical concepts. It is an interesting result that this part of the theory of decision making, namely, the logical theory of the $\mathcal{M}$-functions and the $\mathcal{C}$-functions, can thus be separated from the rest. However, we should note that this logical theory deals only with the abstract, formal aspects of probability, and that the full meaning of (subjective) probability can be understood only in the wider context of decision theory through the connections between probability and the concepts of utility and rational action.

It is important to notice clearly the following distinction. While the *axioms* of inductive logic themselves are formulated in purely logical terms and do not refer to any contingent matters of fact, the *reasons* for our choice of the axioms are not purely logical. For example, when you ask me why I accept the axiom of symmetry (17), then I point out that if X had a Cr_0-function corresponding to an $\mathcal{M}$-function violating (17), then this function Cr_0 would violate R4, and I show that therefore X, in a certain possible knowledge situation, would be led to an unreasonable decision. Thus, in order to give my reasons for the axiom, I move from pure logic to the context of decision theory and speak about beliefs, actions, possible losses, and the like. However, this is not in the field of empirical, but of rational decision theory. Therefore, in giving my reasons, I do not refer to particular empirical results concerning particular agents or particular states of nature and the like. Rather, I refer to a *conceivable* series of observations by X, to conceivable sets of possible acts, of possible states of nature, of possible outcomes of the acts, and the like. These features are characteristic for an analysis of *reasonableness* of a given function Cr_0, in contrast to an investigation of the *successfulness* of the (initial or later) credence function of a given person in the real world. Success depends upon the particular contingent circumstances, rationality does not.

There is a class of axioms of inductive logic which I call *axioms of invariance*. The axiom of symmetry is one of them. Another one says that $\mathcal{M}$

is invariant with respect to any finite permutation of attributes belonging to a family of attributes, e.g., colors, provided these attributes are alike in their logical (including semantical) properties. Still another one says that if E is a proposition about a finite sample from a population, then $\mathcal{M}(E)$ is independent of the size of the population. These and other invariance axioms may be regarded as representing the valid core of the old *principle of indifference* (or principle of insufficient reason). The principle, in its original form, as used by Laplace and other authors in the classical period of the theory of probability, was certainly too strong. It was later correctly criticized by showing that it led to absurd results. However, I believe that the basic idea of the principle is sound. Our task is to restate it by specific restricted axioms.

It seems that most authors on subjective probability do not accept any axioms of invariance. In the case of those authors who take credence as their basic concept, e.g., Ramsey, De Finetti, and Savage, this is inevitable, since the invariance axioms do not hold for general credence functions. In order to obtain a stronger system, it is necessary to take as the basic concept either initial credence or credibility (or other concepts in terms of which these are definable).

When we construct an axiom system for $\mathcal{M}$, then the addition of each new axiom has the effect of excluding certain $\mathcal{M}$-functions. We accept an axiom if we recognize that the $\mathcal{M}$-functions excluded by it correspond to non-rational Cr_0-functions. Even on the basis of all axioms which I would accept at the present time for a simple qualitative language (with one-place predicates only, without physical magnitudes), the number of admissible $\mathcal{M}$-functions, i.e., those which satisfy all accepted axioms, is still infinite; but their class is immensely smaller than that of all coherent $\mathcal{M}$-functions. There will presumably be further axioms, justified in the same way by considerations of rationality. We do not know today whether in this future development the number of admissible $\mathcal{M}$-functions will always remain infinite or will become finite and possibly even be reduced to one. Therefore, at the present time I do not assert that there is only one rational Cr_0-function.

I think that the theory of the $\mathcal{M}$- and $\mathscr{C}$-functions deserves the often misused name of *"inductive logic"*. Earlier I gave my reasons for regarding this theory as a part of logic. The epithet "inductive" seems appropriate because this theory provides the foundation for inductive reasoning (in a wide sense). I agree in this view with John Maynard Keynes and Harold Jeffreys. However, it is important that we recognize clearly the essential form of inductive reasoning. It seems to me that the view of almost all writers on induction in the past and including the great majority of contemporary writers, contains one basic mistake. They regard inductive reasoning as an *inference* leading from some known propositions, called the premises or evidence, to a new proposition, called the conclusion, usually a law or a singular prediction. From this point of view the result of any particular inductive reasoning is the *acceptance* of a new proposition (or its rejection, or its suspension until further evidence is found, as the case may be). This seems to me wrong. On the basis of this view it would be impossible to refute Hume's dictum that there are no rational reasons for induction. Suppose that I find in earlier weather reports that a weather situation like the one we have today has occurred one hundred times and that it was followed each time by rain the next morning. According to the customary view, on the basis of this evidence the "inductive method" entitles me to accept the prediction that it will rain tomorrow morning. (If you demur because the number one hundred is too small, change it to one hundred thousand or any number you like.) I would think instead that inductive reasoning about a proposition should lead, not to acceptance or rejection, but to the assignment of a number to the proposition, viz., its $\mathscr{C}$-value. This difference may perhaps appear slight; in fact, however, it is essential. If, in accordance with the customary view, we accept the prediction, then Hume is certainly right in protesting that we have no rational reason for doing so, since, as everybody will agree, it is still possible that it will not rain tomorrow.

If, on the other hand, we adopt the new view of the nature of inductive reasoning, then the situation is quite different. In this case X does not assert the hypothesis H in question, e.g., the prediction "it will rain tomorrow"; he asserts merely the following statements:

(18) (a) At the present moment T_n, the totality of X's observation results is K_n.

(b) $\mathscr{C}(H|K_n) = 0.8$.

(c) $Cred_X(H|K_n) = 0.8$.

(d) $Cr_{X, T_n}(H) = 0.8$.

(a) is the statement of the evidence at hand, the same as in the first case. But now, instead of accepting H, X asserts the statement (c) of the $Cred$-value for H on his evidence. (c) is the result of X's inductive reasoning. Against this result Hume's objection does not hold, because X can give rational reasons for it. (c) is derived from (b) because X has chosen the function $\mathscr{C}$ as his credibility function. (b) is an analytic statement based on the definition of $\mathscr{C}$. X's choice of $\mathscr{C}$ was guided by the axioms of inductive logic. And for each of the axioms we can give reasons, namely, rationality requirements for credibility functions. Thus $\mathscr{C}$ represents a reasonable credibility function. Finally, X's credence value (d) is derived from (c) by (14).

Now some philosophers, including some of my empiricist friends, would raise the following objection. If the result of inductive reasoning is merely an analytic statement (like (b) or (c)), then induction cannot fulfill its purpose of guiding our practical decisions. As a basis for a decision we need a statement with factual content. If the prediction H itself is not available, then we must use a statement of the *objective* probability of H. In answer to this objection I would first point out that X has a factual basis in his evidence, as stated in (a). And for the determination of a rational decision neither the acceptance of H nor knowledge of the objective probability of H

is needed. The rational subjective probability, i.e., the credence as stated in (d), is sufficient for determining first the rational subjective value of each possible act by (15), and then a rational decision. Thus in our example, in view of (b) X would decide to make a bet on rain tomorrow if it were offered to him at odds of four to one or less, but not more.

The old puzzle of induction consists in the following dilemma. On the one hand we see that inductive reasoning is used by the scientist and the man in the street every day without apparent scruples; and we have the feeling that it is valid and indispensable. On the other hand, once Hume awakens our intellectual conscience, we find no answer to his objection. Who is right, the man of common sense or the critical philosopher? We see that, as so often, both are partially right. Hume's criticism of the customary forms of induction was correct. But still the basic idea of common sense thinking is vindicated: induction, if properly reformulated, can be shown to be valid by rational criteria.

REFERENCES

[1] CARNAP, R. *Logical foundations of probability.* Chicago, 1950.

[2] CARNAP, R. *The continuum of inductive methods.* Chicago, 1952.

[3] CARNAP, R. Inductive logic and rational decisions. (This is an expanded version of the present paper.) To appear as the introductory article in: *Studies in probability and inductive logic,* Vol. I, R. Carnap, ed. Forthcoming.

[4] DAVIDSON, D., and P. SUPPES. *Decision making: An experimental approach.* Stanford, 1957.

[5] DE FINETTI, B. La prévision: ses lois logiques, ses sources subjectives. *Annales de l'Institut Henri Poincaré,* Vol. 7 (1937), pp. 1–68.

[6] GOOD, I. J. *Probability and the weighing of evidence.* London and New York, 1950.

[7] JEFFREYS, H. *Theory of probability.* Oxford (1939), 2nd ed. 1948.

[8] KEMENY, J. Fair bets and inductive probabilities. *Journal of Symbolic Logic,* Vol. 20 (1955), pp. 263–273.

[9] KEYNES, J. M. *A treatise on probability.* London and New York, 1921.

[10] KOLMOGOROFF, A. N. *Foundations of the theory of probability.* New York (1950), 2nd ed. 1956.

[11] KOOPMAN, B. O. The bases of probability. *Bull. Amer. Math. Soc.,* Vol. 46 (1940), pp. 763–774.

[12] LEHMAN, R. S. On confirmation and rational betting. *Journal of Symbolic Logic,* Vol. 20 (1955), pp. 251–262.

[13] MOSTELLER, F. C., and P. NOGEE. An experimental measurement of utility. *Journal of Political Economy,* Vol. 59 (1951), pp. 371–404.

[14] NEUMANN, J. VON, and O. MORGENSTERN. *Theory of games and economic behavior.* Princeton (1944), 2nd ed. 1947.

[15] RAMSEY, F. P. *The foundations of mathematics and other logical essays.* London and New York, 1931.

[16] SAVAGE, L. J. *The foundations of statistics.* New York and London, 1954.

[17] SHIMONY, A. Coherence and the axioms of confirmation. *Journal of Symbolic Logic,* Vol. 20 (1955), pp. 1–28.

[18] WRIGHT, G. H. VON. *The logical problem of induction.* Oxford and New York (1941), 2nd ed., 1957.

STUDIES IN THE LOGIC OF EXPLANATION

CARL G. HEMPEL AND PAUL OPPENHEIM[1]

§1. *Introduction.*

To explain the phenomena in the world of our experience, to answer the question "why?" rather than only the question "what?", is one of the foremost objectives of all rational inquiry; and especially, scientific research in its various branches strives to go beyond a mere description of its subject matter by providing an explanation of the phenomena it investigates. While there is rather general agreement about this chief objective of science, there exists considerable difference of opinion as to the function and the essential characteristics of scientific explanation. In the present essay, an attempt will be made to shed some light on these issues by means of an elementary survey of the basic pattern of scientific explanation and a subsequent more rigorous analysis of the concept of law and of the logical structure of explanatory arguments.

The elementary survey is presented in Part I of this article; Part II contains an analysis of the concept of emergence; in Part III, an attempt is made to exhibit and to clarify in a more rigorous manner some of the peculiar and perplexing logical problems to which the familiar elementary analysis of explanation gives rise. Part IV, finally, is devoted to an examination of the idea of explanatory power of a theory; an explicit definition, and, based on it, a formal theory of this concept are developed for the case of a scientific language of simple logical structure.

PART I. ELEMENTARY SURVEY OF SCIENTIFIC EXPLANATION

§2. *Some illustrations.*

A mercury thermometer is rapidly immersed in hot water; there occurs a temporary drop of the mercury column, which is then followed by a swift rise. How is this phenomenon to be explained? The increase in temperature affects at first only the glass tube of the thermometer; it expands and thus provides a larger space for the mercury inside, whose surface therefore drops. As soon as by heat conduction the rise in temperature reaches the mercury, however, the latter expands, and as its coefficient of expansion is considerably larger than that of

[1] This paper represents the outcome of a series of discussions among the authors; their individual contributions cannot be separated in detail. The technical developments contained in Part IV, however, are due to the first author, who also put the article into its final form.

Some of the ideas presented in Part II were suggested by our common friend, Kurt Grelling, who, together with his wife, became a victim of Nazi terror during the war. Those ideas were developed by Grelling, in a discussion by correspondence with the present authors, of emergence and related concepts. By including at least some of that material, which is indicated in the text, in the present paper, we feel that we are realising the hope expressed by Grelling that his contributions might not entirely fall into oblivion.

We wish to express our thanks to Dr. Rudolf Carnap, Dr. Herbert Feigl, Dr. Nelson Goodman, and Dr. W. V. Quine for stimulating discussions and constructive criticism.

275

glass, a rise of the mercury level results.—This account consists of statements of two kinds. Those of the first kind indicate certain conditions which are realized prior to, or at the same time as, the phenomenon to be explained; we shall refer to them briefly as antecedent conditions. In our illustration, the antecedent conditions include, among others, the fact that the thermometer consists of a glass tube which is partly filled with mercury, and that it is immersed into hot water. The statements of the second kind express certain general laws; in our case, these include the laws of the thermic expansion of mercury and of glass, and a statement about the small thermic conductivity of glass. The two sets of statements, if adequately and completely formulated, explain the phenomenon under consideration: They entail the consequence that the mercury will first drop, then rise. Thus, the event under discussion is explained by subsuming it under general laws, i.e., by showing that it occurred in accordance with those laws, by virtue of the realization of certain specified antecedent conditions.

Consider another illustration. To an observer in a row boat, that part of an oar which is under water appears to be bent upwards. The phenomenon is explained by means of general laws—mainly the law of refraction and the law that water is an optically denser medium than air—and by reference to certain antecedent conditions—especially the facts that part of the oar is in the water, part in the air, and that the oar is practically a straight piece of wood.—Thus, here again, the question "*Why* does the phenomenon happen?" is construed as meaning "according to what general laws, and by virtue of what antecedent conditions does the phenomenon occur?"

So far, we have considered exclusively the explanation of particular events occurring at a certain time and place. But the question "Why?" may be raised also in regard to general laws. Thus, in our last illustration, the question might be asked: Why does the propagation of light conform to the law of refraction? Classical physics answers in terms of the undulatory theory of light, i.e. by stating that the propagation of light is a wave phenomenon of a certain general type, and that all wave phenomena of that type satisfy the law of refraction. Thus, the explanation of a general regularity consists in subsuming it under another, more comprehensive regularity, under a more general law.—Similarly, the validity of Galileo's law for the free fall of bodies near the earth's surface can be explained by deducing it from a more comprehensive set of laws, namely Newton's laws of motion and his law of gravitation, together with some statements about particular facts, namely the mass and the radius of the earth.

§3. The basic pattern of scientific explanation.

From the preceding sample cases let us now abstract some general characteristics of scientific explanation. We divide an explanation into two major constituents, the explanandum and the explanans[2]. By the explanandum, we

---

[2] These two expressions, derived from the Latin *explanare*, were adopted in preference to the perhaps more customary terms "explicandum" and "explicans" in order to reserve the latter for use in the context of explication of meaning, or analysis. On explication in this sense, cf. Carnap, [Concepts], p. 513.—Abbreviated titles in brackets refer to the bibliography at the end of this article.

understand the sentence describing the phenomenon to be explained (not that phenomenon itself); by the explanans, the class of those sentences which are adduced to account for the phenomenon. As was noted before, the explanans falls into two subclasses; one of these contains certain sentences $C_1, C_2, \cdots, C_k$ which state specific antecedent conditions; the other is a set of sentences $L_1, L_2, \cdots L_r$ which represent general laws.

If a proposed explanation is to be sound, its constituents have to satisfy certain conditions of adequacy, which may be divided into logical and empirical conditions. For the following discussion, it will be sufficient to formulate these requirements in a slightly vague manner; in Part III, a more rigorous anlysis and a more precise restatement of these criteria will be presented.

I. *Logical conditions of adequacy.*

(R1) The explanandum must be a logical consequence of the explanans; in other words, the explanandum must be logically deducible from the information contained in the explanans, for otherwise, the explanans would not constitute adequate grounds for the explanandum.

(R2) The explanans must contain general laws, and these must actually be required for the derivation of the explanandum.—We shall not make it a necessary condition for a sound explanation, however, that the explanans must contain at least one statement which is not a law; for, to mention just one reason, we would surely want to consider as an explanation the derivation of the general regularities governing the motion of double stars from the laws of celestial mechanics, even though all the statements in the explanans are general laws.

(R3) The explanans must have empirical content; i.e., it must be capable, at least in principle, of test by experiment or observation.—This condition is implicit in (R1); for since the explanandum is assumed to describe some empirical phenomenon, it follows from (R1) that the explanans entails at least one consequence of empirical character, and this fact confers upon it testability and empirical content. But the point deserves special mention because, as will be seen in §4, certain arguments which have been offered as explanations in the natural and in the social sciences violate this requirement.

II. *Empirical condition of adequacy.*

(R4) The sentences constituting the explanans must be true.

That in a sound explanation, the statements constituting the explanans have to satisfy some condition of factual correctness is obvious. But it might seem more appropriate to stipulate that the explanans has to be highly confirmed by all the relevant evidence available rather than that it should be true. This stipulation however, leads to awkward consequences. Suppose that a certain phenomenon was explained at an earlier stage of science, by means of an explanans which was well supported by the evidence then at hand, but which had been highly disconfirmed by more recent empirical findings. In such a case, we would have to say that originally the explanatory account was a correct explanation, but that it ceased to be one later, when unfavorable evidence was discovered. This does not appear to accord with sound common usage, which directs us to say that on the basis of the limited initial evidence, the truth of the explanans, and thus the soundness of the explanation, had been quite probable, but that the ampler evidence now available made it highly probable that the explanans was not true, and hence that the account in question was not—and had never been—a correct explanation. (A similar point will be made and illustrated, with respect to the requirement of truth for laws, in the beginning of §6.)

Some of the characteristics of an explanation which have been indicated so far may be summarized in the following schema:

$$
\text{Logical deduction}
\left[
\begin{array}{l}
\left\{
\begin{array}{ll}
C_1, C_2, \cdots, C_k & \text{Statements of antecedent} \\
 & \text{conditions} \\
L_1, L_2, \cdots, L_r & \text{General Laws}
\end{array}
\right\} \text{Explanans} \\
\hline
\quad E \qquad \text{Description of the} \\
\qquad\qquad \text{empirical phenomenon} \\
\qquad\qquad \text{to be explained}
\end{array}
\right\} \text{Explanandum}
$$

Let us note here that the same formal analysis, including the four necessary conditions, applies to scientific prediction as well as to explanation. The difference between the two is of a pragmatic character. If E is given, i.e. if we know that the phenomenon described by E has occurred, and a suitable set of statements C_1, C_2, $\cdots$, C_k, L_1, L_2, $\cdots$, L_r is provided afterwards, we speak of an explanation of the phenomenon in question. If the latter statements are given and E is derived prior to the occurrence of the phenomenon it describes, we speak of a prediction. It may be said, therefore, that an explanation is not fully adequate unless its explanans, if taken account of in time, could have served as a basis for predicting the phenomenon under consideration.[2a]—Consequently, whatever will be said in this article concerning the logical characteristics of explanation or prediction will be applicable to either, even if only one of them should be mentioned.

It is this potential predictive force which gives scientific explanation its importance: only to the extent that we are able to explain empirical facts can we attain the major objective of scientific research, namely not merely to record the phenomena of our experience, but to learn from them, by basing upon them theoretical generalizations which enable us to anticipate new occurrences and to control, at least to some extent, the changes in our environment.

Many explanations which are customarily offered, especially in pre-scientific discourse, lack this predictive character, however. Thus, it may be explained

---

[2a] The logical similarity of explanation and prediction, and the fact that one is directed towards past occurrences, the other towards future ones, is well expressed in the terms "postdictability" and "predictability" used by Reichenbach in [Quantum Mechanics], p. 13.

that a car turned over on the road "because" one of its tires blew out while the car was travelling at high speed. Clearly, on the basis of just this information, the accident could not have been predicted, for the explanans provides no explicit general laws by means of which the prediction might be effected, nor does it state adequately the antecedent conditions which would be needed for the prediction.—The same point may be illustrated by reference to W. S. Jevons's view that every explanation consists in pointing out a resemblance between facts, and that in some cases this process may require no reference to laws at all and "may involve nothing more than a single identity, as when we explain the appearance of shooting stars by showing that they are identical with portions of a comet".[1] But clearly, this identity does not provide an explanation of the phenomenon of shooting stars unless we presuppose the laws governing the development of heat and light as the effect of friction. The observation of similarities has explanatory value only if it involves at least tacit reference to general laws.

In some cases, incomplete explanatory arguments of the kind here illustrated suppress parts of the explanans simply as "obvious"; in other cases, they seem to involve the assumption that while the missing parts are not obvious, the incomplete explanans could at least, with appropriate effort, be so supplemented as to make a strict derivation of the explanandum possible. This assumption may be justifiable in some cases, as when we say that a lump of sugar disappeared "because" it was put into hot tea, but it is surely not satisfied in many other cases. Thus, when certain peculiarities in the work of an artist are explained as outgrowths of a specific type of neurosis, this observation may contain significant clues, but in general it does not afford a sufficient basis for a potential prediction of those peculiarities. In cases of this kind, an incomplete explanation may at best be considered as indicating some positive correlation between the antecedent conditions adduced and the type of phenomenon to be explained, and as pointing out a direction in which further research might be carried on in order to complete the explanatory account.

The type of explanation which has been considered here so far is often referred to as causal explanation. If E describes a particular event, then the antecedent circumstances described in the sentences $C_1, C_2, \cdots, C_k$ may be said jointly to "cause" that event, in the sense that there are certain empirical regularities, expressed by the laws $L_1, L_2, \cdots, L_r$, which imply that whenever conditions of the kind indicated by $C_1, C_2, \cdots, C_k$ occur, an event of the kind described in E will take place. Statements such as $L_1, L_2, \cdots, L_r$, which assert general and unexceptional connections between specified characteristics of events, are customarily called causal, or deterministic, laws. They are to be distinguished from the so-called statistical laws which assert that in the long run, an explicitly stated percentage of all cases satisfying a given set of conditions are accompanied by an event of a certain specified kind. Certain cases of scientific explanation involve "subsumption" of the explanandum under a set of laws of which at least some are statistical in character. Analysis of the peculiar logical structure

[1] [Principles], p. 533.

of that type of subsumption involves difficult special problems. The present essay will be restricted to an examination of the causal type of explanation, which has retained its significance in large segments of contemporary science, and even in some areas where a more adequate account calls for reference to statistical laws.[4]

§4. Explanation in the non-physical sciences. Motivational and teleological approaches.

Our characterization of scientific explanation is so far based on a study of cases taken from the physical sciences. But the general principles thus obtained apply also outside this area.[5] Thus, various types of behavior in laboratory animals and in human subjects are explained in psychology by subsumption under laws or even general theories of learning or conditioning; and while frequently, the regularities invoked cannot be stated with the same generality and precision as in physics or chemistry, it is clear, at least, that the general character of those explanations conforms to our earlier characterization.

Let us now consider an illustration involving sociological and economic factors. In the fall of 1946, there occurred at the cotton exchanges of the United States a price drop which was so severe that the exchanges in New York, New Orleans, and Chicago had to suspend their activities temporarily. In an attempt to explain this occurrence, newspapers traced it back to a large-scale speculator in New Orleans who had feared his holdings were too large and had therefore begun to liquidate his stocks; smaller speculators had then followed his example

---

[4] The account given above of the general characteristics of explanation and prediction in science is by no means novel; it merely summarizes and states explicitly some fundamental points which have been recognized by many scientists and methodologists.

Thus, e.g., Mill says:"An individual fact is said to be explained by pointing out its cause, that is, by stating the law or laws of causation of which its production is an instance", and "a law of uniformity in nature is said to be explained when another law or laws are pointed out, of which that law itself is but a case, and from which it could be deduced." ([Logic], Book III, Chapter XII, section 1). Similarly, Jevons, whose general characterization of explanation was critically discussed above, stresses that "the most important process of explanation consists in showing that an observed fact is one case of a general law or tendency." ([Principles], p. 533). Ducasse states the same point as follows: "Explanation essentially consists in the offering of a hypothesis of fact, standing to the fact to be explained as case of antecedent to case of consequent of some already known law of connection." ([Explanation], pp. 150-51). A lucid analysis of the fundamental structure of explanation and prediction was given by Popper in [Forschung], section 12, and, in an improved version, in his work [Society], especially in Chapter 25 and in note 7 referring to that chapter.—For a recent characterization of explanation as subsumption under general theories, cf., for example, Hull's concise discussion in [Principles], chapter I. A clear elementary examination of certain aspects of explanation is given in Hospers, [Explanation], and a concise survey of many of the essentials of scientific explanation which are considered in the first two parts of the present study may be found in Feigl, [Operationism], pp. 284 ff.

[5] On the subject of explanation in the social sciences, especially in history, cf. also the following publications, which may serve to supplement and amplify the brief discussion to be presented here: Hempel, [Laws]; Popper, [Society]; White, [Explanation]; and the articles *Cause* and *Understanding* in Beard and Hook, [Terminology].

in a panic and had thus touched off the critical decline. Without attempting to assess the merits of the argument, let us note that the explanation here suggested again involves statements about antecedent conditions and the assumption of general regularities. The former include the facts that the first speculator had large stocks of cotton, that there were smaller speculators with considerable holdings, that there existed the institution of the cotton exchanges with their specific mode of operation, etc. The general regularities referred to are—as often in semi-popular explanations—not explicitly mentioned; but there is obviously implied some form of the law of supply and demand to account for the drop in cotton prices in terms of the greatly increased supply under conditions of practically unchanged demand; besides, reliance is necessary on certain regularities in the behavior of individuals who are trying to preserve or improve their economic position. Such laws cannot be formulated at present with satisfactory precision and generality, and therefore, the suggested explanation is surely incomplete, but its intention is unmistakably to account for the phenomenon by integrating it into a general pattern of economic and socio-psychological regularities.

We turn to an explanatory argument taken from the field of linguistics.[6] In Northern France, there exist a large variety of words synonymous with the English "bee," whereas in Southern France, essentially only one such word is in existence. For this discrepancy, the explanation has been suggested that in the Latin epoch, the South of France used the word "apicula", the North the word "apis". The latter, because of a process of phonologic decay in Northern France, became the monosyllabic word "é"; and monosyllables tend to be eliminated, especially if they contain few consonantic elements, for they are apt to give rise to misunderstandings. Thus, to avoid confusion, other words were selected. But "apicula", which was reduced to "abelho", remained clear enough and was retained, and finally it even entered into the standard language, in the form "abbeille". While the explanation here described is incomplete in the sense characterized in the previous section, it clearly exhibits reference to specific antecedent conditions as well as to general laws.[7]

While illustrations of this kind tend to support the view that explanation in biology, psychology, and the social sciences has the same structure as in the physical sciences, the opinion is rather widely held that in many instances, the causal type of explanation is essentially inadequate in fields other than physics and chemistry, and especially in the study of purposive behavior. Let us ex-

---

[6] The illustration is taken from Bonfante, [Semantics], section 3.

[7] While in each of the last two illustrations, certain regularities are unquestionably relied upon in the explanatory argument, it is not possible to argue convincingly that the intended laws, which at present cannot all be stated explicitly, are of a causal rather than a statistical character. It is quite possible that most or all of the regularities which will be discovered as sociology develops will be of a statistical type. Cf., on this point, the suggestive observations by Zilsel in [Empiricism] section 8, and [Laws]. This issue does not affect, however, the main point we wish to make here, namely that in the social no less than in the physical sciences, subsumption under general regularities is indispensable for the explanation and the theoretical understanding of any phenomenon.

amine briefly some of the reasons which have been adduced in support of this view.

One of the most familiar among them is the idea that events involving the activities of humans singly or in groups have a peculiar uniqueness and irre-peatability which makes them inaccessible to causal explanation because the latter, which its reliance upon uniformities, presupposes repeatability of the phenomena under consideration. This argument which, incidentally, has also been used in support of the contention that the experimental method is in-applicable in psychology and the social sciences, involves a misunderstanding of the logical character of causal explanation. Every individual event, in the physical sciences no less than in psychology or the social sciences, is unique in the sense that it, with all its peculiar characteristics, does not repeat itself. Nevertheless, individual events may conform to, and thus be explainable by means of, general laws of the causal type. For all that a causal law asserts is that any event of a specified kind, i.e. any event having certain specified char-acteristics, is accompanied by another event which in turn has certain specified characteristics; for example, that in any event involving friction, heat is de-veloped. And all that is needed for the testability and applicability of such laws is the recurrence of events with the antecedent characteristics, i.e. the repeti-tion of those characteristics, but not of their individual instances. Thus, the argument is inconclusive. It gives occasion, however, to emphasize an important point concerning our earlier analysis: When we spoke of the explanation of a single event, the term "event" referred to the occurrence of some more or less complex characteristic in a specific spatio-temporal location or in a certain indi-vidual object, and not to *all* the characteristics of that object, or to all that goes on in that space-time region.

A second argument that should be mentioned here[*] contends that the es-tablishment of scientific generalizations—and thus of explanatory principles—for human behavior is impossible because the reactions of an individual in a given situation depend not only upon that situation, but also upon the previous history of the individual.—But surely, there is no *a priori* reason why generalizations should not be attainable which take into account this dependence of behavior on the past history of the agent. That indeed the given argument "proves" too much, and is therefore a *non sequitur*, is made evident by the existence of certain physical phenomena, such as magnetic hysteresis and elastic fatigue, in which the magnitude of a specific physical effect depends upon the past history of the system involved, and for which nevertheless certain general regularities have been established.

A third argument insists that the explanation of any phenomenon involving purposive behavior calls for reference to motivations and thus for teleological rather than causal analysis. Thus, for example, a fuller statement of the sug-gested explanation for the break in the cotton prices would have to indicate the large-scale speculator's motivations as one of the factors determining the event

[*] Cf., for example, F. H. Knight's presentation of this argument in [Limitations], pp. 251–52.

in question. Thus, we have to refer to goals sought, and this, so the argument runs, introduces a type of explanation alien to the physical sciences. Unquestionably, many of the—frequently incomplete—explanations which are offered for human actions involve reference to goals and motives; but does this make them essentially different from the causal explanations of physics and chemistry? One difference which suggests itself lies in the circumstance that in motivated behavior, the future appears to affect the present in a manner which is not found in the causal explanations of the physical sciences. But clearly, when the action of a person is motivated, say, by the desire to reach a certain objective, then it is not the as yet unrealized future event of attaining that goal which can be said to determine his present behavior, for indeed the goal may never be actually reached; rather—to put it in crude terms—it is (a) his desire, present before the action, to attain that particular objective, and (b) his belief, likewise present before the action, that such and such a course of action is most likely to have the desired effect. The determining motives and beliefs, therefore, have to be classified among the antecedent conditions of a motivational explanation, and there is no formal difference on this account between motivational and causal explanation.

Neither does the fact that motives are not accessible to direct observation by an outside observer constitute an essential difference between the two kinds of explanation; for also the determining factors adduced in physical explanations are very frequently inaccessible to direct observation. This is the case, for instance, when opposite electric charges are adduced in explanation of the mutual attraction of two metal spheres. The presence of those charges, while eluding all direct observation, can be ascertained by various kinds of indirect test, and that is sufficient to guarantee the empirical character of the explanatory statement. Similarly, the presence of certain motivations may be ascertainable only by indirect methods, which may include reference to linguistic utterances of the subject in question, slips of the pen or of the tongue, etc.; but as long as these methods are "operationally determined" with reasonable clarity and precision, there is no essential difference in this respect between motivational explanation and causal explanation in physics.

A potential danger of explanation by motives lies in the fact that the method lends itself to the facile construction of ex-post-facto accounts without predictive force. It is a widespread tendency to "explain" an action by ascribing it to motives conjectured only after the action has taken place. While this procedure is not in itself objectionable, its soundness requires that (1) the motivational assumptions in question be capable of test, and (2) that suitable general laws be available to lend explanatory power to the assumed motives. Disregard of these requirements frequently deprives alleged motivational explanations of their cognitive significance.

The explanation of an action in terms of the motives of the agent is sometimes considered as a special kind of teleological explanation. As was pointed out above, motivational explanation, if adequately formulated, conforms to the conditions for causal explanation, so that the term "teleological" is a misnomer if it is meant to imply either a non-causal character of the explanation or a peculiar determination of the present by the future. If this is borne in mind, however, the term "teleological" may be viewed, in this context, as referring to causal explanations in which some of the antecedent conditions are motives of the agent whose actions are to be explained.[9]

Teleological explanations of this kind have to be distinguished from a much more sweeping type, which has been claimed by certain schools of thought to be indispensable especially in biology. It consists in explaining characteristics of an organism by reference to certain ends or purposes which the characteristics are said to serve. In contradistinction to the cases examined before, the ends are not assumed here to be consciously or subconsciously pursued by the organism in question. Thus, for the phenomenon of mimicry, the explanation is sometimes offered that it serves the purpose of protecting the animals endowed with it from detection by its pursuers and thus tends to preserve the species. —Before teleological hypotheses of this kind can be appraised as to their potential explanatory power, their meaning has to be clarified. If they are intended somehow to express the idea that the purposes they refer to are inherent in the design of the universe, then clearly they are not capable of empirical test and thus violate the requirement (R3) stated in §3. In certain cases, however, assertions about the purposes of biological characteristics may be translatable into statements in non-teleological terminology which assert that those characteristics function in a specific manner which is essential to keeping the organism alive or to preserving the species.[10] An attempt to state precisely what is meant by this latter assertion—or by the similar one that without those characteristics, and other things being equal, the organism or the species would not survive—encounters considerable difficulties. But these need not be discussed here. For even if we assume that biological statements in teleological form can be adequately translated into descriptive statements about the life-preserving function of certain biological characteristics, it is clear that (1) the use of the concept of purpose is not essential in these contexts, since the term "purpose" can be completely eliminated from the statements in question, and (2) teleological assumptions, while now endowed with empirical content, cannot serve as explanatory principles in the customary contexts. Thus, e.g., the fact that a

[9] For a detailed logical analysis of the character and the function of the motivation concept in psychological theory, see Koch, [Motivation].—A stimulating discussion of teleological behavior from the standpoint of contemporary physics and biology is contained in the article [Teleology] by Rosenblueth, Wiener and Bigelow. The authors propose an interpretation of the concept of purpose which is free from metaphysical connotations, and they stress the importance of the concept thus obtained for a behavioristic analysis of machines and living organisms. While our formulations above intentionally use the crude terminology frequently applied in philosophical arguments concerning the applicability of causal explanation to purposive behavior, the analysis presented in the article referred to is couched in behavioristic terms and avoids reference to "motives" and the like.

[10] An analysis of teleological statements in biology along these lines may be found in Woodger, [Principles], especially pp. 432 ff; essentially the same interpretation is advocated by Kaufmann in [Methodology], chapter 8.

given species of butterflies displays a particular kind of coloring cannot be inferred from—and therefore cannot be explained by means of—the statement that this type of coloring has the effect of protecting the butterflies from detection by pursuing birds, nor can the presence of red corpuscles in the human blood be inferred from the statement that those corpuscles have a specific function in assimilating oxygen and that this function is essential for the maintenance of life.

One of the reasons for the perseverance of teleological considerations in biology probably lies in the fruitfulness of the teleological approach as a heuristic device: Biological research which was psychologically motivated by a teleological orientation, by an interest in purposes in nature, has frequently led to important results which can be stated in non-teleological terminology and which increase our scientific knowledge of the causal connections between biological phenomena.

Another aspect that lends appeal to teleological considerations is their anthropomorphic character. A teleological explanation tends to make us feel that we really "understand" the phenomenon in question, because it is accounted for in terms of purposes, with which we are familiar from our own experience of purposive behavior. But it is important to distinguish here understanding in the psychological sense of a feeling of empathic familiarity from understanding in the theoretical, or cognitive, sense of exhibiting the phenomenon to be explained as a special case of some general regularity. The frequent insistence that explanation means the reduction of something unfamiliar to ideas or experiences already familiar to us is indeed misleading. For while some scientific explanations do have this psychological effect, it is by no means universal: The free fall of a physical body may well be said to be a more familiar phenomenon than the law of gravitation, by means of which it can be explained; and surely the basic ideas of the theory of relativity will appear to many to be far less familiar than the phenomena for which the theory accounts.

"Familiarity" of the explicans is not only not necessary for a sound explanation—as we have just tried to show—, but it is not sufficient either. This is shown by the many cases in which a proposed explicans sounds suggestively familiar, but upon closer inspection proves to be a mere metaphor, or an account lacking testability, or a set of statements which includes no general laws and therefore lacks explanatory power. A case in point is the neovitalistic attempt to explain biological phenomena by reference to an entelechy or vital force. The crucial point here is not—as it is sometimes made out to be—that entelechies cannot be seen or otherwise directly observed; for that is true also of gravitational fields, and yet, reference to such fields is essential in the explanation of various physical phenomena. The decisive difference between the two cases is that the physical explanation provides (1) methods of testing, albeit indirectly, assertions about gravitational fields, and (2) general laws concerning the strength of gravitational fields, and the behavior of objects moving in them. Explanations by entelechies satisfy the analogue of neither of these two conditions. Failure to satisfy the first condition represents a violation of (R3); it renders all statements about entelechies inaccessible to empirical test and thus devoid of empirical meaning. Failure to comply with the second condition involves a violation of (R2). It deprives the concept of entelechy of all explanatory import; for explanatory power never resides in a concept, but always in the general laws in which it functions. Therefore, notwithstanding the flavor of familiarity of the metaphor it invokes, the neovitalistic approach cannot provide theoretical understanding.

The preceding observations about familiarity and understanding can be applied, in a similar manner, to the view held by some scholars that the explanation, or the understanding, of human actions requires an empathic understanding of the personalities of the agents[11]. This understanding of another person in terms of one's own psychological functioning may prove a useful heuristic device in the search for general psychological principles which might provide a theoretical explanation; but the existence of empathy on the part of the scientist is neither a necessary nor a sufficient condition for the explanation, or the scientific understanding, of any human action. It is not necessary, for the behavior of psychotics or of people belonging to a culture very different from that of the scientist may sometimes be explainable and predictable in terms of general principles even though the scientist who establishes or applies those principles may not be able to understand his subjects empathically. And empathy is not sufficient to guarantee a sound explanation, for a strong feeling of empathy may exist even in cases where we completely misjudge a given personality. Moreover, as the late Dr. Zilsel has pointed out, empathy leads with ease to incompatible results; thus, when the population of a town has long been subjected to heavy bombing attacks, we can understand, in the empathic sense, that its morale should have broken down completely, but we can understand with the same ease also that it should have developed a defiant spirit of resistance. Arguments of this kind often appear quite convincing; but they are of an *ex post facto* character and lack cognitive significance unless they are supplemented by testable explanatory principles in the form of laws or theories.

Familiarity of the explanans, therefore, no matter whether it is achieved through the use of teleological terminology, through neovitalistic metaphors, or through other means, is no indication of the cognitive import and the predictive force of a proposed explanation. Besides, the extent to which an idea will be considered as familiar varies from person to person and from time to time, and a psychological factor of this kind certainly cannot serve as a standard in assessing the worth of a proposed explanation. The decisive requirement for every sound explanation remains that it subsume the explanandum under general laws.

PART II. ON THE IDEA OF EMERGENCE

§5. *Levels of Explanation. Analysis of Emergence.*

As has been shown above, a phenomenon may often be explained by sets of laws of different degrees of generality. The changing positions of a planet, for example, may be explained by subsumption under Kepler's laws, or by deriva-

---

[11] For a more detailed discussion of this view on the basis of the general principles outlined above, cf. Zilsel, [Empiricism], sections 7 and 8, and Hempel, [Laws], section 6.

tion from the far more comprehensive general law of gravitation in combination with the laws of motion, or finally by deduction from the general theory of relativity, which explains—and slightly modifies—the preceding set of laws. Similarly, the expansion of a gas with rising temperature at constant pressure may be explained by means of the Gas Law or by the more comprehensive kinetic theory of heat. The latter explains the Gas Law, and thus indirectly the phenomenon just mentioned, by means of (1) certain assumptions concerning the micro-behavior of gases (more specifically, the distributions of locations and speeds of the gas molecules) and (2) certain macro-micro principles, which connect such macro-characteristics of a gas as its temperature, pressure and volume with the micro-characteristics just mentioned.

In the sense of these illustrations, a distinction is frequently made between various levels of explanation[11]. Subsumption of a phenomenon under a general law directly connecting observable characteristics represents the first level; higher levels require the use of more or less abstract theoretical constructs which function in the context of some comprehensive theory. As the preceding illustrations show, the concept of higher-level explanation covers procedures of rather different character; one of the most important among them consists in explaining a class of phenomena by means of a theory concerning their micro-structure. The kinetic theory of heat, the atomic theory of matter, the electromagnetic as well as the quantum theory of light, and the gene theory of heredity are examples of this method. It is often felt that only the discovery of a micro-theory affords real scientific understanding of any type of phenomenon, because only it gives us insight into the inner mechanism of the phenomenon, so to speak. Consequently, classes of events for which no micro-theory was available have frequently been viewed as not actually understood; and concern with the theoretical status of phenomena which are unexplained in this sense may be considered as a theoretical root of the doctrine of emergence.

Generally speaking, the concept of emergence has been used to characterize certain phenomena as "novel", and this not merely in the psychological sense of being unexpected[12], but in the theoretical sense of being unexplainable, or unpredictable, on the basis of information concerning the spatial parts or other constituents of the systems in which the phenomena occur, and which in this context are often referred to as wholes. Thus, e.g., such characteristics of water as its transparence and liquidity at room temperature and atmospheric pressure, or its ability to quench thirst have been considered as emergent on the ground that they could not possibly have been predicted from a knowledge of the properties of its chemical constituents, hydrogen and oxygen. The weight of the compound, on the contrary, has been said not to be emergent because it is a mere "resultant" of its components and could have been predicted by simple addition even before the compound had been formed. The conceptions of ex-

---

[11] For a lucid brief exposition of this idea, see Feigl, [Operationism], pp. 284-288.

[12] Concerning the concept of novelty in its logical and psychological meanings, see also Stace, [Novelty].

planation and prediction which underly this idea of emergence call for various critical observations, and for corresponding changes in the concept of emergence.

(1) First, the question whether a given characteristic of a "whole", w, is emergent or not cannot be significantly raised until it has been stated what is to be understood by the parts or constituents of w. The volume of a brick wall, for example, may be inferable by addition from the volumes of its parts if the latter are understood to be the component bricks, but it is not so inferable from the volumes of the molecular components of the wall. Before we can significantly ask whether a characteristic W of an object w is emergent, we shall therefore have to state the intended meaning of the term "part of". This can be done by defining a specific relation Pt and stipulating that those and only those objects which stand in Pt to w count as parts or constituents of w. 'Pt' might be defined as meaning "constituent brick of" (with respect to buildings), or "molecule contained in" (for any physical object), or "chemical element contained in" (with respect to chemical compounds, or with respect to any material object), or "cell of" (with respect to organisms), etc. The term "whole" will be used here without any of its various connotations, merely as referring to any object w to which others stand in the specified relation Pt. In order to emphasize the dependence of the concept of part upon the definition of the relation Pt in each case, we shall sometimes speak of Pt-parts, to refer to parts as determined by the particular relation Pt under consideration.

(2) We turn to a second point of criticism. If a characteristic of a whole is to be qualified as emergent only if its occurrence cannot be inferred from a knowledge of all the properties of its parts, then, as Grelling has pointed out, no whole can have any emergent characteristics. Thus, to illustrate by reference to our earlier example, the properties of hydrogen include that of forming, if suitably combined with oxygen, a compound which is liquid, transparent, etc. Hence the liquidity, transparence, etc. of water can be inferred from certain properties of its chemical constituents. If the concept of emergence is not to be vacuous, therefore, it will be necessary to specify in every case a class G of attributes and to call a characteristic W of an object w emergent relatively to G and Pt if the occurrence of W in w cannot be inferred from a complete characterization of all the Pt-parts with respect to the attributes contained in G, i.e. from a statement which indicates, for every attribute in G, to which of the parts of w it applies. —Evidently, the occurrence of a characteristic may be emergent with respect to one class of attributes and not emergent with respect to another. The classes of attributes which the emergentists have in mind, and which are usually not explicitly indicated, will have to be construed as non-trivial, i.e. as not logically entailing the property of each constituent of forming, together with the other constituents, a whole with the characteristics under investigations.—Some fairly simple cases of emergence in the sense so far specified arise when the class G is restricted to certain simple properties of the parts, to the exclusion of spatial or other relations among them. Thus, the electromotive force of a system of several electric batteries cannot be inferred from the electromotive forces of its

constituents alone without a description, in terms of relational concepts, of the way in which the batteries are connected with each other.[14]

(3) Finally, the predictability of a given characteristic of an object on the basis of specified information concerning its parts will obviously depend on what general laws or theories are available.[15] Thus, the flow of an electric current in a wire connecting a piece of copper and a piece of zinc which are partly immersed in sulfuric acid is unexplainable, on the basis of information concerning any non-trivial set of attributes of copper, zinc and sulfuric acid, and the particular structure of the system under consideration, unless the theory available contains certain general laws concerning the functioning of batteries, or even more comprehensive principles of physical chemistry. If the theory includes such laws, on the other hand, then the occurrence of the current is predictable. Another illustration, which at the same time provides a good example for the point made under (2) above, is afforded by the optical activity of certain substances. The optical activity of sarco-lactic acid, for example, i.e. the fact that in solution it rotates the plane of polarization of plane-polarized light, cannot be predicted on the basis of the chemical characteristics of its constituent elements; rather, certain facts about the relations of the atoms constituting a molecule of sarco-lactic acid have to be known. The essential point is that the molecule in question contains an asymmetric carbon atom, i.e. one that holds four different atoms or groups, and if this piece of relational information is provided, the optical activity of the solution can be predicted provided that furthermore the theory available for the purpose embodies the law that the presence of one asymmetric carbon atom in a molecule implies optical activity of the solution; if the theory does not include this micro-macro law, then the phenomenon is emergent with respect to that theory.

An argument is sometimes advanced to the effect that phenomena such as the

[14] This observation connects the present discussion with a basic issue in Gestalt theory. Thus, e.g., the insistence that "a whole is more than the sum of its parts" may be construed as referring to characteristics of wholes whose prediction requires knowledge of certain structural relations among the parts. For a further examination of this point, see Grelling and Oppenheim, [Gestaltbegriff] and [Functional Whole].

[15] Logical analyses of emergence which make reference to the theories available have been propounded by Grelling and recently, in a very explicit form, by Henle in [Emergence]. In effect, Henle's definition characterizes a phenomenon as emergent if it cannot be predicted, by means of the theories accepted at the time, on the basis of the data available before its occurrence. In this interpretation of emergence, no reference is made to characteristics of parts or constitutents. Henle's concept of predictability differs from the one implicit in our discussion (and made explicit in Part III of this article) in that it implies derivability from the "simplest" hypothesis which can be formed on the basis of the data and theories available at the time. A number of suggestive observations on the idea of emergence and on Henle's analysis of it are contained in Bergmann's article [Emergence].— The idea that the concept of emergence, at least in some of its applications, is meant to refer to unpredictability by means of "simple" laws was advanced also by Grelling in the correspondence mentioned in note (1). Reliance on the notion of simplicity of hypotheses, however, involves considerable difficulties; in fact, no satisfactory definition of that concept is available at present.

flow of the current, or the optical activity, in our last examples, are absolutely emergent at least in the sense that they could not possibly have been predicted before they had been observed for the first time; in other words, that the laws requisite for their prediction could not have been arrived at on the basis of information available before their first observed occurrence.[16] This view is untenable, however. On the strength of data available at a given time, science often establishes generalizations by means of which it can forecast the occurrence of events the like of which have never before been encountered. Thus, generalizations based upon periodicities exhibited by the characteristics of chemical elements then known, enabled Mendeleeff in 1871 to predict the existence of a certain new element and to state correctly various properties of that element as well as of several of its compounds; the element in question, germanium, was not discovered until 1886.—A more recent illustration of the same point is provided by the development of the atomic bomb and the prediction, based on theoretical principles established prior to the event, of its explosion under specified conditions, and of its devastating release of energy.

As Grelling has stressed, the observation that the predictability of the occurrence of any characteristic depends upon the theoretical knowledge available, applies even to those cases in which, in the language of some emergentists, the characteristic of the whole is a mere resultant of the corresponding characteristics of the parts and can be obtained from the latter by addition. Thus, even the weight of a water molecule cannot be derived from the weights of its atomic constituents without the aid of a law which expresses the former as some specific mathematical function of the latter. That this function should be the sum is by no means self-evident; it is an empirical generalization, and at that not a strictly correct one, as relativistic physics has shown.

Failure to realize that the question of the predictability of a phenomenon cannot be significantly raised unless the theories available for the prediction have been specified has encouraged the misconception that certain phenomena have a mysterious quality of absolute unexplainability, and that their emergent status has to be accepted with "natural piety", as F. L. Morgan put it. The observations presented in the preceding discussion strip the idea of emergence of these unfounded connotations: emergence of a characteristic is not an ontological trait inherent in some phenomena; rather it is indicative of the scope of our knowl-

---

[16] C. D. Broad, who in chapter 2 of his book, [Mind], gives a clear presentation and critical discussion of the essentials of emergentism, emphasizes the importance of "laws of composition" in predicting the characteristics of a whole on the basis of those of its parts. (cf. [Mind], pp. 61ff.); but he subscribes to the view characterized above and illustrates it specifically by the assertion that "if we want to know the chemical (and many of the physical) properties of a chemical compound, such as silver-chloride, it is absolutely necessary to study samples of *that particular compound*. . . . The essential point is that it would also be useless to study chemical compounds in general and to compare their properties with those of their elements in the hope of discovering a *general* law of composition by which the properties of *any* chemical compound could be foretold when the properties of its separate elements were known." (Ibid., p. 64)—That an achievement of precisely this sort has been possible on the basis of the periodic system of the elements is pointed out above.

edge at a given time; thus it has no absolute, but a relative character; and what is emergent with respect to the theories available today may lose its emergent status tomorrow.

The preceding considerations suggest the following redefinition of emergence: The occurrence of a characteristic W in an object w is emergent relatively to a theory T, a part relation Pt, and a class G of attributes if that occurrence cannot be deduced by means of T from a characterization of the Pt-parts of w with respect to all the attributes in G.

This formulation explicates the meaning of emergence with respect to *events* of a certain kind, namely the occurrence of some characteristic W in an object w. Frequently, emergence is attributed to *characteristics* rather than to events; this use of the concept of emergence may be interpreted as follows: A characteristic W is emergent relatively to T, Pt, and G if its occurrence in *any* object is emergent in the sense just indicated.

As far as its cognitive content is concerned, the emergentist assertion that the phenomena of life are emergent may now be construed, roughly, as an elliptic formulation of the following statement: Certain specifiable biological phenomena cannot be explained, by means of contemporary physico-chemical theories, on the basis of data concerning the physical and chemical characteristics of the atomic and molecular constituents of organisms. Similarly, the so-called emergent status of mind reduces to the assertion that present-day physical, chemical and biological theories do not suffice to explain all psychological phenomena on the basis of data concerning the physical, chemical, and biological characteristics of the cells or of the molecules or atoms constituting the organisms in question. But in this interpretation, the emergent character of biological and psychological phenomena becomes trivial; for the description of various biological phenomena requires terms which are not contained in the vocabulary of present day physics and chemistry; hence we cannot expect that all specifically biological phenomena are explainable, i.e. deductively inferable, by means of present day physico-chemical theories on the basis of initial conditions which themselves are described in exclusively physico-chemical terms. In order to obtain a less trivial interpretation of the assertion that the phenomena of life are emergent, we have therefore to include in the explanatory theory all those laws known at present which connect the physico-chemical with the biological "level", i.e., which contain, on the one hand, certain physical and chemical terms, including those required for the description of molecular structures, and on the other hand, certain concepts of biology. An analogous observation applies to the case of psychology. If the assertion that life and mind have an emergent status is interpreted in this sense, then its import can be summarized approximately by the statement that no explanation, in terms of micro-structure theories, is available at present for large classes of phenomena studied in biology and psychology.[17]

[17] The following passage from Tolman, [Behavior], may serve to support this interpretation: ". . . 'behavior-acts', though no doubt in complete one-to-one correspondence with the underlying molecular facts of physics and physiology, have, as 'molar' wholes, certain emergent properties of their own. . . . Further, these molar properties of behavior-acts

Assertions of this type, then, appear to represent the rational core of the doctrine of emergence. In its revised form, the idea of emergence no longer carries with it the connotation of absolute unpredictability—a notion which is objectionable not only because it involves and perpetuates certain logical misunderstandings, but also because, not unlike the ideas of neo-vitalism, it encourages an attitude of resignation which is stifling for scientific research. No doubt it is this characteristic, together with its theoretical sterility, which accounts for the rejection, by the majority of contemporary scientists, of the classical absolutistic doctrine of emergence.[18]

PART III. LOGICAL ANALYSIS OF LAW AND EXPLANATION

§6. Problems of the concept of general law.

From our general survey of the characteristics of scientific explanation, we now turn to a closer examination of its logical structure. The explanation of a phenomenon, we noted, consists in its subsumption under laws or under a theory. But what is a law, what is a theory? While the meaning of these concepts seems intuitively clear, an attempt to construct adequate explicit definitions for them encounters considerable difficulties. In the present section, some basic problems of the concept of law will be described and analyzed; in the next section, we intend to propose, on the basis of the suggestions thus obtained, definitions of law and of explanation for a formalized model language of a simple logical structure.

The concept of law will be construed here so as to apply to true statements only. The apparently plausible alternative procedure of requiring high confirmation rather than truth of a law seems to be inadequate: It would lead to a relativized concept of law, which would be expressed by the phrase "sentence S is a law relatively to the evidence E". This does not seem to accord with the meaning customarily assigned to the concept of law in science and in methodological inquiry. Thus, for example, we would not say that Bode's general formula for the distance of the planets from the sun was a law relatively to the astronomical evidence available in the 1770s, when Bode propounded it, and that it ceased to be a law after the discovery of Neptune and the determination of its distance from the sun; rather, we would say that the limited original evidence had given a high probability to the assumption that the formula was a law, whereas more recent additional information reduced that probability so much as to make it practically certain that Bode's formula is not generally true, and hence not a law.[18a]

---

cannot in the present state of our knowledge, i.e., prior to the working-out of many empirical correlations between behavior and its physiological correlates, be known even inferentially from a mere knowledge of the underlying, molecular, facts of physics and physiology." (l. c., pp. 7-8).—In a similar manner, Hull uses the distinction between molar and molecular theories and points out that theories of the latter type are not at present available in psychology. Cf. [Principles], pp. 19ff.; [Variables], p. 275.

[18] This attitude of the scientist is voiced, for example, by Hull in [Principles], pp. 24-28.

[18a] The requirement of truth for laws has the consequence that a given empirical statement S can never be definitely known to be a law; for the sentence affirming the truth of S

Apart from being true, a law will have to satisfy a number of additional conditions. These can be studied independently of the factual requirement of truth, for they refer, as it were, to all logically possible laws, no matter whether factually true or false. Adopting a convenient term proposed by Goodman[17], we will say that a sentence is lawlike if it has all the characteristics of a general law, with the possible exception of truth. Hence, every law is a lawlike sentence, but not conversely.

Our problem of analyzing the concept of law thus reduces to that of explicating the meaning of "lawlike sentence". We shall construe the class of lawlike sentences as including analytic general statements, such as "A rose is a rose", as well as the lawlike sentences of empirical science, which have empirical content.[18] It will not be necessary to require that each lawlike sentence permissible in explanatory contexts be of the second kind; rather, our definition of explanation will be so constructed as to guarantee the factual character of the totality of the laws—though not of every single one of them—which function in an explanation of an empirical fact.

What are the characteristics of lawlike sentences? First of all, lawlike sentences are statements of universal form, such as "All robins' eggs are greenish-blue", "All metals are conductors of electricity", "At constant pressure, any gas expands with increasing temperature". As these examples illustrate, a lawlike sentence usually is not only of universal, but also of conditional form; it makes an assertion to the effect that universally, if a certain set of conditions, C, is realized, then another specified set of conditions, E, is realized as well. The standard form for the symbolic expression of a lawlike sentence is therefore the universal conditional. However, since any conditional statement can be transformed into a non-conditional one, conditional form will not be considered as essential for a lawlike sentence, while universal character will be held indispensable.

But the requirement of universal form is not sufficient to characterize lawlike sentences. Suppose, for example, that a certain basket, b, contains at a certain time t a number of red apples and nothing else.[21] Then the statement

(S_1) Every apple in basket b at time t is red

is both true and of universal form. Yet the sentence does not qualify as a law; we would refuse, for example, to explain by subsumption under it the fact

---

is logically equivalent with S and is therefore capable only of acquiring a more or less high probability, or degree of confirmation, relatively to the experimental evidence available at any given time. On this point, cf. Carnap, [Remarks].—For an excellent non-technical exposition of the semantical concept of truth, which is here applied, the reader is referred to Tarski, [Truth].

[17] [Counterfactuals]. p. 125.

[18] This procedure was suggested by Goodman's approach in [Counterfactuals].—Reichenbach, in a detailed examination of the concept of law, similarly construes his concept of nomological statement as including both analytic and synthetic sentences; cf. [Logic], chapter VIII.

[21] The difficulty illustrated by this example was stated concisely by Langford ([Review]), who referred to it as the problem of distinguishing between universals of fact and causal universals. For further discussion and illustration of this point, see also Chisholm [Conditional], especially pp. 301f.—A systematic analysis of the problem was given by Goodman

that a particular apple chosen at random from the basket is red. What distinguishes S_1 from a lawlike sentence? Two points suggest themselves, which will be considered in turn, namely, finite scope, and reference to a specified object.

First, the sentence S_1 makes, in effect, an assertion about a finite number of objects only, and this seems irreconcilable with the claim to universality which is commonly associated with the notion of law.[12] But are not Kepler's laws considered as lawlike although they refer to a finite set of planets only? And might we not even be willing to consider as lawlike a sentence such as the following?

(S_2) All the sixteen ice cubes in the freezing tray of this refrigerator have a temperature of less than 10 degrees centigrade.

This point might well be granted; but there is an essential difference between S_1 on the one hand and Kepler's laws as well as S_2 on the other: The latter, while finite in scope, are known to be consequences of more comprehensive laws whose scope is not limited, while for S_1 this is not the case.

Adopting a procedure recently suggested by Reichenbach[13], we will therefore distinguish between fundamental and derivative laws. A statement will be called a derivative law if it is of universal character and follows from some fundamental laws. The concept of fundamental law requires further clarification; so far, we may say that fundamental laws, and similarly fundamental lawlike sentences, should satisfy a certain condition of non-limitation of scope.

It would be excessive, however, to deny the status of fundamental lawlike sentence to all statements which, in effect, make an assertion about a finite class of objects only, for that would rule out also a sentence such as "All robins' eggs are greenish-blue", since presumably the class of all robins' eggs—past, present, and future—is finite. But again, there is an essential difference between this sentence and, say, S_1. It requires empirical knowledge to establish the finiteness of the class of robins' eggs, whereas, when the sentence S_1 is construed in a manner which renders it intuitively unlawlike, the terms "basket b" and "apple" are understood so as to imply finiteness of the class of apples in the basket at time t. Thus, so to speak, the meaning of its constitutive terms alone—without additional factual information—entails that S_1 has a finite scope.—Fundamental laws, then, will have to be construed so as to satisfy what we have called a condition of non-limited scope; our formulation of that condition however, which refers to what is entailed by "the meaning" of certain expressions, is too vague and will have to be revised later. Let us note in passing that the stipulation here envisaged would bar from the class of fundamental lawlike sentences also such undesirable candidates as "All uranic objects are spherical", where "uranic" means the property

---

in [Counterfactuals], especially part III.—While not concerned with the specific point under discussion, the detailed examination of counterfactual conditionals and their relation to laws of nature, in Chapter VIII of Lewis's work [Analysis], contains important observations on several of the issues raised in the present section.

[12] The view that laws should be construed as not being limited to a finite domain has been expressed, among others, by Popper ([Forschung], section 13) and by Reichenbach ([Logic], p. 369).

[13] [Logic], p. 361.—Our terminology as well as the definitions to be proposed later for the two types of law do not coincide with Reichenbach's, however.

of being the planet Uranus; indeed, while this sentence has universal form, it fails to satisfy the condition of non-limited scope.

In our search for a general characterization of lawlike sentences, we now turn to a second clue which is provided by the sentence S_1. In addition to violating the condition of non-limited scope, this sentence has the peculiarity of making reference to a particular object, the basket b; and this, too, seems to violate the universal character of a law.[14] The restriction which seems indicated here, should however again be applied to fundamental lawlike sentences only; for a true general statement about the free fall of physical bodies on the moon, while referring to a particular object, would still constitute a law, albeit a derivative one.

It seems reasonable to stipulate, therefore, that a fundamental lawlike sentence must be of universal form and must contain no essential—i.e., uneliminable —occurrences of designations for particular objects. But this is not sufficient; indeed, just at this point, a particularly serious difficulty presents itself. Consider the sentence

(S_3) Everything that is either an apple in basket b at time t or a sample of ferric oxide is red.

If we use a special expression, say "x is ferple", as synonymous with "x is either an apple in b at t or a sample of ferric oxide", then the content of S_3 can be expressed in the form

(S_4) Everything that is ferple is red.

The statement thus obtained is of universal form and contains no designations of particular objects, and it also satisfies the condition of non-limited scope; yet clearly, S_4 can qualify as a fundamental lawlike sentence no more than can S_3.

As long as "ferple" is a defined term of our language, the difficulty can readily be met by stipulating that after elimination of defined terms, a fundamental lawlike sentence must not contain essential occurrences of designations for particular objects. But this way out is of no avail when "ferple", or another term of the kind illustrated by it, is a primitive predicate of the language under consideration. This reflection indicates that certain restrictions have to be imposed upon those predicates—i.e., terms for properties or relations,—which may occur in fundamental lawlike sentences.[15]

---

[14] In physics, the idea that a law should not refer to any particular object has found its expression in the maxim that the general laws of physics should contain no reference to specific space-time points, and that spatio-temporal coordinates should occur in them only in the form of differences or differentials.

[15] The point illustrated by the sentences S_3 and S_4 above was made by Goodman, who has also emphasized the need to impose certain restrictions upon the predicates whose occurrence is to be permissible in lawlike sentences. These predicates are essentially the same as those which Goodman calls projectible. Goodman has suggested that the problems of establishing precise criteria for projectibility, of interpreting counterfactual conditionals, and of defining the concept of law are so intimately related as to be virtually aspects of a single problem. (Cf. his articles [Query] and [Counterfactuals].) One suggestion for an analysis of projectibility has recently been made by Carnap in [Application]. Goodman's note [Infirmities] contains critical observations on Carnap's proposals.

More specifically, the idea suggests itself of permitting a predicate in a fundamental lawlike sentence only if it is purely universal, or, as we shall say, purely qualitative, in character; in other words, if a statement of its meaning does not require reference to any one particular object or spatio-temporal location. Thus, the terms "soft", "green", "warmer than", "as long as", "liquid", "electrically charged", "female", "father of".are purely qualitative predicates, while "taller than the Eiffel Tower", "medieval", "lunar", "arctic", "Ming" are not.[14]

Exclusion from fundamental lawlike sentences of predicates which are not purely qualitative would at the same time ensure satisfaction of the condition of non-limited scope; for the meaning of a purely qualitative predicate does not require a finite extension; and indeed, all the sentences considered above which violate the condition of non-limited scope make explicit or implicit reference to specific objects.

The stipulation just proposed suffers, however, from the vagueness of the concept of purely qualitative predicate. The question whether indication of the meaning of a given predicate in English does or does not require reference to some one specific object does not always permit an unequivocal answer since English as a natural language does not provide explicit definitions or other clear explications of meaning for its terms. It seems therefore reasonable to attempt definition of the concept of law not with respect to English or any other natural language, but rather with respect to a formalized language—let us call it a model language, L,—which is governed by a well-determined system of logical rules, and in which every term either is characterized as primitive or is introduced by an explicit definition in terms of the primitives.

This reference to a well-determined system is customary in logical research and is indeed quite natural in the context of any attempt to develop precise criteria for certain logical distinctions. But it does not by itself suffice to overcome the specific difficulty under discussion. For while it is now readily possible to characterize as not purely qualitative all those among the defined predicates in L whose definiens contains an essential occurrence of some individual name, our problem remains open for the primitives of the language, whose meanings are not determined by definitions within the language, but rather by semantical rules of interpretation. For we want to permit the interpretation of the primitives of L by means of such attributes as blue, hard, solid, warmer, but

[14] That laws, in addition to being of universal form, must contain only purely universal predicates was clearly argued by Popper ([Forschung], sections 14, 15).—Our alternative expression "purely qualitative predicate" was chosen in analogy to Carnap's term "purely qualitative property" (cf. [Application]).—The above characterization of purely universal predicates seems preferable to a simpler and perhaps more customary one, to the effect that a statement of the meaning of the predicate must require no reference to particular objects. For this formulation might be too exclusive since it could be argued that stating the meaning of such purely qualitative terms as "blue" or "hot" requires illustrative reference to some particular object which has the quality in question. The essential point is that no one specific object has to be chosen; any one in the logically unlimited set of blue or of hot objects will do. In explicating the meaning of "taller than the Eiffel Tower", "being an apple in basket b at time t", "medieval", etc., however, reference has to be made to one specific object or to some one in a limited set of objects.

not by the properties of being a descendant of Napoleon, or an arctic animal, or a Greek statue; and the difficulty is precisely that of stating rigorous criteria for the distinction between the permissible and the non-permissible interpretations. Thus the problem of setting up an adequate definition for purely qualitative attributes now arises again; namely for the concepts of the metalanguage in which the semantical interpretation of the primitives is formulated. We may postpone an encounter with the difficulty by presupposing formalization of the semantical meta-language, the meta-meta-language, and so forth; but somewhere, we will have to stop at a non-formalized meta-language, and for it a characterization of purely qualitative predicates will be needed and will present much the same problems as non-formalized English, with which we began. The characterization of a purely qualitative predicate as one whose meaning can be made explicit without reference to any one particular object points to the intended meaning but does not explicate it precisely, and the problem of an adequate definition of purely qualitative predicates remains open.

There can be little doubt, however, that there exists a large number of property and relation terms which would be rather generally recognized as purely qualitative in the sense here pointed out, and as permissible in the formulation of fundamental lawlike sentences; some examples have been given above, and the list could be readily enlarged. When we speak of purely qualitative predicates, we shall henceforth have in mind predicates of this kind.

In the following section, a model language L of a rather simple logical structure will be described, whose primitives will be assumed to be qualitative in the sense just indicated. For this language, the concepts of law and explanation will then be defined in a manner which takes into account the general observations set forth in the present section.

§7. *Definition of law and explanation for a model language.*

Concerning the syntax of our model language L, we make the following assumptions:

L has the syntactical structure of the lower functional calculus without identity sign. In addition to the signs of alternation (disjunction), conjunction, and implication (conditional), and the symbols of universal and existential quantification with respect to individual variables, the vocabulary of L contains individual constants ('a', 'b', $\cdots$), individual variables ('x', 'y', $\cdots$), and predicates of any desired finite degree; the latter may include, in particular, predicates of degree 1 ('P', 'Q', $\cdots$), which express properties of individuals, and predicates of degree 2 ('R', 'S', $\cdots$), which express dyadic relations among individuals.

For simplicity, we assume that all predicates are primitive, i.e., undefined in L, or else that before the criteria subsequently to be developed are applied to a sentence, all defined predicates which it contains are eliminated in favor of primitives.

The syntactical rules for the formation of sentences and for logical inference in L are those of the lower functional calculus. No sentence may contain free variables, so that generality is always expressed by universal quantification.

For later reference, we now define, in purely syntactical terms, a number of auxiliary concepts. In the following definitions, S is always understood to be a sentence in L.

(*7.1a*) S is formally true (formally false) in L if S (the denial of S) can be proved in L, i.e. by means of the formal rules of logical inference for L. If two sentences are mutually derivable from each other in L, they will be called equivalent.

(*7.1b*) S is said to be a singular, or alternatively, a molecular sentence if S contains no variables. A singular sentence which contains no statement connectives is also called atomic. Illustrations: The sentences '$R(a, b) \supset (P(a) \cdot \sim Q(a))$', '$\sim Q(a)$', '$(R(a, b)$', '$P(a)$' are all singular, or molecular; the last two are atomic.

(*7.1c*) S is said to be a generalized sentence if it consists of one or more quantifiers followed by an expression which contains no quantifiers. S is said to be of universal form if it is a generalized sentence and all the quantifiers occurring in it are universal. S is called purely generalized (purely universal) if S is a generalized sentence (is of universal form) and contains no individual constants. S is said to be essentially universal if it is of universal form and not equivalent to a singular sentence. S is called essentially generalized if it is not equivalent to a singular sentence.

Illustrations: '$(x)(P(x) \supset Q(x))$', '$(x)R(a, x)$', '$(x)(P(x) \vee P(a))$', '$(x)(P(x)\vee \sim P(x))$', '$(Ex)(P(x)\cdot \sim Q(x))$', '$(Ex)(y)(R(a, x)\cdot S(a, y))$' are all generalized sentences; the first four are of universal form, the first and fourth are purely universal; the first and second are essentially universal, the third being equivalent to the singular sentence '$P(a)$', and the fourth to '$P(a) \vee \sim P(a)$'. All sentences except the third and fourth are essentially generalized.

Concerning the semantical interpretation of L, we lay down the following two stipulations:

(*7.2a*) The primitive predicates of L are all purely qualitative.

(*7.2b*) The universe of discourse of L, i.e., the domain of objects covered by the quantifiers, consists of all physical objects, or of all spatio-temporal locations.

A linguistic framework of the kind here characterized is not sufficient for the formulation of scientific theories since it contains no functors and does not provide the means for dealing with real numbers. Besides, the question is open at present whether a constitution system can be constructed in which all of the concepts of empirical science are reduced, by chains of explicit definitions, to a basis of primitives of a purely qualitative character. Nevertheless, we consider it worthwhile to study the problems at hand for the simplified type of language just described because the analysis of law and explanation is far from trivial even for our model language L, and because that analysis sheds light on the logical character of the concepts under investigation also in their application to more complex contexts.

In accordance with the considerations developed in section 6, we now define:

(*7.3a*) S is a fundamental lawlike sentence in L if S is purely universal; S is a fundamental law in L if S is purely universal and true.

(*7.3b*) S is a derivative law in L if (1) S is essentially, but not purely, universal and (2) there exists a set of fundamental laws in L which has S as a consequence.

(*7.3c*) S is a law in L if it is a fundamental or a derivative law in L.

The fundamental laws as here defined obviously include, besides general statements of empirical character, all those statements of purely universal form which are true on purely logical grounds; i.e. those which are formally true in L, such as '$(x)(P(x)\mathbf{v} \sim P(x))$', and those whose truth derives exclusively from the interpretation given to its constituents, as is the case with '$(x)(P(x) \supset Q(x))$', if 'P' is interpreted as meaning the property of being a father, and 'Q' that of being male.—The derivative laws, on the other hand, include neither of these categories; indeed, no fundamental law is also a derivative one.

As the primitives of L are purely qualitative, all the statements of universal form in L also satisfy the requirement of non-limited scope, and thus it is readily seen that the concept of law as defined above satisfies all the conditions suggested in section 6.[17]

The explanation of a phenomenon may involve generalized sentences which are not of universal form. We shall use the term "theory" to refer to such sentences, and we define this term by the following chain of definitions:

(7.4a) S is a fundamental theory if S is purely generalized and true.

(7.4b) S is a derivative theory in L if (1) S is essentially, but not purely, generalized and (2) there exists a set of fundamental theories in L which has S as a consequence.

(7.4c) S is a theory in L if it is a fundamental or a derivative theory in L.

By virtue of the above definitions, every law is also a theory, and every theory is true.

With the help of the concepts thus defined, we will now reformulate more precisely our earlier characterization of scientific explanation with specific reference to our model language L. It will be convenient to state our criteria for a sound explanation in the form of a definition for the expression "the ordered couple of sentences, (T, C), constitutes an explanans for the sentence E." Our analysis will be restricted to the explanation of particular events, i.e., to the case where the explanandum, E, is a singular sentence.[18]

---

[17] As defined above, fundamental laws include universal conditional statements with vacuous antecedents, such as "All mermaids are brunettes". This point does not appear to lead to undesirable consequences in the definition of explanation to be proposed later.— For an illuminating analysis of universal conditionals with vacuous antecedents, see Chapter VIII in Reichenbach's [Logic].

[18] This is not a matter of free choice: The precise rational reconstruction of explanation as applied to general regularities presents peculiar problems for which we can offer no solution at present. The core of the difficulty can be indicated briefly by reference to an example: Kepler's laws, K, may be conjoined with Boyle's law, B, to a stronger law $K.B$; but derivation of K from the latter would not be considered as an explanation of the regularities stated in Kepler's laws; rather, it would be viewed as representing, in effect, a pointless "explanation" of Kepler's laws by themselves. The derivation of Kepler's laws from Newton's laws of motion and of gravitation, on the other hand, would be recognized as a genuine explanation in terms of more comprehensive regularities, or so-called higher-level laws. The problem therefore arises of setting up clear-cut criteria for the distinction of levels of explanation or for a comparison of generalized sentences as to their comprehensiveness. The establishment of adequate criteria for this purpose is as yet an open problem.

In analogy to the concept of lawlike sentence, which need not satisfy a requirement of truth, we will first introduce an auxiliary concept of potential explanans,
which is not subject to a requirement of truth; the notion of explanans will then
be defined with the help of this auxiliary concept.—The considerations presented in Part I suggest the following initial stipulations:

(7.5)   An ordered couple of sentences, (T, C), constitutes a potential explanans
for a singular sentence E only if
   (1) T is essentially generalized and C is singular
   (2) E is derivable in L from T and C jointly, but not from C alone.

(7.6)   An ordered couple of sentences, (T, C), constitutes an explanans for a
singular sentence E if and only if
   (1) (T, C) is a potential explanans for E
   (2) T is a theory and C is true.

(7.6) is an explicit definition of explanation in terms of the concept of potential
explanation.[29]   On the other hand, (7.5) is not suggested as a definition, but as
a statement of necessary conditions of potential explanation.   These conditions
will presently be shown not be sufficient, and additional requirements will be
discussed by which (7.5) has to be supplemented in order to provide a definition
of potential explanation.

Before we turn to this point, some remarks are called for concerning the formulation of (7.5).   The analysis presented in Part I suggests that an explanans for
a singular sentence consists of a class of generalized sentences and a class of
singular ones.   In (7.5), the elements of each of these classes separately are
assumed to be conjoined to one sentence.   This provision will simplify our
formulations, and in the case of generalized sentences, it serves an additional
purpose: A class of essentially generalized sentences may be equivalent to a
singular sentence; thus, the class {'$P(a)\mathbf{v}(x)Q(x)$', '$P(a)\mathbf{v} \sim (x)Q(x)$'} is equivalent with the sentence '$P(a)$'.   Since scientific explanation makes essential use
of generalized sentences, sets of laws of this kind have to be ruled out; this is
achieved above by combining all the generalized sentences in the explanans into
one conjunction, T, and stipulating that T has to be essential generalized.
—Again, since scientific explanation makes essential use of generalized sentences,
E must not be a consequence of C alone: The law of gravitation, combined with
the singular sentence "Mary is blonde and blue-eyed" does not constitute an
explanans for "Mary is blonde".   The last stipulation in (7.5) introduces the
requisite restriction and thus prohibits complete self-explanation of the explanandum, i.e., the derivation of E from some singular sentence which has E
as a consequence.—The same restriction also dispenses with the need for a
special requirement to the effect that T has to have factual content if (T, C) is
to be a potential explanans for an empirical sentence E.   For if E is factual,
then, since E is a consequence of T and C jointly, but not of C alone, T must be
factual, too.

[29] It is necessary to stipulate, in (7.6) (2), that T be a theory rather than merely that
T be true, for as was shown in section 6, the generalized sentences occurring in an explanans
have to constitute a theory, and not every essentially generalized sentence which is true
is actually a theory, i.e., a consequence of a set of purely generalized true sentences.

Our stipulations in (7.5) do not preclude, however, what might be termed partial self-explanation of the explanandum. Consider the sentences $T_1 = $ '$(x)(P(x) \supset Q(x))$', $C_1 = $ '$R(a, b) \cdot P(a) \cdot U(b)$', $E_1' = $ '$Q(a) \cdot R(a, b)$'. They satisfy all the requirements laid down in (7.5), but it seems counterintuitive to say that (T_1, C_1) potentially explains E_1, because the occurrence of the component '$R(a, b)$' of C_1 in the sentence E_1 amounts to a partial explanation of the explanandum by itself. Is it not possible to rule out, by an additional stipulation, all those cases in which E shares part of its content with C, i.e. where C and E have a common consequence which is not formally true in L? This stipulation would be tantamount to the requirement that C and E have to be exhaustive alternatives in the sense that their alternation is formally true, for the content which any two sentences have in common is expressed by their alternation. The proposed restriction, however, would be very severe. For if E does not share even part of its content with C, then C is altogether unnecessary for the derivation of E from T and C, i.e., E can be inferred from T alone. Therefore, in every potential explanation in which the singular component of the explanans is not dispensable, the explanandum is partly explained by itself. Take, for example, the potential explanation of $E_2 = $ '$Q(a)$' by $T_2 = $ '$(x)(P(x) \supset Q(x))$' and $C_2 = $ '$P(a)$', which satisfies (7.5), and which surely is intuitively unobjectionable. Its three components may be equivalently expressed by the following sentences: $T_2' = $ '$(x)(\sim P(x) \mathbf{v} Q(x))$'; $C_2' = $ '$(P(a) \mathbf{v} Q(a)) \cdot (P(a) \mathbf{v} \sim Q(a))$'; $E_2' = $ '$(P(a) \mathbf{v} Q(a)) \cdot (\sim P(a) \mathbf{v} Q(a))$'. This reformulation shows that part of the content of the explanandum is contained in the content of the singular component of the explanans and is, in this sense, explained by itself.

Our analysis has reached a point here where the customary intuitive idea of explanation becomes too vague to provide further guidance for rational reconstruction. Indeed, the last illustration strongly suggests that there may be no sharp boundary line which separates the intuitively permissible from the counterintuitive types of partial self-explanation; for even the potential explanation just considered, which is acceptable in its original formulation, might be judged unacceptable on intuitive grounds when transformed into the equivalent version given above.

The point illustrated by the last example is stated more explicitly in the following theorem, which we formulate here without proof.

(7.7) *Theorem.* Let (T, C) be a potential explanans for the singular sentence E. Then there exist three singular sentences, E_1, E_2, and C_1 in L such that E is equivalent to the conjunction $E_1 \cdot E_2$, C is equivalent to the conjunction $C_1 \cdot E_1$, and E_2 can be derived in L from T alone.[19]

In more intuitive terms, this means that if we represent the deductive structure

---

[19] In the formulation of the above theorem and subsequently, statement connective symbols are used not only as signs *in* L, but also autonomously in speaking *about* compound expressions of L. Thus, when 'S' and 'T' are names or name variables for sentences in L, their conjunction and disjunction will be designated by '$S.T$' and 'SvT', respectively; the conditional which has S as antecedent and T as consequent will be designated by '$S \supset T$', and the denial of S by '$\sim S$'. (Incidentally, this convention has already been used, tacitly, at one place in note 28).

of the given potential explanation by the schema $\{T, C\} \to E$, then this schema can be restated in the form $\{T, C_1 \cdot E_1\} \to E_1 \cdot E_2$, where E_2 follows from T alone, so that C_1 is entirely unnecessary as a premise; hence, the deductive schema under consideration can be reduced to $\{T, E_1\} \to E_1 \cdot E_2$, which can be decomposed into the two deductive schemata $\{T\} \to E_2$ and $\{E_1\} \to E_1$. The former of these might be called a purely theoretical explanation of E_2 by T, the latter a complete self-explanation of E_1. Theorem (7.7) shows, in other words, that every explanation whose explanandum is a singular sentence can be decomposed into a purely theoretical explanation and a complete self-explanation; and any explanation of this kind in which the singular constituent of the explanans is not completely unnecessary involves a partial self-explanation of the explanandum.[11]

To prohibit partial self-explanation altogether would therefore mean limitation of explanation to purely theoretical explanation. This measure seems too severely restrictive. On the other hand, an attempt to delimit, by some special rule, the permissible degree of self-explanation does not appear to be warranted because, as we saw, customary usage provides no guidance for such a delimitation, and because no systematic advantage seems to be gained by drawing some arbitrary dividing line. For these reasons, we refrain from laying down stipulations prohibiting partial self-explanation.

The conditions laid down in (7.5) fail to preclude yet another unacceptable type of explanatory argument, which is closely related to complete self-explanation, and which will have to be ruled out by an additional stipulation. The point is, briefly, that if we were to accept (7.5) as a definition, rather than merely as a statement of necessary conditions, for potential explanation, then, as a consequence of (7.6), any given particular fact could be explained by means of any true lawlike sentence whatsoever. More explicitly, if E is a true singular sentence—say, "Mt. Everest is snowcapped",—and T is a law—say, "All metals are good conductors of heat",—then there exists always a true singular sentence C such that E is derivable from T and C, but not from C alone; in other words, such that (7.5) is satisfied. Indeed, let T_s be some arbitrarily chosen particular instance of T, such as "If the Eiffel Tower is metal, it is a good conductor of heat". Now since E is true, so is the conditional $T_s \supset E$, and if the latter is chosen as the sentence C, then T, C, E satisfy the conditions laid down in (7.5).

In order to isolate the distinctive characteristic of this specious type of explanation, let us examine an especially simple case of the objectionable kind.

---

[11] The characteristic here referred to as partial self-explanation has to be distinguished from what is sometimes called the circularity of scientific explanation. The latter phrase has been used to cover two entirely different ideas. (a) One of these is the contention that the explanatory principles adduced in accounting for a specific phenomenon are inferred from that phenomenon, so that the entire explanatory process is circular. This belief is false, since general laws cannot be inferred from singular sentences. (b) It has also been argued that in a sound explanation the content of the explanandum is contained in that of the explanans. That is correct since the explanandum is a logical consequence of the explanans; but this peculiarity does not make scientific explanation trivially circular since the general laws occurring in the explanans go far beyond the content of the specific explanandum. For a fuller discussion of the circularity objection, see Feigl, [Operationism], pp. 286 ff, where this issue is dealt with very clearly.

Let $T_1 = $ '$(x)P(x)$' and $E_1 = $ '$R(a, b)$'; then the sentence $C_1 = $ '$P(a) \supset R(a, b)$' is formed in accordance with the preceding instructions, and T_1, C_1, E_1 satisfy the conditions (7.5). Yet, as the preceding example illustrates, we would not say that (T_1, C_1) constitutes a potential explanans for E_1. The rationale for the verdict may be stated as follows: If the theory T_1 on which the explanation rests, is actually true, then the sentence C_1, which can also be put into the form '$\sim P(a) \vee R(a, b)$', can be verified, or shown to be true, only by verifying '$R(a, b)$', i.e., E_1. In this broader sense, E_1 is here explained by itself. And indeed, the peculiarity just pointed out clearly deprives the proposed potential explanation for E_1 of the predictive import which, as was noted in Part I, is essential for scientific explanation: E_1 could not possibly be predicted on the basis of T_1 and C_1 since the truth of C_1 cannot be ascertained in any manner which does not include verification of E_1. (7.5) should therefore be supplemented by a stipulation to the effect that if (T, C) is to be a potential explanans for E, then the assumption that T is true must not imply that verification of C necessitates verification of E.[11]

How can this idea be stated more precisely? Study of an illustration will suggest a definition of verification for molecular sentences. The sentence $M = $ '$(\sim P(a) \cdot Q(a)) \vee R(a, b)$' may be verified in two different ways, either by ascertaining the truth of the two sentences '$\sim P(a)$' and '$Q(a)$', which jointly have M as a consequence, or by establishing the truth of the sentence '$R(a, b)$', which, again, has M as a consequence. Let us say that S is a basic sentence in L if S is either an atomic sentence or the denial of an atomic sentence in L. Verification of a molecular sentence S may then be defined generally as establishment of the truth of some class of basic sentences which has S as a consequence. Hence, the intended additional stipulation may be restated: The assumption that T is true must not imply that every class of true basic sentences which has C as a consequence also has E as a consequence.

As brief reflection shows, this stipulation may be expressed in the following form, which avoids reference to truth: T must be compatible in L with at least one class of basic sentences which has C but not E as a consequence; or, equivalently: There must exist at least one class of basic sentences which has C, but neither $\sim T$ nor E as a consequence in L.

If this requirement is met, then surely E cannot be a consequence of C, for otherwise there could be no class of basic sentences which has C but not E as a consequence; hence, supplementation of (7.5) by the new condition renders the second stipulation in (7.5) (2) superfluous.—We now define potential explanation as follows:

(7.8) An ordered couple of sentences, (T, C), constitutes a potential explanans for a singular sentence E if and only if the following conditions are satisfied:

(1) T is essentially generalized and C is singular

[11] It is important to distinguish clearly between the following two cases: (a) If T is true then C cannot be true without E being true; and (b) If T is true, C cannot be verified without E being verified.—Condition (a) must be satisfied by any potential explanation; the much more restictive condition (b) must not be satisfied if (T, C) is to be a potential explanans for E.

(2) E is derivable in L from T and C jointly

(3) T is compatible with at least one class of basic sentences which has C but not E as a consequence.

The definition of the concept of explanans by means of that of potential explanans as formulated in (7.6) remains unchanged.

In terms of our concept of explanans, we can give the following interpretation to the frequently used phrase "this fact is explainable by means of that theory":

(7.9)   A singular sentence E is explainable by a theory T if there exists a singular sentence C such that (T, C) constitutes an explanans for E.

The concept of causal explanation, which has been examined here, is capable of various generalizations. One of these consists in permitting T to include statistical laws. This requires, however, a previous strengthening of the means of expression available in L, or the use of a complex theoretical apparatus in the metalanguage.—On the other hand, and independently of the admission of statistical laws among the explanatory principles, we may replace the strictly deductive requirement that E has to be a consequence of T and C jointly by the more liberal inductive one that E has to have a high degree of confirmation relatively to the conjunction of T and C. Both of these extensions of the concept of explanation open important prospects and raise a variety of new problems. In the present essay, however, these issues will not be further pursued.

PART IV. THE SYSTEMATIC POWER OF A THEORY

§8. Explication of the concept of systematic power.

Scientific laws and theories have the function of establishing systematic connections among the data of our experience, so as to make possible the derivation of some of those data from others. According as, at the time of the derivation, the derived data are, or are not yet, known to have occurred, the derivation is referred to as explanation or as prediction. Now it seems sometimes possible to compare different theories, at least in an intuitive manner, in regard to their explanatory or predictive powers: Some theories seem powerful in the sense of permitting the derivation of many data from a small amount of initial information, others seem less powerful, demanding comparatively more initial data, or yielding fewer results. Is it possible to give a precise interpretation to comparisons of this kind by defining, in a completely general manner, a numerical measure for the explanatory or predictive power of a theory? In the present section, we shall develop such a definition and examine some of its implications; in the following section, the definition will be expanded and a general theory of the concept under consideration will be outlined.

Since explanation and prediction have the same logical structure, namely that of a deductive systematization, we shall use the neutral term "systematic power" to refer to the intended concept. As is suggested by the preceding intuitive characterization, the systematic power of a theory T will be reflected in the ratio of the amount of information derivable by means of T to the amount of initial information required for that derivation. This ratio will obviously depend on the particular set of data, or of information, to which T is applied, and we shall therefore relativize our concept accordingly. Our aim, then, is to construct a definition for $s(T, K)$, the systematic power of a theory T with respect to a finite class K of data, or the degree to which T deductively systematizes the information contained in K.

Our concepts will be constructed again with specific reference to the language L. Any singular sentence in L will be said to express a potential datum, and K will accordingly be construed as a finite class of singular sentences[22]. T will be construed in a much broader sense than in the preceding sections; it may be any sentence in L, no matter whether essentially generalized or not. This liberal convention is adopted in the interest of the generality and simplicity of the definitions and theorems now to be developed.

To obtain values between 0 and 1 inclusive, we might now try to identify $s(T, K)$ with the percentage of those sentences in K which are derivable from the remainder by means of T. Thus, if $K_1 = \{'P(a)', 'Q(a)', '\sim P(b)', '\sim Q(b)', 'Q(c)', '\sim P(d)'\}$, and $T_1 = '(x)(P(x) \supset Q(x))'$, then exactly the second and third sentence in K_1 are derivable by means of T_1 from the remainder, in fact from the first and fourth sentence. We might therefore consider setting $s(T_1, K_1) = 2/6 = 1/3$. But then, for the class $K_2 = \{'P(a) \cdot Q(a)', '\sim P(b) \cdot \sim Q(b)', 'Q(c)', '\sim P(d)'\}$, the same T_1 would have the s-value 0, although K_2 contains exactly the same information as K_1; again, for yet another formulation of that information, namely, $K_3 = \{'P(a) \cdot \sim Q(b)', 'Q(a) \cdot \sim P(b)', 'Q(c)', '\sim P(d)'\}$, T_1 would have the s-value 1/4, and so on. But what we seek is a measure of the degree to which a given theory deductively systematizes a given body of factual information, i.e., a certain content, irrespective of the particular structure and grouping of the sentences in which that content happens to be expressed. We shall therefore make use of a method which represents the contents of any singular sentence or class of singular sentences as composed of certain uniquely determined smallest bits of information. By applying our general idea to these bits, we shall obtain a measure for the systematic power of T in K which is independent of the way in which the content of K is formulated. The sentences expressing those smallest bits of information will be called minimal sentences, and an exact formulation of the proposed procedure will be made possible by an explicit definition of this auxiliary concept. To this point we now turn.

If, as will be assumed here, the vocabulary of L contains fixed finite numbers of individual constants and of predicate constants, then only a certain finite number, say n, of different atomic sentences can be formulated in L. By a minimal

---

[22] As this stipulation shows, the term "datum" is here understood as covering actual as well as potential data. The convention that any singular sentence expresses a potential datum is plausible especially if the primitive predicates of L refer to attributes whose presence absence in specific instances can be ascertained by direct observation. In this case, each singular sentence in L may be considered as expressing a potential datum, in the sense of describing a logically possible state of affairs whose existence might be ascertained by direct observation.—The assumption that the primitives of L express directly observable attributes is, however, not essential for the definition and the formal theory of systematic power set forth in sections 8 and 9.

sentence in L, we will understand a disjunction of any number k $(0 \leq k \leq n)$ of different atomic sentences and the denials of the $n-k$ remaining ones. Clearly, n atomic sentences determine 2^n minimal sentences. Thus, if a language L_1 contains exactly one individual constant, 'a', and exactly two primitive predicates, 'P' and 'Q', both of degree 1, then L_1 contains two atomic sentences, '$P(a)$' and '$Q(a)$', and four minimal sentences, namely, '$P(a)vQ(a)$', '$P(a)v \sim Q(a)$', '$\sim P(a)vQ(a)$', '$\sim P(a)v \sim Q(a)$'. If another language, L_2, contains in addition to the vocabulary of L_1 a second individual constant, 'b', and a predicate 'R' of degree 2, then L_2 contains eight atomic sentences and 256 minimal sentences, such as '$P(a)v P(b)v \sim Q(a)v Q(b)v R(a, a)v R(a, b)v \sim R(b, a)v \sim R(b, b)$'.

The term "minimal sentence" is to indicate that the statements in question are the singular sentences of smallest non-zero content in L, which means that every singular sentence in L which follows from a minimal sentence is either equivalent with that minimal sentence or formally true in L. However, minimal sentences do have consequences other than themselves which are not formally true in L, but these are not of singular form; '$(Ex)(P(x)vQ(x))$' is such a consequence of '$P(a)vQ(a)$' in L_1 above.

Furthermore, no two minimal sentences have any consequence in common which is not formally true in L; in other words, the contents of any two minimal sentences are mutually exclusive.

By virtue of the principles of the sentential calculus, every singular sentence which is not formally true in L can be transformed into a conjunction of uniquely determined minimal sentences; this conjunction will be called the minimal normal form of the sentence. Thus, e.g., in the language L_1 referred to above, the sentences '$P(a)$' and '$Q(a)$' have the minimal normal forms '$P(a)vQ(a)) \cdot (P(a)v \sim Q(a))$', and '$(P(a)vQ(a)) \cdot (\sim P(a)vQ(a))$', respectively; in L_2, the same sentences have minimal normal forms consisting of 128 conjoined minimal sentences each.—If a sentence is formally true in L, its content is zero, and it cannot be represented by a conjunction of minimal sentences. It will be convenient, however, to say that the minimal normal form of a formally true sentence in L is the vacuous conjunction of minimal sentences, which does not contain a single term.

As a consequence of the principle just mentioned, any class of singular sentences which are not all formally true can be represented by a sentence in minimal normal form. The basic idea outlined above for the explication of the concept of systematic power can now be expressed by the following definition:

(8.1) Let T be any sentence in L, and K any finite class of singular sentences in L which are not all formally true. If K' is the class of minimal sentences which occur in the minimal normal form of K, consider all divisions of K' into two mutually exclusive subclasses, K_1' and K_2', such that every sentence in K_2' is derivable from K_1' by means of T. Each division of this kind determines a ratio $n(K_2')/n(K')$, i.e. the number of minimal sentences in K_2' divided by the total number of minimal sentences in K'. Among the values of these ratios, there must be a largest one; $s(T, K)$ is to equal that maximum ratio. (Note that if all the elements of K were formally true, $n(K')$ would be 0 and the above ratio would not be defined.)

Illustration: Let L_1 contain only one individual constant, 'a', and only two predicates, 'P' and 'Q', both of degree 1. In L_1, let $T = $ '$(x)(P(x) \supset Q(x))$', $K = \{$'$P(a)$', '$Q(a)$'$\}$. Then we have $K' = \{$'$P(a)\mathbf{v}Q(a)$', '$P(a)\mathbf{v} \sim Q(a)$', '$\sim P(a)\mathbf{v}Q(a)$'$\}$. From the subclass K'_1 consisting of the first two elements of K'—which together are equivalent to '$P(a)$'—we can derive, by means of T, the sentence '$Q(a)$', and from it, by pure logic, the third element of K'; it constitutes the only element of K'_2. No "better" systematization is possible, hence $s(T, K) = 1/3$.

Our definition leaves open, and is independent of, the question whether for a given K' there might not exist different divisions each of which would yield the maximum value for $n(K'_2)/n(K')$. Actually, this can never happen: there exists always exactly one optimal subdivision of a given K'. This fact is a corollary of a general theorem, to which we now turn. It will be noticed that in the last illustration, K'_2 can be derived from T alone, without the use of K'_1 as a premise; indeed, '$\sim P(a)\mathbf{v}Q(a)$' is but a substitution instance of the sentence '$(x)(\sim P(x)\mathbf{v}Q(x))$', which is equivalent to T. The theorem now to be formulated, which might appear surprising at first, shows that this observation applies analogously in all other cases.

(8.2) *Theorem.* Let T be any sentence, K' a class of minimal sentences, and K'_2 a subclass of K' such that every sentence in K'_2 is derivable by means of T from the class $K - K'_2$; then every sentence in K'_2 is derivable from T alone.

The proof, in outline, is as follows: Since the contents of any two different minimal sentences are mutually exclusive, so must be the contents of K'_1 and K'_2, which have not a single minimal sentence in common. But since the sentences of K'_2 follow from K'_1 and T jointly, they must therefore follow from T alone.

We note the following consequences of our theorem:

(8.2a) *Theorem.* In any class K' of minimal sentences, the largest subclass which is derivable from the remainder by means of a sentence T is identical with the class of those elements in K' which are derivable from T alone.

(8.2b) *Theorem.* Let T be any sentence, K a class of singular sentences which are not all formally true, K' the equivalent class of minimal sentences, and K'_i the class of those among the latter which are derivable from T alone. Then the concept s defined in (8.1) satisfies the following equation:

$$s(T, K) = n(K'_i)/n(K')$$

§9. *Systematic power and logical probability of a theory. Generalization of the concept of systematic power.*

The concept of systematic power is closely related to that of degree of confirmation, or logical probability, of a theory. A study of this relationship will shed new light on the proposed definition of s, will suggest certain ways of generalizing it, and will finally lead to a general theory of systematic power which is formally analogous to that of logical probability.

The concept of logical probability, or degree of confirmation, is the central concept of inductive logic. Recently, different explicit definitions for this concept have been proposed, for languages of a structure similar to that of our model language, by Carnap[34] and by Helmer, Hempel, and Oppenheim[35].

While the definition of s proposed in the preceding section rests on the concept of minimal sentence, the basic concept in the construction of a measure for logical probability is that of state description, or, as we shall also say, of maximal sentence. A maximal sentence is the dual[36] of a minimal sentence in L; it is a conjunction of k ($0 \leqq k \leqq n$) different atomic sentences and of the denials of the remaining n-k atomic sentences. In a language with n atomic sentences, there exist 2^n state descriptions. Thus, e.g., the language L_1 repeatedly mentioned in §8 contains the following four maximal sentences: '$P(a) \cdot Q(a)$', '$P(a) \cdot \sim Q(a)$', '$\sim P(a) \cdot Q(a)$', '$\sim P(a) \cdot \sim Q(a)$'.

The term "maximal sentence" is to indicate that the sentences in question are the singular sentences of maximum non-universal content in L, which means that every singular sentence in L which has a maximal sentence as a consequence is either equivalent with that maximal sentence or formally false in L.

As we saw, every singular sentence can be represented in a conjunctive, or minimal, normal form, i.e., as a conjunction of certain uniquely determined minimal sentences; similarly, every singular sentence can be expressed also in a disjunctive, or maximal, normal form, i.e. as a disjunction of certain uniquely determined maximal sentences. In the language L_1, for example, '$P(a)$' has the minimal normal form '$(P(a) \mathbf{v} Q(a)) \cdot (P(a) \mathbf{v} \sim Q(a))$' and the maximal normal form '$(P(a) \cdot Q(a)) \mathbf{v} (P(a) \cdot \sim Q(a))$'; the sentence '$P(a) \supset Q(a)$' has the minimal normal form '$\sim P(a) \mathbf{v} Q(a)$' and the maximal normal form '$(P(a) \cdot Q(a)) \mathbf{v} (\sim P(a) \cdot Q(a)) \mathbf{v} (\sim P(a) \cdot \sim Q(a))$'; the minimal normal form of a formally true sentence is the vacuous conjunction, while its maximal normal form is the disjunction of all four state descriptions in L_1. The minimal normal form of any formally false sentence is the conjunction of all four minimal sentences in L_1, while its maximal normal form is the vacuous disjunction, as we shall say.

The minimal normal form of a singular sentence is well suited as an indicator of its content, for it represents the sentence as a conjunction of standard components whose contents are minimal and mutually exclusive. The maximal normal form of a sentence is suited as an indicator of its range, that is, intuitively speaking, of the variety of its different possible realizations, or of the variety of

[34] Cf. especially [Inductive Logic], [Concepts], [Application].

[35] See Helmer and Oppenheim, [Probability]; Hempel and Oppenheim, [Degree].—Certain general aspects of the relationship between the confirmation of a theory and its predictive or systematic success are examined in Hempel, [Studies], Part II, sections 7 and 8. The definition of s developed in the present essay establishes a quantitative counterpart of what, in that paper, is characterized, in non-numerical terms, as the prediction criterion of confirmation.

[36] For a definition and discussion of this concept, cf. Church, [Logic], p. 172.

those possible states of the world which, if realized, would make the statement true. Indeed, each maximal sentence may be considered as describing, as completely as is possible in L, one possible state of the world; and the state descriptions constituting the maximal normal form of a given singular sentence simply list those among the possible states which would make the sentence true.

Just as the contents of any two different minimal sentences, so also the ranges of any two maximal sentences are mutually exclusive: No possible state of the world can make two different maximal sentences true because any two maximal sentences are obviously incompatible with each other.[17]

Range and content of a sentence vary inversely. The more a sentence asserts, the smaller the variety of its possible realizations, and conversely. This relationship is reflected in the fact that the larger the number of constituents in the minimal normal form of a singular sentence, the smaller the number of constituents in its maximal normal form, and conversely. In fact, if the minimal normal form of a singular sentence U contains m_U of the $m = 2^n$ minimal sentences in L, then its maximal normal form contains $l_U = m - m_U$ of the m maximal sentences in L. This is illustrated by our last four examples, where $m = 4$, and $m_U = 2, 1, 0, 4$ respectively.

The preceding observations suggest that the content of any singular sentence U might be measured by the corresponding number m_U or by some magnitude proportional to it. Now it will prove convenient to restrict the values of the content measure function to the interval from 0 to 1, inclusive; and therefore, we define a measure, $g_1(U)$, for the content of any singular sentence in L by the formula

(9.1)
$$g_1(U) = m_U/m$$

To any finite class K of singular sentences, we assign, as a measure $g_1(K)$ of its content, the value $g_1(S)$, where S is the conjunction of the elements of K.

By virtue of this definition, the equation in theorem (8.2b) may be rewritten:

$$s(T, K) = g_1(K_t')/g_1(K')$$

Here, K_t' is the class of all those minimal sentences in K' which are consequences of T. In the special case where T is a singular sentence, K_t' is therefore equivalent with $T \lor S$, where S is the conjunction of all the elements of K'. Hence, the preceding equation may then be transformed into

(9.2)
$$s(T, S) = g_1(T \lor S)/g_1(S)$$

This formula holds when T and S are singular sentences, and S is not formally true. It bears a striking resemblance to the general schema for the definition of the logical probability of T in regard to S:

(9.3)
$$p(T, S) = r(T \cdot S)/r(S)$$

[17] A more detailed discussion of the concept of range may be found in Carnap, [Inductive Logic], section 2, and in Carnap, [Semantics], sections 18 and 19, where the relation of range and content is examined at length.

Here, $r(U)$ is, for any sentence U in L, a measure of the range of U, T is any sentence in L, and S any sentence in L with $r(S) \neq 0$.

The several specific definitions which have been proposed for the concept of logical probability accord essentially with the pattern exhibited by (9.3)[38], but they differ in their choice of a specific measure function for ranges, i.e. in their definition of r.   One idea which comes to mind is to assign, to any singular sentence U whose maximal normal form contains l_U maximal sentences, the range measure

$$(9.4) \qquad\qquad r_1(U) = l_U/m$$

which obviously is defined in strict analogy to the content measure g_1 for singular sentences as introduced in (9.1).   For every singular sentence U, the two measures add up to unity:

$$(9.5) \qquad\qquad r_1(U) + g_1(U) = (l_U + m_U)/m = 1$$

As Carnap has shown, however, the range measure r_1 confers upon the corresponding concept of logical probability, i.e., upon the concept p_1 defined by means of it according to the schema (9.3), certain characteristics which are incompatible with the intended meaning of logical probability[39]; and Carnap as well as Helmer jointly with the present authors have suggested certain alternative measure functions for ranges, which lead to more satisfactory concepts of probability or of degree of confirmation.   While we need not enter into details here, the following general remarks seem indicated to prepare the subsequent discussion.

The function r_1 measures the range of a singular sentence essentially by counting the number of maximal sentences in its maximal normal form; it thus gives equal weight to all maximal sentences (definition (9.1) deals analogously with minimal sentences).   The alternative definitions just referred to are based on a different procedure.   Carnap, in particular, lays down a rule which assigns a specific weight, i.e. a specific value of r, to each maximal sentence, but these weights are not the same for all maximal sentences.   He then defines the range measure of any other singular sentence as the sum of the measures of its constituent maximal sentences.   In terms of the function thus obtained—let us call it r_2—Carnap defines the corresponding concept of logical probability, which we shall call p_2, for singular sentences T, S in accordance with the schema (9.3): $p_2(T, S) = r_2(T. S)/r_2(S)$.   The definitions of r_2 and p_2 are then extended, by means of certain limiting processes, to the cases where T and S are no longer both singular.[40]

[38] In Carnap's theory of logical probability, $p(T, S)$ is defined, for certain cases, as the limit which the function $r(T. S)/r(S)$ assumes under specified conditions (cf. Carnap, [Inductive Logic], p. 75); but we shall refrain here from considering this generalization of that type of definition which is represented by (9.3).

[39] [Inductive Logic], pp. 80–81.

[40] The alternative approach suggested by Helmer and the present authors involves use of a range measure function r_I which depends in a specified manner on the empirical information I available; hence, the range measure of any sentence U is determined only if a sentence

Now it can readily be seen that just as the function r_1 defined in (9.5) is but but one among an infinity of possible range measures, so the analogous function g_1 defined in (9.1) is but one among an infinity of possible content measures; and just as each range measure may serve to define, according to the schema (9.3), a corresponding measure of logical probability, so each content measure function may serve to define, by means of the schema illustrated by (9.2), a corresponding measure of systematic power. The method which suggests itself here for obtaining alternative content measure functions is to choose some range measure r other than r_1 and then to *define* a corresponding content measure g in terms of it by means of the formula

$$(9.6) \qquad\qquad g(U) = 1 - r(U)$$

so that g and r satisfy the analogue to (9.5) by definition. The function g thus defined will lead in turn, via a definition analogous to (9.2), to a corresponding concept s. Let us now consider this procedure a little more closely.

We assume that a function r is given which satisfies the customary requirements for range measures, namely:

(9.7) 1. $r(U)$ is uniquely determined for *all* sentences U in L.

2. $0 \leq r(U) \leq 1$ for every sentence U in L.

3. $r(U) = 1$ if the sentence U is formally true in L and thus has universal range.

4. $r(U_1 \vee U_2) = r(U_1) + r(U_2)$ for any two sentences U_1, U_2 whose ranges are mutually exclusive, i.e., whose conjunction is formally false.

In terms of the given range measure let the corresponding content measure g be defined by means of (9.6). Then g can readily be shown to satisfy the following conditions:

(9.8) 1. $g(U)$ is uniquely determined for *all* sentences U in L.

2. $0 \leq g(U) \leq 1$ for every sentence U in L.

3. $g(U) = 1$ if the sentence U is formally false in L and thus has universal content.

I, expressing the available empirical information, is given. In terms of this range measure function, the concept of degree of confirmation, dc, can be defined by means of a formula similar to (9.3). The value of $dc(T, S)$ is not defined, however, in certain cases where S is generalized, as has been pointed out by McKinsey (cf. [Review]); also, the concept dc does not satisfy all the theorems of elementary probability theory (cf. the discussion of this point in the first two articles mentioned in note (35)); therefore, the degree of confirmation of a theory relatively to a given evidence is not a probability in the strict sense of the word. On the other hand, the definition of dc here referred to has certain methodologically desirable features, and it might therefore be of interest to construct a related concept of systematic power by means of the range measure function r_1. In the present paper, however, this question will not be pursued.

4. $g(U_1 \cdot U_2) = g(U_1) + g(U_2)$ for any two sentences U_1, U_2 whose contents are mutually exclusive, i.e., whose disjunction is formally true.

In analogy to (9.2), we next define, by means of g, a corresponding function s:

(9.9) $$s(T, S) = g(T \lor S)/g(S)$$

This function is determined for every sentence T, and for every sentence S with $g(S) \neq 0$, whereas the definition of systematic power given in §8 was restricted to those cases where S is singular and not formally true. Finally, our range measure r determines a corresponding probability function by virtue of the definition

(9.10) $$p(T, S) = r(T \cdot S)/r(S)$$

This formula determines the function p for any sentence T, and for any sentence S with $r(S) \neq 0$.

In this manner, every range measure r which satisfies (9.7) determines uniquely a corresponding content measure g which satisfies (9.8), a corresponding function s, defined by (9.9), and a corresponding function p, defined by (9.10). As a consequence of (9.7) and (9.10), the function p can be shown to satisfy the elementary laws of probability theory, especially those listed in (9.12) below; and by virtue of these, it is finally possible to establish a very simple relationship which obtains, for any given range measure r, between the corresponding concepts $p(T, S)$ and $s(T, S)$. Indeed, we have

(9.11) $$\begin{aligned} s(T, S) &= g(T \lor S)/g(S) \\ &= (1 - r(T \lor S))/(1 - r(S)) \\ &= r(\sim (T \lor S))/r(\sim S) \\ &= r(\sim T \cdot \sim S)/r(\sim S) \\ &= p(\sim T, \sim S) \end{aligned}$$

We now list, without proof, some theorems concerning p and s which follow from our assumptions and definitions; they hold in all cases where the values of p and s referred to exist, i.e., where the r-value of the second arguments of p, and the g-value of the second arguments of s, is not 0.

(9.12) (1) a. $0 \leqq p(T, S) \leqq 1$
        b. $0 \leqq s(T, S) \leqq 1$
     (2) a. $p(\sim T, S) = 1 - p(T, S)$
        b. $s(\sim T, S) = 1 - s(T, S)$
     (3) a. $p(T_1 \lor T_2, S) = p(T_1, S) + p(T_2, S) - p(T_1 \cdot T_2, S)$
        b. $s(T_1 \cdot T_2, S) = s(T_1, S) + s(T_2, S) - s(T_1 \lor T_2, S)$
     (4) a. $p(T_1 \cdot T_2, S) = p(T_1, S) \cdot p(T_2, T_1 \cdot S)$
        b. $s(T_1 \lor T_2, S) = s(T_1, S) \cdot s(T_2, T_1 \lor S)$

In the above grouping, these theorems exemplify the relationship of dual correspondence which obtains between p and s. A general characterization of this correspondence is given in the following theorem, which can be proved on the basis of (9.11), and which is stated here in a slightly informal manner in order to avoid the tedium of lengthy formulations.

(9.13) Dualism theorem. From any demonstrable general formula expressing an equality or an inequality concerning p, a demonstrable formula concerning s is obtained if 'p' is replaced, throughout, by 's', and '$\cdot$' and 'v' are exchanged for each other. The same exchange, and replacement of 's' by 'p', conversely transforms any theorem expressing an equality or an inequality concerning s into a theorem about p.

We began our analysis of the systematic power of a theory in regard to a class of data by interpreting this concept, in §8, as a measure of the optimum ratio of those among the given data which are derivable from the remainder by means of the theory. Systematic elaboration of this idea has led to the definition, in the present section, of a more general concept of systematic power, which proved to be the dual counterpart of the concept of logical probability. This extension of our original interpretation yields a simpler and more comprehensive theory than would have been attainable on the basis of our initial definition.

But the theory of systematic power, in its narrower as well as in its generalized version, is, just like the theory of logical probability, purely formal in character, and a significant application of either theory in epistemology or the methodology of science requires the solution of certain fundamental problems which concern the logical structure of the language of science and the interpretation of its concepts. One urgent desideratum here is the further elucidation of the requirement of purely qualitative primitives in the language of science; another crucial problem is that of choosing, among an infinity of formal possibilities, an adequate range measure r. The complexity and difficulty of the issues which arise in these contexts has been brought to light by recent investigations[41]; it can only be hoped that recent developments in formal theory will soon be followed by progress in solving those open problems and thus clarifying the conditions for a sound application of the theories of logical probability and of systematic power

Queens College, Flushing, N. Y.
Princeton, N. J.

BIBLIOGRAPHY

Throughout the article, the abbreviated titles in brackets are used for reference

Beard, Charles A., and Hook, Sidney. [Terminology] Problems of terminology in historical writing. Chapter IV of Theory and practice in historical study: A report of the Committee on Historiography. Social Science Research Council, New York, 1946.

Bergmann, Gustav. [Emergence] Holism, historicism, and emergence. *Philosophy of Science*, vol. 11 (1944), pp. 209–221.

Bonfante, G. [Semantics] Semantics, language. An article in P. L. Harriman, ed., The encyclopedia of psychology. Philosophical Library, New York, 1946.

Broad, C. D. [Mind] The mind and its place in nature. New York, 1925.

---

[41] Cf. especially Goodman, [Query], [Counterfactuals], [Infirmities], and Carnap, [Application]. See also notes (21) and (25).

Carnap, Rudolf. [Semantics] Introduction to semantics. Harvard University Press, 1942.

————. [Inductive Logic] On inductive logic. *Philosophy of science*, vol 12 (1945), pp. 72–97.

————. [Concepts] The two concepts of probability. *Philosophy and phenomenological research*, vol. 5 (1945), pp. 513–532.

————. [Remarks] Remarks on induction and truth. *Philosophy and phenomenological research*, vol. 6 (1946), pp. 590–602.

————. [Application] On the application of inductive logic. *Philosophy and phenomenological research*, vol. 8 (1947), pp. 133–147.

Chisholm, Roderick M. [Conditional] The contrary-to-fact conditional. *Mind*, vol. 55 (1946), pp. 289–307.

Church, Alonzo. [Logic] Logic, formal. An article in Dagobert D. Runes, ed. The dictionary of philosophy. Philosophical Library, New York, 1942.

Ducasse, C. J. [Explanation] Explanation, mechanism, and teleology. *The journal of philosophy*, vol. 22 (1925), pp. 150–155.

Feigl, Herbert. [Operationism] Operationism and scientific method. *Psychological review*, vol. 52 (1945), pp. 250–259 and 284–288.

Goodman, Nelson. [Query] A query on confirmation. *The journal of philosophy*, vol. 43 (1946), pp. 383–385.

————. [Counterfactuals]. The problem of counterfactual conditionals. *The journal of philosophy*, vol. 44 (1947), pp. 113–128.

————. [Infirmities] On infirmities of confirmation theory. *Philosophy and phenomenological research*, vol. 8 (1947), pp. 149–151.

Grelling, Kurt and Oppenheim, Paul. [Gestaltbegriff] Der Gestaltbegriff im Lichte der neuen Logik. *Erkenntnis*, vol. 7 (1937–38), pp. 211–225 and 357–359.

Grelling, Kurt and Oppenheim, Paul. [Functional Whole] Logical Analysis of "Gestalt" as "Functional whole". Preprinted for distribution at Fifth Internat. Congress for the Unity of Science, Cambridge, Mass., 1939.

Helmer, Olaf and Oppenheim, Paul. [Probability] A syntactical definition of probability and of degree of confirmation. *The journal of symbolic logic*, vol. 10 (1945), pp. 25–60.

Hempel, Carl G. [Laws] The function of general laws in history. *The journal of philosophy*, vol. 39 (1942), pp. 35–48.

————. [Studies] Studies in the logic of confirmation. *Mind*, vol. 54 (1945); Part I: pp. 1–26, Part II: pp. 97–121.

Hempel, Carl G. and Oppenheim, Paul. [Degree] A definition of "degree of confirmation". *Philosophy of science*, vol. 12 (1945), pp. 98–115.

Henle, Paul. [Emergence] The status of emergence. *The journal of philosophy*, vol. 39 (1942), pp. 486–493.

Hospers, John. [Explanation] On explanation. *The journal of philosophy*, vol. 43 (1946), pp. 337–356.

Hull, Clark L. [Variables] The problem of intervening variables in molar behavior theory. *Psychological review*, vol. 50 (1943), pp. 273–291.

————. [Principles] Principles of behavior. New York, 1943.

Jevons, W. Stanley. [Principles] The principles of science. London, 1924. (1st ed. 1874).

Kaufmann, Felix. [Methodology] Methodology of the social sciences. New York, 1944.

Knight, Frank H. [Limitations] The limitations of scientific method in economics. In Tugwell, R., ed., The trend of economics. New York, 1924.

Koch, Sigmund. [Motivation] The logical character of the motivation concept. *Psychological review*, vol. 48 (1941). Part I: pp. 15–38, Part II: pp. 127–154.

Langford, C. H. [Review] Review in *The journal of symbolic logic*, vol. 6 (1941), pp. 67–68.

Lewis, C. I. [Analysis] An analysis of knowledge and valuation. La Salle, Ill., 1946.

McKinsey, J. C. C. [Review] Review of Helmer and Oppenheim, [Probability]. *Mathematical reviews*, vol. 7 (1946), p. 45.

Mill, John Stuart. [Logic] A system of Logic.

Morgan, C. Lloyd. Emergent evolution, New York, 1923.

———. The emergence of novelty. New York, 1933.

Popper, Karl. [Forschung] Logik der Forschung. Wien, 1935.

———. [Society] The open society and its enemies. London, 1945.

Reichenbach, Hans. [Logic] Elements of symbolic logic. New York, 1947.

———. [Quantum mechanics] Philosophic foundations of quantum mechanics. University of California Press, 1944.

Rosenblueth, A., Wiener, N., and Bigelow, J. [Teleology] Behavior, Purpose, and Teleology. Philosophy of science, vol. 10 (1943), pp. 18-24.

Stace, W. T. [Novelty] Novelty, indeterminism and emergence. Philosophical review, vol. 48 (1939), pp. 296-310.

Tarski, Alfred. [Truth] The semantical conception of truth, and the foundations of semantics. Philosophy and phenomenological research, vol. 4 (1944), pp. 341-376.

Tolman, Edward Chase. [Behavior] Purposive behavior in animals and men. New York 1932.

White, Morton G. [Explanation] Historical explanation. Mind, vol. 52 (1943), pp. 212-229.

Woodger, J. H. [Principles] Biological principles. New York, 1929.

Zilsel, Edgar. [Empiricism] Problems of empiricism. In International encyclopedia of unified science, vol. II, no. 8. The University of Chicago Press, 1941.

———. [Laws] Physics and the problem of historico-sociological laws. Philosophy of science, vol. 8 (1941), pp. 567-579.

315

The Meaning of Reduction in the Natural Sciences

ERNEST NAGEL

Professor of Philosophy, Columbia University

4

$\mathcal{T}$he science of mechanics was the first branch of mathematical physics to assume the form of a comprehensive theory. The success of that theory in explaining and bringing into systematic relation a large variety of phenomena was for a long time unprecedented; and the belief entertained by many eminent scientists and philosophers, sometimes supported by a priori arguments, that all the processes of nature would eventually fall within the scope of its principles was repeatedly confirmed by the absorption of several sectors of physics into mechanics. However, it is now common knowledge that classical mechanics no longer occupies the position of the "universal" physical science once claimed for it; for since the latter part of the nineteenth century the difficulties facing the extension of mechanics to various further domains of physical inquiry have come to be acknowledged as insuperable, and rival candidates for the office of a universal physical science have been proposed. Moreover, with some exceptions, no serious students today believe that some particular physical theory can be established on a priori grounds as the universal or fundamental theory of natural processes; and to many thinkers it is even an open question whether the ideal of a comprehensive theory which would thoroughly integrate all domains of natural science is realizable. Nevertheless, the phenomenon of a relatively autonomous branch of science becoming absorbed by, or "reduced" to, some other discipline is an undeniable and recurrent feature of the history of modern science, and there is no reason to suppose that such reduction will not continue to take place in the future.

It is with this phenomenon that the present paper is concerned. The successful reduction of one science to another, as well as the failures in effecting such a reduction in a number of notable cases, have been occasions, exploited by both practicing scientists and professional as well as lay philosophers, for far-reaching reinterpretations of the nature and limits of knowledge, science, and the allegedly ultimate constitution of things in general. These interpretations take various forms. Discoveries concerning the physics and physiology of perception have been frequently used to support the conclusion that the findings of physics are incompatible with so-called common sense or naïve realism (the belief that things encountered in normal experience do possess the traits which are manifest to controlled observation); and elaborate epistemologies have been proposed for resolving the paradox that, in spite of this presumed incompatibility, science takes its point of departure from, and finds its evidence in, such common-sense knowledge. The successful reduction of thermodynamics to statistical mechanics in the nineteenth century, and the more recent expansion of electrical theories of matter, have been taken to show that spatial displacements are the only form of intelligible and genuine change; that the qualitative and behavioral diversities noted in ordinary experience are "unreal" and illusory; or, conversely, that the "mysterious world" discovered by microscopic physics is but an insubstantial symbol which expresses a pervasive spiritual reality not alien to human values. On the other hand, the failure to explain electro-dynamical phenomena in terms of the principles of me-chanics, and the general decline of mechanics as the uni-versal science of nature in contemporary physics, has been hailed as evidence for the bankruptcy of classical science, for the necessity of instituting an "organismic" point of view and "organismic" categories of explanation in the study of all natural phenomena, and for a variety of meta-physical doctrines concerning levels of being, emergence, and creative novelty.

I do not believe that these speculative interpretations of the assumed facts of science are warranted by the evi-dence. On the contrary, I believe that the problems to which they are addressed are generated by misconstruing the statements of the natural sciences and reading them in senses not in accordance with the meanings that actual usage in scientific contexts establishes for those statements. However, it is not my present aim to examine the detailed arguments which lead to the adoption of views such as those just briefly indicated. I wish instead to consider what is done when one science is reduced to another, and to suggest that an important source of much dubious com-mentary on the nature and the interrelations of the sci-ences lies in the failure to recognize the conditions which must be fulfilled when such a reduction is effected. It is a commonplace that linguistic expressions, associated with established habits or rules of usage in one set of homogeneous contexts, frequently come to be used in other contexts on the assumption of definite analogies or continuities between the several domains. But judging from the practice of many philosophers and scientists, it is still not a commonplace that when the range of application of expressions is thus extended, these expressions may undergo critical changes in meaning, and that unless care is exercised in interpreting them so that specific contexts of relevant usage are noted, serious misunderstandings and spurious problems are bound to arise. In any event, misconceptions having their basis in just such careless handling of language seem to me to accompany much traditional and current discussion of the significance of scientific reduction. The present essay is an attempt to indicate some quite familiar and yet frequently neglected distinctions that are pertinent to the analysis of this recurrent phenomenon in the development of the natural sciences.

Before turning to my actual theme, it will be useful to distinguish a type of reduction in the history of science which generally, though certainly not always, is unaccompanied by serious misapprehensions. I have in mind the normal expansion of some body of theory, initially proposed for a certain extensive domain of phenomena, so that laws which previously may have been found to hold in a narrow sector of that domain, or in some other domain homogenous in a readily identifiable sense with the first, are shown to be derivable from that theory when suitably specialized. For example, Galileo's *Two New Sciences* was a contribution to the physics of freely falling terrestrial bodies; but when Newton showed that his own general theory of mechanics and gravitation, when supplemented by appropriate boundary conditions, entailed Galileo's laws, the latter were incorporated into the Newtonian theory as a special case. Were we to regard this branch of inquiry cultivated by Galileo as a distinctive science, the subsequent facts could be described by saying that Galileo's special discipline was reduced to the science of Newton. However, although it is possible to distinguish the subject matters of the Newtonian and the (initially distinct) Galilean sciences (for example, the latter was concerned solely with terrestrial phenomena, while the former included celestial ones), these subject matters are in an obvious sense homogenous and continuous; for it is the motions of bodies and the determinants of such motions that are under investigation in each case, and in each case inquiry is directed toward discovering relations between physical traits that are the common concern of both disciplines. Stated more formally, the point is that no descriptive terms appear in the formulations of the Galilean science which do not occur essentially and with approximately the same meanings in the statements of Newtonian mechanics. The history of science is replete with illustrations of reductions of this type, but I shall ignore them in what follows, because the logical issues involved in them do not appear to generate typical forms of philosophic puzzlement or to stimulate fundamental reinterpretations of the nature of knowledge.

The situation seems to be quite different, however, in those cases of reduction in which a subject matter possessing certain distinctive properties is apparently assimilated with another that supposedly does not manifest those traits; and acute intellectual discomfort is often experienced in those instances of reduction in which the science that suffers reduction is concerned with so-called "macroscopic" or "molar" phenomena, while the science to which the reduction is effected employs a theory that postulates some "microscopic" structure for molar physical systems. Thus, consider the following example. Most adults, if provided with ordinary mercury thermometers, are able to determine with reasonable accuracy the temperatures of various bodies, and understand what is meant by such statements as that the temperature of a glass of milk is 10° C. Accordingly, such individuals know how to use the word "temperature," at any rate within a broad context, though doubtless a large fraction of them would be incapable of stating adequately the tacit rules governing such usage, or of explicating the meaning of the word to the satisfaction of someone schooled in thermodynamics. However, if such an individual were to use the word so that its application was always associated with the behavior of a mercury column in a glass tube when the latter was placed in proximity to the body whose temperature was in question, he might be at a loss to construe the sense of such a statement as that the temperature of a certain substance at its melting point is several thousand degrees high; and he might protest that since at such alleged "high temperatures" ordinary thermometers would be vaporized, the statement had no definite meaning for him. But a slight study of physics would readily remove this source of puzzlement. The puzzled individual would discover that the word "temperature" is associated with a more inclusive set of rules of usage than he had originally supposed, and that in its extended usage it refers to a physical state of a body, which may be manifested in other ways than in the volume expansion of a mercury column—for example, in changes in electrical resistance, or in the generation of electric currents. Accordingly, once the laws are understood which connect the behavior of ordinary thermometers with the behavior of bolometers, pyrometers, and other overtly identifiable recording instruments, the grounds for the more inclusive usage of the term "temperature" become intelligible. This wider use of the word, then, rarely appears to cover a mystery, any more than does the extension of the word from its uses in contexts of direct experience of hot and cold to contexts in which the mercury thermometer replaces the human organism as a test body.

Suppose, however, that the layman for whom the word "temperature" thus acquires a more generalized meaning than he originally associated with it now pursues his study of physics into the kinetic theory of matter. Here he discovers that the temperature of a body is simply the mean kinetic energy of the molecules constituting the body. But this bit of information usually produces renewed perplexity, and, indeed, in an especially acute form. For the layman is assured by the best authorities that while on the one hand individual molecules possess no temperatures, nevertheless the meaning of the word "temperature" must by definition be taken as identical with the meaning of such expressions as "energy of molecular motions." And questions that are typical of a familiar philosophical tradition now seem both relevant and inescapable. If the meaning of "temperature" is the same as that of "kinetic energy of molecular motion," what are we talking about when milk is said to have a temperature of 10° C.? Surely not the kinetic energies of the molecular constituents of the liquid, for the uninstructed layman is able to understand what is thus said without possessing any notions about the molecular composition of milk. Perhaps, then, the familiar distinctions between hot and cold, between various temperatures as specified in terms of the behavior of identifiable instruments, are distinctions which refer to a domain of illusion. Perhaps, also, the temperatures that are measured in ordinary experience as well as in laboratories are merely indications of some fundamental underlying reality which is inherently incapable of being characterized by such expressions as "temperature" understood in its customary sense. Or should we perhaps regard temperature as an emergent trait, not present on lower levels of physical reality? But if this is the correct way of viewing the matter, does a theory that is about such lower levels ever really explain emergent traits such as temperature? It would be easy to enlarge the list of such queries, but those cited suffice to suggest the general character of the instances of reduction which provoke them. To avoid repeated circumlocution, and for lack of better labels, let me refer to a science to which another is reduced as the "primary science," and to the science which suffers such reduction as the "secondary science." Philosophical problems of the sort indicated, then, seem to be generated when the subject matter of the primary science is "qualitatively discontinuous" or "in-homogenous" with the subject matter of the secondary science—or, to put the matter perhaps more clearly, when the statements of the secondary science contain descriptive terms that do not occur in the theories of the primary science.

It is reductions of this type that I wish to consider. And since the reduction of thermodynamics to mechanics, more exactly, to statistical mechanics and the kinetic theory of matter, is both a typical and a relatively familiar and simple example of this type, I propose to center my discussion around this illustration.

I will first briefly recall some well-known historical facts. The study of thermal phenomena goes back in modern times to Galileo and his circle, and during the subsequent three centuries a large number of laws were established dealing with special phases of the thermal behavior of bodies—laws which were eventually exhibited as systematically interrelated on the basis of a small num-

ber of general principles. Thermodynamics, as this science
came to be called, employed concepts, distinctions, and
general laws which were also used in mechanics—for
example, the notions of volume, weight, mass, and pressure, and laws such as the principle of the lever and
Hooke's Law. Nevertheless, it was regarded as a science
relatively autonomous with respect to mechanics, because
it made use of such distinctive notions as temperature,
heat, and entropy, and because it assumed laws and principles which were not corollaries of the fundamental assumptions of mechanics. Accordingly, though many propositions of mechanics were constantly employed in the
exploration of thermal phenomena, thermodynamics was
generally assumed for a long time to be a special discipline,
plainly distinguishable from mechanics and not simply a
chapter of it. In this respect, the relation of thermodynamics to mechanics was considered analogous to the relation
between mechanics and physical geometry: mechanics was
held to be distinguishable from physical geometry, even
though geometrical propositions were employed in the
formulation of mechanical laws and in the construction of
instruments used to test these laws. Indeed, thermodynamics is still frequently expounded as a physical theory
that is autonomous in the indicated sense with respect to
mechanics; and in such expositions the findings of the science are presented in such a manner that the propositions
asserted can be understood and verified in terms of explanations and procedures which do not assume the reducibility of thermodynamics to some other theory. However,

experimental work early in the nineteenth century on the mechanical equivalent of heat stimulated theoretical inquiry to find a more intimate connection between thermal and mechanical phenomena than the bare facts seemed to assert. And when Maxwell and Boltzmann were finally able to "derive" the Boyle-Charles Law from assumptions apparently statable in terms of mechanics concerning the molecular constitution of ideal gases, and especially when the entropy principle was shown to be capable of interpretation as a statistical law concerning the aggregate mechanical behavior of molecules, thermodynamics was widely believed to have lost its autonomy and to have been reduced to mechanics.

Just how is this reduction effected, and what is the argument which apparently makes possible the derivation of statements containing such terms as "temperature," "heat," and "entropy" from a set of theoretical assumptions that do not use or mention them? It is not possible, without producing a treatise on the subject, to exhibit the complete argument. I shall therefore fix my attention on a small fragment of the complicated analysis, the derivation of the Boyle-Charles Law for ideal gases from the assumptions of the kinetic theory of matter.

Suppressing most of the details that do not contribute directly to the clarification of the main issues, a simplified form of the derivation is in outline as follows. Assume an ideal gas to occupy a volume V. The gas is taken to be composed of a large number of molecules possessing equal mass and size, each perfectly elastic and with dimensions that are negligible when compared with the average distance between them. The molecules are further supposed to be in constant relative motion, and subject only to forces of impact between themselves and the walls of the containing volume, also taken to be perfectly elastic. Accordingly, the motions of the molecules are assumed to be analyzable in terms of the principles of Newtonian mechanics. The problem now is to determine the relation of the pressure which the molecules exert on the walls of their container to other aspects of their motion.

However, since the instantaneous co-ordinates of state of the individual molecules are not actually ascertainable, the usual mathematical procedure of classical mechanics cannot be applied; and in order to make headway with the problem, a further assumption must be introduced— an assumption which is a statistical one concerning the positions and momenta of the molecules. This statistical assumption takes the following form. Suppose that the volume V of the gas is subdivided into a very large number of smaller volumes whose dimensions are equal but nevertheless are large compared with the diameters of the molecules; suppose also that the maximum range of velocity of the molecules is divided into a large number of equal intervals of velocity; and associate with each small volume all possible velocity intervals, calling each complex obtained by associating a volume with a velocity interval a "phase-cell." The statistical assumption then is that the probability of a molecule's occupying an assigned phase-cell is the same for all molecules and phase-cells, and that (subject to certain qualifications which need not be mentioned here) the probabilities that any pair of molecules will occupy the same phase-cell are independent. From this set of assumptions it is now possible to deduce that the pressure p which the molecules exert on the walls of the container is related in a definite way to the mean kinetic energy E of the molecules of the gas, and that in fact $p = 2E/3V$, or $pV = 2E/3$. But a comparison of this equation with the Boyle-Charles Law (according to which $pV = kT$, where k is constant for a given mass of gas and T its absolute temperature), suggests that the latter could be deduced from the assumptions mentioned, *if* temperature were "identified" with the mean kinetic energy of molecular motions. Accordingly, let us adopt this "identification" in the form of the hypothesis that $2E/3 = kT$ (i.e., that the absolute temperature of an ideal gas is proportional to the mean kinetic energy of the molecules which are assumed to constitute it). The Boyle-Charles Law is then a logical consequence of the general principles of mechanics, when these are supplemented by a statistical postulate on the motions of molecules constituting a gas, a hypothesis on the connection between temperature and kinetic energy, and various further assumptions that have been indicated.

If the derivation of the Boyle-Charles Law is used as a basis for generalization, what are the essential requirements for reducing one science to another? The following comments fall into two groups, the first dealing with

matters that are primarily of a formal nature, the second
with questions of an empirical character.

1. In the first place, the derivation requires that all the
assertions, postulates, or hypotheses of each of the sciences
involved in the reduction are available in the form of
explicit statements, whose meanings are assumed to be
fixed in terms of procedures and rules of usage appropriate to each discipline. Moreover, the statements within
each science fall into a number of fairly distinct groups
when a classification is introduced on the basis of the
logical functions the statements possess in the discipline.
The following schematic list, though not exhaustive, indicates what I believe to be the more important groupings.

(a) In a highly developed science such as mechanics
there usually is a class T of statements which constitute
the fundamental theory of the discipline and thus serve
as principles of explanation and as partial premises in
most deductions undertaken in the science, e.g., the principles of Newtonian mechanics. In a given exposition of
the science, these statements are logically primitive, in
the sense that they are not derived from any other class
of statements in the science. Whether this class of statements is best conceived as a set of leading principles,
empirical rules of inference, or methodological rules of
analysis, rather than as premises in the usual sense of the
word, is a question that can be ignored here.

(b) A science which contains a fundamental theory will
also contain a class of statements or theorems which are
logically derivable from T. These theorems in all but

trivial cases are usually of a conditional form, and their consequents are derivable from T only if the latter is supplemented by various special assumptions which appear as the antecedents in the theorems. Two subdivisions of this class of special assumptions may be distinguished. (i) There is the group of assumptions which serve as general hypotheses concerning a variety of conditions to which the fundamental theory may be applied. Thus, one such assumption in the application of Newtonian principles to the study of gases is that of a physical system composed of a large number of point-masses, with forces of impact as the only forces present. An alternative assumption might be that of a physical system consisting of bodies with non-negligible diameters subject to gravitational forces. (ii) And there is also the group of assumptions which specify the detailed boundary or initial conditions for the application of the theory. Thus, in the above example the initial conditions are stated as a statistical assumption concerning the position and velocities of gas molecules.

(c) Finally, every positive science will contain a large class of singular statements which formulate procedures and the outcome of observations relevant for the conduct of inquiry in the science; and it will usually also make use of general laws which its fundamental theory does not pretend to explain but which are simply borrowed from some other special discipline. Call the first group of these statements "observation statements," and the second group "borrowed statements." Observation statements may on occasion serve as specifications of the initial conditions for the application of the theory, or they may state the predicted consequences of the theory when other such statements are used to supplement the latter as initial conditions. Accordingly, observation statements will normally have members in common with the class of statements of boundary and initial conditions, though in general these two classes will not coincide. Indeed, many observation statements will describe instruments required for testing general assumptions of the science, and in doing so may make use of general laws and hence of expressions referring to distinctions that fall within the province of some other specialized discipline. For example, if Newtonian assumptions are employed in the study of celestial phenomena, telescopes may be required to test these assumptions; but the description of telescopes, and the interpretation of the observations that are obtained through their use, generally involves the use of expressions that refer to distinctions studied primarily in theoretical optics rather than in Newtonian mechanics.

2. This brings me to my second formal observation. The statements of a science, to whichever of the above classes they may belong, can be analyzed as linguistic structures compounded out of more elementary expressions in accordance with tacit or explicit rules of construction. These elementary expressions E are of various sorts, but they may be assumed to have fairly definite meanings fixed by habit or explicit rules of usage. Some of them are the familiar expressions of logic, arithmetic, and perhaps

higher mathematics; but most of them will usually be so-called "descriptive" terms or combinations of terms which signify allegedly empirical objects, traits, and processes.

Though there may be serious difficulties both theoretical and practical in distinguishing descriptive expressions from others, let us suppose that the distinction can be carried through in some fashion, and let us consider the class of descriptive expressions in E. Many of the descriptive expressions of a science are taken over from the language of ordinary affairs and retain their customary, everyday meanings; others, however, may be specific to the science, and may, moreover, have meanings which preclude their application to matters of familiar experience. Thus the statements constituting the fundamental theory of a science, as well as many of the special assumptions which are used to supplement the theory in various ways, normally contain several descriptive expressions of this latter sort.

Now it is generally possible to explicate the meanings of many descriptive expressions in E with the help of other such expressions, though of course logical expressions will play a role in the explication. Let us refer to those descriptive expressions with the help of which the meanings of all other such expressions may be explicated —whether the explication is given in the form of conventional explicit definition or through the use of different and more complicated logical techniques—as the "primitive expressions" of the science. (Expressions that are primitive in this sense may be primitive only in some specific context of analysis and not in another. But this point, though not without importance for a general theory of definition, does not affect the present discussion.)

However, the explication of the meaning of an expression may have either of two objectives, and accordingly it is useful to distinguish between two classes of primitive expression. (a) The explication may aim at specifying the meaning of an expression in terms familiar from everyday usage; and in consequence, the primitives employed may be restricted to those expressions which refer to matters of common observation, laboratory procedure, and other forms of overt behavior. Call such primitives "experiential primitives," even if no sharp line may be drawn between expressions that are experiential and those that are not. For example, the meaning of the word "temperature" is often specified by means of statements describing the volume expansion of liquids and gases, or the behavior of other readily observable bodies; and in this instance the primitives employed in the explication are experiential ones.

(b) On the other hand, an explication may aim at specifying the meaning of an expression by exhibiting its relation to the meanings of expressions used in formulating the fundamental theory or the various supplementary assumptions of the science. And in consequence, the primitives employed may in fact contain no expression which refers to matters accessible to direct observation. Call such primitives the "theoretical primitives" of the science. For example, the meaning of the word "temperature" is sometimes specified with the help of statements describing the Carnot cycle of heat transformations, statements which contain expressions like "perfect nonconductor," "infinite heat-reservoir," and "infinitely slow volume expansion," that have no manifest reference to anything that is observable. Again, the explication of the expression "center of mass," as customarily given in treatises on mechanics, involves the use of other expressions that are basic in formulating the principles of mechanics, though they do not all refer to directly observable characteristics of bodies.

It is not necessary to decide, for the purpose of the present discussion, whether the meanings of all theoretical primitives of a science are explicable with the help of its experiential primitives. And though the class of theoretical primitives of a discipline and the class of its experiential primitives may have expressions in common, the two do not in general coincide.

3. I come to my third comment of a formal nature. A comparison of the statements belonging to the primary science involved in a reduction with those belonging to the secondary science shows that in general the two sciences share a number of common statements and expressions, the fixed meanings of these expressions being the same for both sciences. Statements certifiable in logic and demonstrative mathematics are obvious examples of such common expressions, but, in addition to them, the two sciences will frequently share statements and other expressions which have a descriptive or empirical content. For example, many

propositions that fall within the field of mechanics, such
as the law of the lever, also enjoy important uses in thermodynamics, as one of the borrowed statements of the latter
science; and thermodynamics also employs such expressions as "volume," "weight," and "pressure" in senses which
coincide with the meanings of these words in mechanics.
On the other hand, the secondary science prior to its reduction generally contains statements and expressions not occurring in the primary science, except possibly as members
of the class of observation and borrowed statements. For
example, theoretical mechanics in its classical form contains neither the Boyle-Charles Law nor the word "temperature," though both of these occur in thermodynamics,
and though the word may on occasion be employed in
statements which describe the conditions of application of
the first principles of mechanics.

Now it is of the utmost importance to observe that expressions peculiar to a science will possess meanings that
are fixed by its *own* procedures, and that are therefore intelligible in terms of its own rules of usage, whether or not
the science has been or will be reduced to some other discipline. In many cases, to be sure, the meanings of some
expressions in a science can be explicated with the help of
thóse occurring in another, and, indeed, even with the help
of the theoretical primitives of the latter. For example, it
is usually assumed that an analytical equivalence can be
exhibited between the word "pressure" as employed in thermodynamics and other expressions belonging to the class
of theoretical primitives in the science of mechanics. But

it obviously does *not* follow that every expression used in a sense that is specified in a given science must or need be explicable in terms of the primitives, whether theoretical or experiential, of another discipline.

Let us finally consider what is formally required for the reduction of one science to another. The objective of the reduction is to show that the laws or general principles of the secondary science are simply logical consequences of the assumptions of the primary science. However, if these laws contain expressions that do not occur in the assumptions of the primary science, a logical derivation is clearly impossible. Accordingly, a necessary condition for the derivation is the explicit formulation of suitable relations between such expressions in the secondary science and expressions occurring in the premises of the primary discipline.

Now it may be possible to explicate the meaning of an expression occurring in a law of the secondary science in terms of the experiential primitives of the primary one, especially if, as is perhaps normally the case, the experiential predicates of the two sciences are the same. But this possibility is not in general sufficient for the purposes of reduction, since the problem here is to establish a certain kind of connection between expressions that occur in the secondary science but not in the premises of the primary discipline and expressions that do appear in these premises, especially those expressions of the latter class in terms of which the fundamental theory of the primary science is formulated. For though the uses of each of two expressions may be specifiable with the help of a common set of experiential primitives, it by no means follows that one of the expressions must be definable in terms of the other. The words "uncle" and "grandfather," for instance, are each definable in terms of "male" and "parent," but "uncle" is not definable in terms of "grandfather." Accordingly, a crucial step in reduction consists in establishing a proper kind of relation—that is, one which will make possible the indicated logical derivation—between expressions occurring in the laws of the secondary science and the theoretical primitives of the primary science.

There appear to be just two general ways of doing this. One is to show that an expression in question is logically related, either by synonymity or entailment, to some expression in the premises of the primary science. In consequence, the meaning of the expression in the secondary science, as fixed by the usage established in this discipline, must be explicable in terms of the theoretical primitives of the primary science. The other way is to adopt a material or physical hypothesis according to which the occurrence of the properties designated by some expression in the premises of the primary science is a sufficient, or a necessary and sufficient, condition for the occurrence of the properties designated by the expression of the secondary discipline. But in this case the meaning of the expression in the secondary science, as fixed by the established usages of the latter, is not declared to be analytically related to the established meaning of the corresponding expression in the primary science. In consequence, the indicated hypothesis cannot be asserted on the strength of purely logical considerations, but is at best a contingent truth requiring support from empirical data.

Let us now assume that the word "temperature" is the only expression that occurs in the Boyle-Charles Law which does not also occur in the various premises of mechanics and the kinetic theory of gases from which the law is to be derived. Accordingly, if the deduction is to be possible, an additional assumption must be introduced—the assumption that temperature is proportional to the mean kinetic energy of the gas molecules. How is this assumption to be understood, and in particular what sort of considerations support the indicated connection between the word "temperature" and the expression "mean kinetic energy"? But it is clear that in the sense in which "temperature" is used in thermodynamics, the word is neither synonymous with "mean kinetic energy" nor is its meaning entailed by the meaning of the latter expression. For it is surely not by analyzing the meaning of "temperature," in its thermodynamical sense, that the additional assumption required for deducing the Boyle-Charles Law from the premises of mechanics can be established. This additional assumption is evidently an empirical hypothesis, which postulates a determinate factual connection between two properties of physical systems that are in principle independently identifiable—between temperature as specified in thermodynamics on the one hand and the state of having a certain mean kinetic energy on the other; and if the hypothesis is true, it is at best only contingently true.

One objection to this last claim must be briefly considered. It is well known that though an expression may possess a certain fixed meaning at one stage in the development of inquiry, the redefinition of expressions is a recurrent feature in the history of the sciences. Accordingly, so the objection runs, while in an earlier usage the word "temperature" possessed a meaning which was specified by the procedures of thermometry and classical thermodynamics, it is now so used that temperature is "identical by definition" with molecular motion. The deduction of the Boyle-Charles Law does not therefore require that the premises of mechanics be supplemented with a contingent physical hypothesis but simply makes use of this definitional identity. This objection seems to me to illustrate the curious double talk of which highly competent scientists are sometimes guilty, to the detriment of essential clarity. It is obviously possible to so redefine the word "temperature" that it becomes synonymous with "mean kinetic energy of molecules." But it should be no less obvious that on this redefined usage, the word has a different meaning from the one associated with it on the basis of the usage customary in thermometry and thermodynamics, and in consequence a different meaning from the one associated with it in the Boyle-Charles Law. If, then, thermodynamics is to be reduced to mechanics, it is temperature in the sense specified in the former science which must be shown to be connected with mean kinetic energy. Accordingly, if the word "temperature" is redefined as proposed, the hypothesis must be adopted that the state of bodies described by the word "temperature" in its thermodynamical meaning is also correctly characterized by the word "temperature" in its redefined and different sense. But then this hypothesis is one which does not hold simply by definition. And unless it is adopted, it is not the Boyle-Charles Law which is derived from the premises of mechanics; what is derived is a sentence with a physical and syntactical structure similar to the law, but with a sense that is entirely different from what the law asserts.

I now turn to my second set of comments, those concerned with matters that are not primarily formal.

1. Thus far, I have been arguing the doubtless obvious point that the reduction of one science to another is not possible unless the various expressions occurring in the laws of the former also appear in the premises of the latter. But it is perhaps equally evident that these premises must satisfy further conditions if a proposed reduction is to count as an important scientific achievement. For if the premises of an alleged primary science could be selected quite arbitrarily, subject only to the formal requirements that have been mentioned thus far, the logical deduction of the laws of a secondary science from such premises selected *ad hoc* would in most cases represent only a trivial scientific accomplishment. And in point of fact, an essential condition that is normally imposed upon the assumptions of the primary science is that they be supported by empirical evidence possessing some measure of adequacy. The issues raised by this requirement, and especially the problems

connected with the notion of adequate evidence, cannot be discussed in the present paper, and in any case are not pertinent exclusively to the analysis of reduction. However, a few brief reminders bearing on this requirement that are especially relevant to the reduction of thermodynamics to mechanics may contribute something to the present analysis.

It is well known that the general assumption according to which physical bodies in different states of aggregation are systems of molecules is confirmed by a large number of well-established experimental facts of chemistry and of molar physics, facts which are not primarily about thermal properties of bodies. Accordingly, the adoption of this hypothesis for the special task of accounting for the thermal behavior of gases is in line with the normal strategy of the natural sciences to extend the use of ideas fruitful in one set of inquiries into related domains. Similarly, the fundamental principles of mechanics, which serve as partial premises in the reduction of thermodynamics to mechanics, are supported by evidence drawn from many fields of study distinct from the study of gases. The assumption that these principles characterize the behavior of the hypothetical molecular constituents of a gas thus involves what is essentially the extrapolation of a theory from domains in which the theory has been well confirmed to another domain whose relevant features are postulated to be homogenous with those of the former domains. But in addition to all this, it is especially noteworthy that the combined set of assumptions employed in the reduction of thermodynamics to mechanics, including the special hypothesis on the connection of temperature and kinetic energy, make it possible to bring into systematic relations a large number of propositions on the behavior of gases as well as of other bodies, propositions whose factual dependence on one another might otherwise not have become evident. Many of these propositions were known to be in approximate agreement with experimental facts long before the reduction was effected, but some of them, certainly, were discovered only subsequently to the reduction, and partly as a consequence of the stimulus to inquiry which the reduction supplied.

This last point needs to be stressed. It is fairly safe to maintain that the mere deduction of the Boyle-Charles Law from the assumptions of mechanics does not provide critical evidence for those assumptions, and especially for the assumption on the connection between temperature and kinetic energy, for prior to the reduction this law was already known to hold, at least approximately, for most gases far removed from their points of liquefaction. And though the adoption of those assumptions does effect, in consequence of the mere deduction of the law, a unification of physical knowledge, the unification is obtained on the basis of what to many practicing scientists seems an *ad hoc* postulation. The crucial evidence for those assumptions, and therefore for the scientific importance of the reduction, appears to come from two related lines of inquiry: the deduction from these assumptions of hitherto unknown connections between observable phenomena, or of propositions which are in better agreement with experimental findings than any that had been previously accepted; and secondly, the evaluation, from data of observation, of various constants or parameters that appear in the assumptions, with the proviso that there is good agreement between the values of a constant calculated from data obtained from independent lines of inquiry. For example, though the Boyle-Charles Law holds approximately for ideal gases, most gases under all but exceptional circumstances do not behave in accordance with it. On the other hand, if some of the assumptions used in the deduction of the law from mechanics are modified in a manner not radically altering their main features—specifically, if molecules are assumed to have diameters that are not negligible in comparison with the mean distances separating them, and if cohesive forces between molecules are also postulated—the proposition known as Van der Waal's equation can be derived, which is in much closer approximation to the actual behavior of most gases than is the Boyle-Charles Law. Again, to illustrate the second type of evidence generally accepted as critical for the importance of the reduction of thermodynamics to mechanics, one of the assumptions involved in that reduction is that under conditions of standard pressure and temperature equal volumes of a gas contain an equal number of molecules, quite irrespective of the chemical nature of the gas. Now the number of molecules contained in a liter of a gas (Avogadro's number) can be calculated on the basis of data obtained from observations, though to be sure only if these data are interpreted in a specified manner; and it turns out that alternative ways of calculating this number yield estimates that are in good agreement with one another, even when the measurements which serve as the basis of the calculations are obtained from the study of quite different materials—e.g., Brownian movements and crystal structure, as well as thermal phenomena.

2. These admittedly sketchy remarks on the character of the empirical evidence which supports the assumptions of a primary science merely hint at the complex considerations that are actually involved in judging whether a proposed reduction of one science to another is a significant advance in the organization of knowledge or whether it is simply a formal logical exercise. However, these remarks will perhaps help make plain that even though a science continues to be distinguished from other branches of inquiry on the basis of the general character of its fundamental theory, it may with the progress of inquiry modify or supplement the details of many of its subordinate and yet still quite general assumptions.

And this brings me to my next comment. For if this last point is well taken, it is clear that the question whether a given science is reducible to another needs to be made more explicit by the introduction of a definite date. No practicing physicist will take seriously the claim that, say, electrodynamics is reducible to mechanics—even if the claim were accompanied by a formal deduction of the equations of electrodynamics from a set of assumptions that by common consent are taken to fall within mechanics—unless these assumptions are warranted by independ-

ent evidence available at the time the claim is made. It is
thus one thing to say that thermodynamics is reducible to
mechanics when the latter includes among its assumptions
certain hypotheses on the behavior of molecules, and quite
a different thing to claim that the reduction is possible to a
science of mechanics that does not countenance such hypotheses. More specifically, thermodynamics can be reduced
to a mechanics that postdates 1866, but it is not reducible
to a mechanics as this science was conceived in 1700. Similarly, a certain part of chemistry is reducible to a post-1925
physical theory, though not to the physical theory of a
hundred years ago.

In consequence, much traditional and recent controversy
over the interrelations of the various special sciences and
concerning the supposed limits of the explanatory power of
physical theory can be regarded as a debate over what at
a given time is the most promising line of research and
scientific advance. Thus, biologists who insist upon the
importance of an "organismic" theory of biological behavior and who reject "machine-theories" of living structures may be construed as maintaining, though by no means
always clearly, that in the present state of physical and biological theory it is advantageous to conduct their inquiries
without abandoning distinctions peculiar to biology in
favor of modes of analysis typical of modern physics. On
the other hand, the mechanists in biology can be understood as recommending, though often in the language of a
dogmatically held ultimate philosophy, a general line of
attack on biological problems which in their opinion would

advance the solution of these problems and at the same time hasten the assimilation of biology to physics—even if the physics to which biology may eventually be reduced may differ from the present science of physics in important though unspecified respects. However this may be, if the controversy over the scope of physics is conceived in this manner, no major philosophical or logical issue appears to be raised by it, though subsidiary questions involved in the controversy may require logical clarification. If one takes sides in the debate, one is primarily venturing a prediction, on what are often only highly conjectural grounds, as to what will be the most fertile avenue of exploration in a given subject matter at a given stage of the development of several sciences. On the other hand, when such controversies overlook the fact that the reduction of one science to another involves a tacit reference to a date, they assume the character of typically irresoluble debates over what are alleged to be metaphysical ultimates; and differences and similarities between departments of inquiry that may possess only a temporary autonomy with respect to one another come to be cited as evidence for some immutably final account of the inherent nature of things.

3. These last remarks have prepared the way for my final comment. Unlike the present discussion, which views the reduction of one science to another in terms of the logical connections between certain empirically confirmed statements of the two sciences, analyses of reduction and of the relations between sciences in general frequently approach these questions in terms of the possibility or im-

possibility of deducing the properties of one subject matter from the properties of another. Thus, a contemporary
writer argues that because "a headache is not an arrangement or rearrangement of particles in one's cranium" and
our sensation of violet is not a change in the optic nerve,
psychology is demonstrably an autonomous discipline;
and accordingly, though the mind is said to be connected
with physical processes, "it cannot be reduced to those
processes, nor can it be explained by the laws of those
processes." Another recent writer, in presenting the case
for the occurrence of "genuine novelties" in the inorganic
realm, warns that "it is an error to assume that *all* the
properties of a compound can be deduced solely from the
nature of its elements." And a third influential contemporary author asserts that the characteristic behavior of a
chemical whole or compound, such as water, "*could* not,
even in theory, be deduced from the most complete knowledge of the behavior of its components, taken separately
or in other combinations, and of their properties and arrangements in this whole."

Such an approach to the question almost invariably
transforms what is eminently a logical and empirical problem, capable in principle of being resolved with the help
of familiar scientific methods and techniques, into a speculative issue that becomes the concern of an obscure and
inconclusive dialectic. And in any case, formulations such
as those just cited are highly misleading, in so far as they
imply that the reduction of one science to another deprives any properties known to occur of a status in exist-

ence, or in so far as they suggest that the reducibility of one science to another can be asserted or denied without reference to the specific theories actually employed in a primary science for specifying the so-called "natures" of its ostensible elements.

It is clearly a slipshod formulation, and at best an elliptic one, which talks about the "deduction" of properties from one another—as if in the reduction of one science to another one were engaged in the black magic of extracting one set of phenomena from others incommensurably different from the first. Once such an image is associated with the facts of scientific reduction, the temptation is perhaps irresistible to read these facts as if in consequence some characters of things were "unreal" and the number of "genuine" properties in existence were being diminished. And it is simply naïveté to suppose that the natures of the various hypothetical objects assumed in physics and chemistry can be ascertained once and for all and by way of a direct inspection of those objects, so that in consequence it is possible to establish for all time what can or cannot be deduced from those natures. To the extent that one bases one's account of these matters on the study of scientific procedure, rather than on the frequently loose talk of scientists, it is plain that just as the fundamental nature of electricity is stated by Maxwell's equation, so the natures of molecules and atoms and of the properties of these postulated objects are always specified by a more or less explicitly articulated theory or set of general statements.

It follows that whether a given set of properties or behavioral traits of macroscopic objects can be explained by or reduced to the properties and behavioral traits of atoms and molecules is in part a function of the theory that is adopted for specifying the natures of the latter. Accordingly, while the deduction of the properties studied by one science from those of another may not be possible if the latter discipline postulates certain properties for its elements in terms of one theory, the reduction may be quite feasible when a different theory is adopted for specifying the natures of the elements of the primary science. Thus, to repeat in the present context a point already made, if the nature of molecules is stipulated in terms of the theoretical primitives and assumptions of classical mechanics, the reduction of thermodynamics to mechanics is possible only if an additional hypothesis is introduced connecting temperature and kinetic energy. But as has been seen, the impossibility of the reduction without some such special hypothesis follows from purely formal considerations, and not from some alleged ontological hiatus between the microscopic and the macroscopic, the mechanical and the thermodynamical. Laplace was thus demonstrably in error when he imagined a divine intelligence that could foretell the future in every detail on the basis of knowing simply the instantaneous positions and momenta of all material particles as well as the magnitudes and directions of the forces acting between them. At any rate, Laplace was in error if his divine intelligence is assumed to draw inferences in accordance with the canons of logic, and is therefore assumed to be incapable of the blunder of asserting a statement as a conclusion if it contains expressions not occurring in the premises.

The question whether genuine novelties occur in nature when elements combine to form complex structures is clearly ambiguous. It can be construed as asking whether properties may not occur from time to time which have never before appeared anywhere in the cosmos. And it can also be understood as asking whether properties exhibited by various bodies assumed to be complex are in some cases at least different from and irreducible to the properties of their constituents. The question in the first sense clearly raises a problem in history which requires to be resolved with the help of the normal methods of historical inquiry; and the considerations raised in the present paper are not directly relevant to it. But the question in the second sense does call for a brief comment at this place. For the issue whether the properties of complexes are novel, in the nontemporal sense of the word, in relation to the properties of their elements, appears to be identical with the issue whether statements about the former are reducible to a primary science which deals with the latter. And if this is so, then the question whether the reduction is possible—and whether the properties alleged to be novel are indeed as thus described—cannot be discussed without reference to the specific theory which formulates the nature of the elements and of their properties. Failure to observe that novelty is a relational characteristic of properties with respect to a definite theory, and the supposition that on the contrary certain properties of compounds are inherently novel relative to the properties of the elements, irrespective of any theory which may be used to specify these elements and their properties, are among the chief sources for the widespread tendency to convert the analytic truths of logic into the dogmas of a footless ontology.

The chief burden of this paper, accordingly, is that the reducibility or irreducibility of a science is not an absolute characteristic of it. If the laws of chemistry—e.g., the law that under certain specified conditions, hydrogen and oxygen combine to form a stable compound, which in turn exhibits certain modes of behavior in the presence of other chemical substances—cannot be systematically deduced from one theory of atomic structure, they may be deducible from an alternate set of assumptions concerning the natures of chemical elements. Indeed, although not so long ago such a deduction was regarded as impossible— as it indeed was impossible from the then accepted physcal theories of the atom—the reduction of various parts of chemistry to the quantum theory of atomic structure now appears to be making steady if slow headway, and only the stupendous mathematical difficulties involved in making the relevant deductions from the quantum-theoretical assumptions seem to stand in the way of carrying the task through to completion. At the same time, the reduction of chemical law to contemporary physical theory does not wipe out, or transform into a mere appearance, the distinctions and the types of behavior which chemistry recognizes. Similarly, if and when the detailed physical, chemical, and physiological conditions for the occurrence of headaches are ascertained, headaches will not thereby be shown to be nonexistent or illusory. On the contrary, if in consequence of such discoveries a portion of psychology will have been reduced to another science or to a combination of other sciences, all that will have happened is that the occurrence of headaches will have been explained. But the explanation will be of essentially the same sort as those obtainable in other domains of positive science. It will not consist in establishing a logically necessary connection between the occurrence of headaches and the occurrence of traits specified by physics, chemistry, or physiology; nor will it consist in establishing the synonymity of the term "headache" with expressions defined with the help of the theoretical primitives of these disciplines. It will consist, so the history and the procedures of the sciences seem to indicate, in stating the conditions, specified in terms of these primitives, which as a matter of contingent fact do occur when a determinate psychological phenomenon takes place.

The nature of theoretical concepts
and the role of models in an advanced science

by R. B. Braithwaite

The function of every science is to establish laws—true hypotheses—which cover the behaviour of the observable things and events which are the subject-matter of the science thereby enabling the scientist both to connect together his knowledge of particular events and to enable him to predict what events will happen under certain circumstances. If these circumstances can be produced at will, knowledge of the general laws will enable to some extent the applied scientist to control the course of nature and to construct machines or other artefacts which will behave in known ways.

An advanced science like physics is not content only with establishing lowest-level generalisations covering physical events: it aims at, and has been largely successful in, subsuming its lowest-level generalisations under higher-level hypotheses, and thus in organising its hypotheses into a hierarchical deductive system—a scientific theory—in which a hypothesis at a lower level is shown to be deducible from a set of hypotheses at a higher level. Notable examples of this are Maxwell's subsumption of the laws of optics under his very general electromagnetic equations and Einstein's subsumption of gravitational laws under his General Theory of Relativity.

The concepts which enter into the higher-level hypotheses of an advanced science are usually concepts (e.g. electric-field vector, electron, Schrödinger wave-function) which are not directly observable things or properties as are those which appear in the lowest-level generalisations of the science: instead they are ' theoretical concepts ' which appear at the beginning of the deductive theory but which are eliminated in the course of the deduction.   These theoretical concepts present a problem to the philosopher of science: namely, what is their epistemological status? They are clearly in some way empirical concepts—an electron or a Schrödinger wave-function is not an object of pure mathematics like a prime number; but an electron is not observable in the sense in which a flash of light or the pointer reading on a measuring scale is observable. Nevertheless the truth of propositions about electrons is tested by the observable behaviour of measuring instruments; and the question is therefore in what manner an electron or other theoretical concept is an empirical concept. .

An answer to this question, an answer implicit in the writings of many philosophers of science such as Ernst Mach and Karl Pearson, was given explicity by Bertrand Russell in his doctrine of ' logical constructions '. " The supreme maxim in scientific philosophising is this: Wherever possible, logical constructions are to be substituted for inferred entities. " (*Mysticism and Logic and other essays*, 1918, p. 155). According to the logical construction view electrons, for example, are logical constructions out of the observed events and objects by which their presence can be detected; this is equivalent to saying that the word " electron " can be explicitly defined in terms of such observations.   On this view every sentence containing the word " electron " is translatable, without loss of meaning, into a sentence in which there only occur words which denote entities (events, objects, properties, relations) which are directly observable.   It is the business of a philosopher of science to show how these translations are to be made, and thus to show how the theoretical terms of a science can be explicitly defined by means of observable entities.   A philosopher of physics should be able to make this translation in the case of the word " electron ", and thus be able to exhibit the way in which electrons are logical constructions out of observable entities.   Russell's programme of logical construction is similar to the ' operationalist ' programme proposed by P. W. Bridgman, according to which theoretical terms must be defined by means of the empirical ' operations ' involved in their measurement (*The Logic of Modern Physics*, 1927).

This ' logical construction ' view of the nature of theoretical concepts was criticised by F. P. Ramsey in some notes which he wrote in 1929, a few months before his death at the age of 26, and which were published posthumously (*The Foundations of Mathematics and other logical essays*, 1931, pp. 212 ff.). Ramsey developed his criticism by constructing a simple example of a scientific theory; I have been able to construct (in my *Scientific Explanation*, 1953, Chapter III) even simpler examples which show precisely the defects of the logical construction view.

The point displayed by both Ramsey's and my examples is that, although it is always possible to define the theoretical terms occurring in the highest-level hypotheses of a theory by means of the terms denoting directly observable entities which occur in the lowest-level generalisations which the theory was propounded to explain, such a definition will prevent the theory from being expanded into a wider theory capable of explaining new lowest-level generalisations which may subsequently be established. To treat theoretical concepts as logical constructions out of observable entities would be to *ossify* the scientific theory in which they occur: the theory would be adequate to cover systematically a particular set of lowest-level generalisations already established, but there would be no hope of extending the theory to explain more generalisations than it was originally designed to explain. A scientific theory to be capable (like all good scientific theories) of this sort of growth must give more freedom of play to its theoretical concepts than the logical construction view will allow them to have.

What then, if the logical construction view is inadequate, is the epistemological status of theoretical concepts? A way of answering this question, essentially that of Ramsey, is to say that the status of the theoretical concept electron is given by specifying its place in the deductive system of contemporary physics in the following way: there is a property E (called " being an electron ") which is such that from certain higher-level hypotheses which are true and which are about the property electron there follow certain lowest-level generalisations which are empirically testable. According to this answer, nothing is asserted about the ' nature ' of the property E in itself; all that is asserted is that there are instances of E, namely electrons. To say that electrons exist is to assert the truth of the physical theory in which there occurs the concept of being an electron.

There is, however, another way of answering the status-question which is open to a philosopher of science who, with knowledge of the work of logicians such as Carnap done since Ramsey's death, would wish to make a sharper distinction than was made by Ramsey between a scientific theory arranged as a *deductive system* and the *calculus* (or language) representing the deductive system. So let us consider, not the nature of the theoretical concept electron in the deductively-arranged physical theory, but the role played by the term " electron " (or other synonymous symbol) in the calculus representing the theory. This calculus consists of a series of formulae (or sentences) arranged in such a way that all the formulae, except for a small number of formulae (called " initial formulae "), are derived from these initial formulae in accordance with the rules of the calculus. The calculus will be interpreted to represent the physical deductive theory by taking the highest-level hypotheses of the theory to be represented by initial formulae, and lower-level hypotheses and generalisations to be represented by derived formulae in the calculus. For this interpretation to be possible it is necessary that the rules of the calculus should correspond to the logical and mathematical principles of deduction used in making the deductions within the deductive theory.

When the deductive system to be represented by the calculus is a pure one, i.e. a system containing only logically necessary propositions (e.g. a deductive system of arithmetic starting with Peano's axioms), the calculus is interpreted all in a piece. Meanings are attached to the ' primitive ' symbols occurring in the initial formulae representing the axioms (e.g. to " number ", " successor of ", " zero " in Peano's arithmetical calculus), and meanings are attached to all the other symbols used in the calculus by the use of formulae of the calculus interpreted as explicit definitions of these other symbols in terms of the primitive symbols. (These definitions may, of course, be 'contextual definition ', i.e. definitions of a whole formula rather than of a separate term.) The meanings attached to the symbols do not depend upon the order of the formulae in the calculus representing the pure deductive system : the meaning, for example, of " prime number " does not depend upon the order in which theorems about prime numbers are deduced from the axioms of arithmetic.

The situation is quite different in the case in which the deductive system to be represented by the calculus is a scientific theory containing empirical hypotheses at different levels. Here, on my view, the calculus is not interpreted all in a piece: meanings are first attached to the symbols denoting the directly observable properties and relations which occur in the directly testable lowest-level generalisations of the theory. Meaning is then attached to the symbols which are to denote the theoretical concepts of the theory merely by virtue of the fact that they occur in the initial formulae of the calculus representing the theory. The formulae of the calculus are arranged in an order corresponding to the deductive arrangement of the hypotheses of the theory, with the initial formulae corresponding to the highest-level hypotheses. The initial formulae are interpreted as representing propositions from which directly testable propositions logically follow, and the symbols (e.g. " electrons ") which occur in these initial formulae are interpreted as being essential parts of these formulae. No direct meaning is attached to the term " electron ": it is given a meaning indirectly by the function it plays in the calculus which is interpreted as representing the physical theory.

On this view the status of the concept electron, and the question " Do electrons really exist? ", can only be discussed in terms of the role played by the word " electron " in the exposition of physical theory. This view resembles Russell's ' logical construction ' view in that in both cases what is in question is the meaning of the word or other symbol. But whereas Russell would say that the word is given a meaning by being *explicitly* defined by means of words denoting observable entities, I would say that it is given a meaning by showing its place in a calculus representing a scientific theory. This may be called giving an *implicit* or a *contextual* definition of the theoretical terms by means of words denoting observable entities occurring in the formulae of the calculus which represent the directly testable lowest-level generalisations of the theory; and my account may be taken as an elucidation of such indirect or contextual definition. (It will give a wider sense of " contextual definition " than the usual one, for it is the whole interpreted calculus and not only one sentence (or type of sentence) which will form the ' context ' for the purpose of the contextual definition.)

A direction of attention upon the calculus representing a deductive scientific theory will also throw light upon the use of a *model* in thinking about the scientific theory. Suppose that the calculus which is interpreted as representing the theory can also be interpreted as a deductive system in such a way that a direct meaning is given to all the symbols occurring in its initial formulae. In this second interpretation of the calculus the propositions represented will contain no ' theoretical ' concepts, and the calculus will be interpreted all in a piece—as in the case of a calculus representing a pure mathematical, deductive system. Thus this interpretation does not present the epistemological difficulty presented by the original interpretation of the calculus as representing the scientific theory. The deductive system which is this second interpretation of the calculus may be regarded as a model for the theory. A theory and a model for it have the same formal structure, since they are both represented by one and the same calculus. There is a one-one correlation between the propositions of the theory and those of the model, and the deductive arrangement of the propositions in the theory corresponds to that of the correlated propositions in the model (and vice versa). But the epistemological structure of the theory and of a model for it are different: in order to give the model the calculus is directly interpreted all in a piece, whereas to give the theory it is derived formulae of the calculus which are directly interpreted, the earlier formulae being interpreted indirectly by virtue of their place in the calculus.

The fact that a model has the same logical structure as, but a simpler epistemological structure than, the theory for which it is a model explains the use of models in thinking about a theory in an advanced science. For thinking about the model will, for many purposes, serve as a substitute for thinking explicitly about the calculus of which both theory and model are interpretations, since the model is a quite straightforward interpretation of the calculus. So to think of the model instead of the calculus in connection with the theory avoids the self-consciousness required in thinking at the same time of a theory and of the language in which it is expounded, and thus allows of a philosophically unsophisticated approach to an understanding of the logical structure of a scientific deductive theory.

The dangers of thinking of a theory by way of thinking of a model for it are, first, that we may, if we are not careful, suppose that there are concepts involved in the theory which correspond to *all* the properties of the objects in the model; we may think, for example, that the electrons in an atom have all the spatial properties of the balls in a ' solar system ' model of an atom. A second—more subtle—danger is that it may well happen that some of the propositions in the model which are the interpretations of the initial formulae of the calculus are logically necessary propositions (indeed they may all be logically necessary propositions, in which case the model is a pure mathematical model). We may then be tempted illicitly to transfer the logical necessity of these propositions in the model on to the correlated propositions in the theory, and thus to suppose that some or all of the highest-level hypotheses of the theory are logically necessary instead of contingent. In using models we must never forget that we are engaging in *as-if* thinking: the theoretical concepts in a scientific theory behave as if they were elements in the model, but only in certain respects. To forget the limitations is to misuse the valuable aid to thought provided by the model.

The topics of this paper are discussed at greater length in Chapters III and IV of my book *Scientific Explanation* (Cambridge: at the University Press: 1953).

King's College Cambridge.

RISTO HILPINEN

CARNAP'S NEW SYSTEM OF INDUCTIVE LOGIC*

I

During his life Carnap published two extensive, systematic works on inductive logic: *Logical Foundations of Probability* [4] and *The Continuum of Inductive Methods* [5]. These works report Carnap's research in induction and probability in the 1940's and early 1950's. After the publication of the *Continuum* in 1952 Carnap continued his research in inductive logic, and in the course of this research both his theory of inductive probability and his philosophy of induction underwent development. Most of Carnap's publications since 1952 have dealt with the philosophy and methodology of induction rather than inductive logic proper.[1] A part of his recent research in inductive logic will be published in an extensive work 'The Basic System of Inductive Logic', forthcoming in *Studies in Inductive Logic and Probability*, I–II (edited by Rudolf Carnap and Richard Jeffrey).[2]

This paper discusses the development of Carnap's system of inductive logic and his philosophy of induction during the past twenty years. Our exposition and discussion of Carnap's inductive logic is based mainly on 'The Basic System of Inductive Logic' [13B–C] and partly on an earlier (unpublished) version of the same work, 'An Axiom System of Inductive Logic' [13A]. The theory presented in the 'Basic System' will be termed 'Carnap's New System' or the 'Basic System', and that presented in *Logical Foundations of Probability* and the *Continuum* will be called 'Carnap's Old System'. Sections II–III of this paper discuss Carnap's philosophy of induction, Sections IV–VII contain an exposition of the main features of the 'Basic System', and the concluding Sections VIII–IX discuss the philosophical implications of Carnap's New System.

* An earlier version of this paper was read in a Symposium on Carnap's Philosophy at Stanford University, Stanford, 1970. This work has been supported by a U.S. Department of State Grant No. 70-066-A.

361

In *Logical Foundations of Probability* Carnap explains the notion of logical probability from three different points of view.[3] First, a conditional logical probability $c(h, e)$ can be understood as a measure of the degree to which the hypothesis h is confirmed or supported by the data e. This interpretation may be termed the *degree of confirmation*-explication or briefly the *inductivist* conception of logical probability and inductive logic. Secondly, logical probabilities can be regarded as (rational) degrees of belief or as 'fair betting quotients': according to this explication, $c(h, e) = r$ means that a bet on h with a betting quotient r is a fair bet if e describes the total knowledge concerning h available to the bettors.[4] This viewpoint will be termed the *belief-explication* of logical probability. Thirdly, logical probabilities can in certain cases be used as estimates of statistical probabilities (relative frequencies in the long run).[5]

The inductivist conception of logical probability is associated with the classical empiricist philosophy of science, in particular, with the problem of explicating how general theories or hypotheses are justified (or supported) by observational data. The belief-explication (or betting quotient-explication), on the other hand, is closely related to utility theory and the foundations of decision-making under uncertainty. (We might also call it the 'decision-theoretic conception of logical probability'.)[6] These explications are not mutually exclusive; they merely illustrate different aspects and fields of application of logical probability and inductive logic.[7]

In the early 1940's Carnap was mainly interested in the classical inductivist application of inductive logic; in fact, originally he used the expression 'degree of confirmation' as a technical term for logical probability (in order to distinguish it from the statistical concept of probability).[8] Carnap assumed that the degree to which an evidential statement e confirms a hypothesis h is expressible by a conditional logical probability $c(h, e) = r$, where the value r depends only on certain logical (or semantical) relations between h and e.

This conception has been criticized by several philosophers, especially by Ernest Nagel and Karl R. Popper.[9] Nagel and Popper have argued that no adequate *explicatum* of the presystematic notion of confirmation can have the same formal properties as the concept of conditional probability. In many cases 'confirmation' seems to mean *increase* of probability rather than just conditional probability; this is indicated, for instance, by the fact that we speak of *dis*confirmation as well as confirmation. This ambiguity in the intuitive notion of confirmation was pointed out by Carnap in the Preface to the Second Edition of *Logical Foundations of Probability*; here Carnap noted that 'confirmation' can mean either the *firmness* or the *increase of firmness* of a hypothesis (relative to evidential data).[10] Popper's criticism of Carnap's theory seems to be based partly on a confusion between these two notions.[11] In this later work Carnap preferred the belief-explication and emphasized the application of inductive logic to decision problems.[12]

This change of interest was perhaps not motivated merely by the ambiguity in the presystematic notion of confirmation. Even if the degree of confirmation cannot be expressed as a conditional probability (of a hypothesis relative to data), it is plausible to assume that it can be defined in probabilistic terms, and the theory of logical probability can thus serve as a foundation for the theory of confirmation. Most formal definitions of support proposed in recent literature are in fact based on the concept of probability.[13] In the case of Carnap's inductive logic, the emphasis on the decision-theoretic viewpoint may also be motivated by considerations of a different type. The inductivist conception is primarily relevant to situations in which we are interested in whether – and to what degree – some general hypothesis or theory is supported by empirical data. However, it has turned out to be very difficult to evaluate the degree of support of general theories in terms of conditional probabilities; for instance, in Carnap's Old System all genuinely universal hypotheses with factual content have zero probability relative to any type of observational evidence.[14] This difficulty is not specific to Carnap's inductive logic; the customary statistical methodology is subject to similar problems.[15] The belief-explication is related to decision-problems, and in this context the problem of assigning probabilities to general theories does not automatically arise. In practical decision problems it is normally not necessary to consider the probabilities of universal generalizations. Thus, quite apart from the ambiguity involved in the inductivist interpretation, Carnap's inductive logic seems to be more appropriate to applications related to the belief-explication than to discussion of the classical philosophical problems of inductive inference, among which questions related to general theories loom large.[16]

III

Carnap's conception of inductive probability is a variant of the personalist or Bayesian conception of probability. However, there are important differences between Carnap's theory and the customary Bayesian approach. First, Carnap defines probabilities as relative to a formalized language or a conceptual system with a fully specified logical structure. In Carnap's Old System probabilities are assigned to the *sentences* of a given formalized language $\mathscr{L}$. In the New System they are assigned to *propositions*, and propositions are identified with sets of models. However, the fields of sets (or propositions) considered in the New System are defined in terms of the atomic sentences of a formalized language; thus probabilities are relative to languages in Carnap's New System, too. Preoccupation with formalized languages or conceptual systems is characteristic of Carnap's philosophy of science in general; this feature of his probability theory can perhaps be seen as a consequence of his general philosophical methodology. Inductive logic is an extension of semantics and can best be studied by similar methods. However, this interest in conceptual systems is also related to another aspect of Carnap's theory. Carnap's probability theory is stronger than the standard Bayesian theory: he discusses and accepts axioms which are not among the axioms of standard probability theory. His theory of inductive logic has always been in an unfinished state; even in his last works he often says that his inductive logic is still "in the initial phase of the whole construction".[17] Carnap did not present a completed axiom system, but he was constantly attempting to find new axioms and conditions for logical probabilities. In the 'Basic System' the standard axioms are termed *basic axioms* or *axioms of coherence*; the c-functions satisfying these axioms are termed *regular* or *coherent* c-functions. Carnap points out that for any pair of contingent and logically independent sentences h, e and any real number r ($0 < r < 1$), there exists a coherent c-function c such that $c(h, e) = r$. This shows, according to Carnap, that quite unacceptable and unreasonable c-functions are consistent with the basic axioms. In his paper 'On Inductive Logic' Carnap says that the systems of standard axioms

restrict themselves to the first part of inductive logic, which, although fundamental and important, constitutes only a very small and weak section of the whole of inductive logic. The weakness of this part shows itself in the fact that it does not determine the value of c for any pair h, e except in some special cases where the value is 0 or 1. The theorems of this part tell us merely how to calculate further values of c if some values are given. Thus it is clear that this part alone is quite useless for application and must be supplemented by additional axioms.[18]

The main objective of Carnap's research in inductive logic from the early 1940's to recent years has been the formulation of such additional axioms and specification of a more restricted class of probability measures which would correspond to the intuitive notion of reasonable belief better than the class defined by the basic axioms alone.

Originally, in the 1940's, Carnap seemed to believe that one can find, for a given formalized language $\mathcal{L}$, a unique c-function c which can be regarded as *the* explicatum of 'rational degree of belief' or 'degree of confirmation' (this can be seen, for instance, from the paragraph cited above). Certain passages in 'On Inductive Logic' and in *Logical Foundations of Probability* indicate that at a certain time Carnap assumed that he had in fact found this special measure of confirmation; this confirmation function was the function c^*.[19] However, in the *Continuum of Inductive Methods* he defined a comprehensive class of c-functions, all of which satisfy the basic axioms and certain plausible additional requirements. In this class c^* does not seem to be the only reasonable or acceptable c-function; the main distinguishing feature of c^* is only the relative simplicity of its definition. Carnap did not claim any more that there exists a unique 'correct' c-function. In the *Continuum* the adequacy of an inductive method in a given universe is characterized in terms of its 'success' in this universe; this success is measured by the mean square error of the corresponding c-estimate function in the universe in question. Different c-functions are appropriate in different circumstances, and the choice of an inductive method (i.e., the choice of a c-function) should thus be dependent on empirical considerations.[20]

Later Carnap rejected this viewpoint, and assumed that the choice of a c-function is, at least partly, a subjective and personal decision which depends on certain personality traits of the individual concerned, and no objectively founded choice is possible. One of the relevant personality traits was termed by Carnap "inductive inertia"; this trait is reflected by the parameter λ (the characteristic parameter of the λ-system). The larger λ is, the stronger is the inductive inertia of the person in question, and the slower he modifies his *a priori* beliefs on the basis of observational evidence. Carnap calls this the "personalist point of view", but the term "psychologism" may be more appropriate. (This view must be distinguished from Savage's "personalism".)[21]

In the 'Basic System' Carnap returned to a more 'objectivist' and rationalist position and argued again that the choice of an inductive method should depend, not on observed facts and subjective factors, but only on *a priori* considerations.[22] However, he did not maintain that there exists only one acceptable *c*-function. Now he expressed the following view of the nature of inductive logic:

The person X wishes to assign rational credence values to unknown propositions on the basis of the observations he has made. It is the purpose of inductive logic to help him to do this in a rational way; or, more precisely, to give him some general rules, each of which warns him against certain unreasonable steps. The rules do not in general lead him to specific values; they leave some freedom of choice within certain limits. What he chooses is not a credence value for a given proposition but rather certain features of a general policy for determining credence values.[23]

Carnap's aim was to find methods of determining, for each knowledge situation, the 'correct' or 'acceptable' probability values (or intervals of probability values) of various propositions on the basis of the conceptual structure of the situation. The standard probability theory is not sufficient for this purpose: it says that if the probabilities of certain propositions are given, then the probabilities of certain other propositions are determined in accordance with the axioms of probability, but it cannot justify the total set of credence-values used in a given situation. According to Carnap's original program, the theory of logical probability is a "strong extension" of the standard theory in which the structure of the probability space (or sample space) determines a unique measure or a restricted class of probability measures,[24] and the structure of the probability space is determined by the structure of language. The latter assumption was rejected in Carnap's later work.[25] The theory presented in the 'Basic System' did not lead to a unique measure, but even in his last works Carnap was optimistic about the possibility of obtaining – at least in certain special cases – positive results in this direction and restricting the range of acceptable measures on the basis of *a priori* considerations. However, Carnap wanted to keep an "open mind" in this respect:

Suppose that at some point in the context of a given problem, say, the choice of a parameter value, we find that we have a free choice within certain boundaries, and that at the moment we cannot think of any additional rationality requirement which would constrict the boundaries. Then it would certainly be imprudent to assert that the present range of choice will always remain open. How could we deny the possibility that we shall find tomorrow an additional requirement? But it would also be unwise to regard it as certain that such an additional requirement will be found, or even to predict that by the discovery of further requirements the range will shrink down to one point.[26]

In the 'Basic System' Carnap distinguishes the 'subjectivist' and the 'objectivist' approach to inductive logic as follows: Subjectivism emphasizes the existing freedom of choice, whereas objectivism tends to stress the existence of limitations. Carnap was interested in finding new rationality requirements, and this led to an objectivist tendency in his inductive logic.[27] His success (or lack of success) in actually restricting the range of acceptable probability measures by new axioms is not the philosophically most significant aspect of his work, however. Carnap developed an interesting theory about certain conceptual factors that influence rational probability assignments, and this aspect of his work is of fundamental importance to the philosophy of induction. It has inspired philosophers who have studied other dimensions of inductive reasoning by similar methods, for instance the work on inductive generalization by Jaakko Hintikka[28] and Juhani Pietarinen.[29]

IV

In Carnap's Old System the functions m (measure function, absolute probability) and c (conditional probability) are defined on the sentences of a formalized language $\mathscr{L}$. In the 'Basic System' probabilities are defined on *events* or *propositions*; propositions are defined as sets of models (of a formalized language $\mathscr{L}$). Instead of the functions m and c applied to sentences, Carnap now uses the corresponding functions $\mathscr{M}$ and $\mathscr{C}$ applied to sets of models. Following Carnap's notation, we shall denote propositions by E, H, A, B, etc.; the absolute probability of H and the probability of H relative to E are expressed by $\mathscr{M}(H)$ and $\mathscr{C}(H \mid E)$, respectively.

In Carnap's New System probabilities are defined as measures of sets, but they are nevertheless relative to a formalized language $\mathscr{L}$, in particular, to the descriptive constants of $\mathscr{L}$. For each type of language discussed in the 'Basic System', Carnap defines the set of *atomic propositions* on $\mathscr{L}$, $\mathscr{E}^{at}$. Let $Z_\mathscr{L}$ (briefly Z) be the space of models of $\mathscr{L}$. The class of molecular propositions on $\mathscr{L}$, $\mathscr{E}_\mathscr{L}^{mol}$, is the field of sets generated by $\mathscr{E}_\mathscr{L}^{at}$ on Z, and the class of propositions on $\mathscr{L}$, $\mathscr{E}_\mathscr{L}$, is the σ-field generated by $\mathscr{E}_\mathscr{L}^{at}$ on Z. The functions $\mathscr{M}$ and $\mathscr{C}$ are defined for subfields of $\mathscr{E}_\mathscr{L}$.[30]

For the most part Carnap discusses only languages with a relatively simple structure which are termed *monadic predicate languages*. The New System is developed in detail only for such simple languages; in this respect it is similar to the Old System. The descriptive symbols of these languages include individual constants (names of individuals) $a_1, a_2, ...,$ and monadic predicates which denote attributes of individuals. The attributes denoted by the primitive predicates (primitive attributes) are classified into families; the attributes of a single family belong to the same general kind or modality. For instance, colors and shapes are modalities; these are qualitative modalities, but a monadic language $\mathscr{L}$ can, according to Carnap, also contain predicates for quantitative modalities or numerical magnitudes such as age, weight, length, etc. It is assumed that the class of families $\mathscr{F} = \{F^1, F^2, ..., F^n\}$ in $\mathscr{L}$ is always finite, and each family $F^m \in \mathscr{F}$ contains a finite or denumerable number of primitive attributes $P_1^m, P_2^m, ..., P_j^m, ...$[31]

The attributes denoted by the predicates of a family form an *attribute space*. The points of an attribute space represent the most elementary or specific properties of the modality in question, but the attributes in F^m are less specific; they correspond to finite regions of the attribute space U^m. These regions form a finite or countable partition of U^m, $X_1^m, X_2^m, ...,$ where each X_j^m corresponds to P_j^m. The distance between any two points in U^m reflects the similarity between the corresponding qualities: the more two qualities P an P' resemble each other, the nearer each other their representative points are in U^m. A quantitative property of individuals is based on a measurable magnitude G^m, the points of the corresponding attribute space represent the possible numerical values of G^m. Such an attribute space U^m can thus be termed a *value space*. In this case different attributes of F^m correspond to parts of a countable partition of the value space; for instance, if U^m is an interval of the set of real numbers, the attributes of F^m correspond to a partition of subintervals $X_1^m, X_2^m...,$ $X_j^m, ...$ The magnitude G^m itself is not denoted by any sign of a monadic predicate language $\mathscr{L}$, the language $\mathscr{L}$ contains only the predicate symbols P_j^m.[32]

In addition to the basic probability axioms, Carnap accepts a number of axioms of *invariance* for $\mathscr{M}$ and $\mathscr{C}$. According to the axiom of symmetry for individuals, the $\mathscr{M}$- and $\mathscr{C}$-values are invariant with respect to permutations of individuals,[33] and according to the axiom of invariance for families of predicates, the $\mathcal{M}$- and $\mathcal{C}$-values are invariant with respect to the introduction of new families of predicates into the language $\mathcal{L}$. If we wish to determine the numerical value of $\mathcal{C}(H \mid E)$ for given propositions H and E, then, by virtue of the axiom of invariance for families, it suffices to consider only a sublanguage $\mathcal{L}'$ of $\mathcal{L}$ containing only those families of which some attributes are involved in H or E. Conversely, if $\mathcal{C}(H \mid E)$ is based on $\mathcal{L}'$, this language can be extended to a new language $\mathcal{L}$ with new families of predicates without revising the $\mathcal{C}$-values based on $\mathcal{L}'$.[34]

In the 'Basic System' Carnap discusses mainly a case in which $\mathcal{L}$ contains only one finite family F with k attributes $P_1, P_2, ..., P_k$. According to the axiom of invariance for families, the results concerning this language hold for any family of attributes in languages containing any number of families. In the next sections I shall describe the main aspects of Carnap's theory for a single family of predicates.

Carnap's Old System (as presented in *Logical Foundations of Probability* and *The Continuum of Inductive Methods*) does not concern a single family of predicates, but a language $\mathcal{L}$ with several families F^m, each of which contains two attributes, P^m and $\sim P^m$. However, the Old System can easily be reinterpreted in such a way that it becomes comparable to the New System: the Q-predicates defined in terms of the predicates P^m form a partition; let us assume that the number of Q-predicates in $\mathcal{L}$ is k.[35] We shall reinterpret the Q-predicates as the primitive predicates P_j of a one-family language and ignore the original two-attribute families F^m; thus all results and formulae of the Old System become applicable to a single family with k primitive attributes. This reinterpretation will be presupposed in the comparisons made below.[36]

V

Let *Ind* be a denumerable set of individuals $\{a_1, a_2, ...\}$ with an index set D and let D' be the index set of $Ind' \subset Ind$. The class of all atomic propositions with an individual index i is termed $\mathcal{E}_i^{at}$, and $\mathcal{E}_{D'}^{at}$ is the class of atomic propositions with individual indices $i \in D'$; thus

$$\mathcal{E}_{D'}^{at} = \bigcup_{i \in D'} \mathcal{E}_i^{at}.$$

E is termed a *sample proposition* for Ind' if and only if there is a class $\mathcal{A} \subset \mathcal{E}_{D'}^{at}$ such that for every $i \in D'$, $\mathcal{A}$ contains exactly one atomic proposition from $\mathscr{E}_i^{at}$, and $E = \bigcap \mathscr{A}$ (the intersection of all propositions in $\mathscr{A}$). Thus a sample proposition for Ind' assigns an elementary attribute to every member of Ind'. Let E_s be a sample proposition for a set Ind_s; the index set D_s of Ind_s consists of the first s indices in D. Thus E_s specifies the primitive attributes of the individuals $a_1, a_2, ..., a_s$. For each attribute index j, s_j members of Ind_s have the attribute P_j. The k-tuple s $= \langle s_1, ..., s_k \rangle$ is termed the k-tuple of E_s.[37]

In the sequel I shall employ Carnap's notation: s is any k-tuple $\langle s_1, ...,$ $s_j, ..., s_k \rangle$, where $\sum_j s_j = s$. If s $= \langle s_1, ..., s_j, ..., s_k \rangle$, s^j is the k-tuple $\langle s_1, ...,$ $s_j + 1, ..., s_k \rangle$ with the sum $s + 1$. The k-tuple $\langle 0, ..., 0 \rangle$ with $s = 0$ is termed s$_0$.[38]

If two sample propositions have the same k-tuple, they can be transformed into each other by permutations of individuals, and by virtue of the axiom of symmetry for individuals, they have the same $\mathscr{M}$-value. Thus the measure function $\mathscr{M}$ can be represented by its *representative function MI* which determines the prior probability of each sample proposition as a function of its k-tuple. In a similar way, a symmetric confirmation function can be represented by functions which take numbers (i.e., k-tuples of numbers) as arguments: let s be the k-tuple of E_s and let H_j be the atomic proposition $P_j a_{s+1}$; the functions $C_1, ..., C_k$ are termed the *representative C_j-functions* for $\mathscr{C}$ if and only if

$$C_j(s) = \mathscr{C}(H_j \mid E_s)$$

for every j and s. C is the representative C-function for $\mathscr{C}$ if and only if $C(s) = \langle C_1(s), ..., C_k(s) \rangle$, where each C_j is a representative C_j-function for $\mathscr{C}$. The subfunctions of C_j and C restricted to k-tuples with the sum s are denoted by C_j^s and C^s.[39]

In Carnap's Old System, the probability of a hypothesis H depends on the *logical width* of the predicates involved in H. If P_j is an elementary attribute in a family of k attributes, the (relative) logical width of P_j is, according to the Old System, always $1/k$, and the a priori probability of a hypothesis $P_j a_1$ is also $1/k$, i.e.

(1) $C_j(s_0) = \mathscr{M}(P_j a_1) = 1/k.$

In the New System (1) does not always hold. $\mathscr{M}(P_j a_1)$ is dependent on the width of P_j, but the relative width of an elementary attribute is not always $1/k$. The number k indicates the *size* of a family (the number of different basic attributes); in the New System width is not measured in terms of size, but by a separate width function.

The $\mathcal{M}$- and $\mathcal{C}$-values of hypotheses involving attributes of a family F depend on the structure of the corresponding attribute space U. The primitive attributes of F correspond to disjoint and connected regions of the space U. In the construction of his inductive logic, Carnap accepts the following *methodological conjecture*:

(C)   Only those properties of P_j (relations between P_j and P_h) are essential for inductive logic, i.e., for the determination of values of $\mathcal{C}$ for propositions involving P_j (and P_h), which are reflected in the topological and metric properties of the corresponding region X_j (in the topological and metric relations between X_j and X_h).[40]

More specifically, Carnap suggests that it suffices to take as *basic magnitudes* relevant to the determination of $\mathcal{M}$- and $\mathcal{C}$-values the *width* of a basic region X_j and the *distance* between two disjoint regions X_j and X_h.[41]

The concept of *distance* between points in an attribute space is based on judgments of similarity. If X_1, X_2 and X_3 are regions in an attribute space U, the distance between X_1 and X_2 is shorter than that between X_1 and X_3 if and only if X_1 is judged to be more similar to X_2 than to X_3. In some cases it is possible to define a quantitative distance function over the attribute space, that is, a function d such that if u, v and w are points in U, $d(u, v) = 0$ if and only if $u = v$; $d(u, v) = d(v, u)$; and $d(u, v) + d(v, w) \geq d(u, w)$.[42] If the attribute space is the value space of a measurable magnitude, the distance function can be defined in terms of the accepted scale for the magnitude in question (for instance, spatial length and temperature are such magnitudes).

The width of an attribute P_j is, roughly speaking, the size or volume of the region X_j. If a distance function d for U is available, a *width function* w can be defined in terms of d. For instance, if U is a 1-dimensional continuous space, it can be mapped in a 1-to-1 correspondence onto a subset of real numbers by means a coordinate x such that if the coordinates of u and $v \in U$ are x_1 and x_2 respectively, $d(u, v) = x_2 - x_1$. In this case we can take as the width function w the Lebesgue measure on the set of real numbers. Each connected region $X \subset U$ corresponds to an interval in the set of real numbers; $w(X)$ is simply the length of this interval. If U is an n-dimensional space, the width function can be defined in a similar way in terms of an n-dimensional coordinate system as an n-dimensional Lebesgue measure function.[43]

Given a width function w, a normalized width function is defined by

(2) $w'(X_j) = w(X_j)/w(U)$.

The corresponding normalized distance function can be defined by

(3) $d'(u, v) = d(u, v)/w(U)$

if U is a one-dimensional space; in the case of an n-dimensional space

$$d'(u, v) = \frac{d(u, v)}{w(U)^{1/n}}.$$

If F is a family of k qualitative attributes for which no natural metric is available, but only a comparative serial order, Carnap suggests the following choice of w and d: (i) For each of the k primitive attributes, $w(P_j) = 1/k$; (ii) for any two adjacent attributes in the series, $d(P_j, P_h) = 1/k$.[44]

VI

In the 'Basic System' the dependence of $\mathcal{M}$- and $\mathcal{C}$-values on the widths of attributes and distances between attributes is expressed in terms of two new parameters (or types of parameters), γ and η. If j is an attribute index,

(4) $\gamma_j = C_j^0(s_0) = \mathcal{M}(P_j a_1)$.

According to Carnap, γ_j is equal to the relative width of the attribute P_j. If no quantitative concept of width is available, but there is no (a priori) reason for expecting the occurrence of P_j more than that of P_h, we should, according to Carnap, take $\gamma_j = \gamma_h$, and if P_j has a greater width than P_h, we should take $\gamma_j > \gamma_h$. If w is a quantitative width function,

(5) $\gamma_j = aw(P_j)$,

where a is the normalizing constant $1/\sum_{h=1}^k w(P_h)$. If w is a normalized width function, $\gamma_j = w(P_j)$.[45] The γ-values for molecular attributes are determined in an obvious way on the basis of γ-values for primitive attributes.

Carnap's early work, for instance, the system of the *Continuum*, concerns only a case in which $\gamma_j = 1/k$ for every attribute P_j. In this case the $\mathcal{M}$- and $\mathcal{C}$-values satisfy the condition of attribute symmetry. In his early work Carnap took this condition as an axiom for all $\mathscr{C}$-functions.[46] In the 'Basic System' the condition of attribute symmetry is satisfied only by certain special families F.

According to (5), the *a priori* probability of $P_j a_1$ is equal to the relative width of P_j; thus Carnap accepts an even *a priori* distribution over the attribute space. In the case of very simple attribute spaces, for instance, in those considered in Carnap's Old System, an even distribution seems natural, but in more complex cases it is often implausible or even impossible. For instance, if U is the value space of a real-valued quantitative magnitude G, and the set of possible values of G is not bounded, the η-values cannot be determined in terms of a normalized distance function (3).[47] Carnap discusses mainly finite families with bounded value spaces, but in many cases (e.g. if the value space can be mapped onto a subset of real numbers) the choice of the boundaries seems partly arbitrary, and this very arbitrariness indicates that an even *a priori* distribution over the whole space is not acceptable. Moreover, in many cases it seems natural to regard countable and bounded 'observational' families as approximations to continuous, unbounded 'theoretical' families; this requires that the corresponding probability distributions be mutually consistent.

If j and h are two attribute indices,

(6) $\qquad \eta_{jh} = C_j^1(s_0^h)/C_j^0(s_0)$
$\qquad\qquad = \mathscr{C}(P_j a_2 \mid P_h a_1)/\mathscr{M}(P_j a_2)$;

hence

(7) $\qquad \mathscr{C}(P_j a_2 \mid P_h a_1) = \eta_{jh} \mathscr{M}(P_j a_2)$.[48]

Thus the parameter η_{jh} indicates how the knowledge that some individual has the property P_h affects the credibility of the proposition that some other (as yet unobserved) individual has the property P_j.

It is natural to regard this influence as dependent on the *similarity* – that is, the distance – between the attributes P_j and P_h. The parameter η can be termed an *analogy parameter*. (7) implies that observation of an individual with the attribute P_h is *positively relevant* to the proposition $P_j a_2$ if $\eta_{jh} > 1$; if $\eta_{jh} < 1$, it is negatively relevant to $P_j a_2$. In the 'Basic System' Carnap accepts *an axiom of instantial relevance* according to which

(8) $\qquad \mathscr{C}(H_j' \mid E_s \cap H_j) > \mathscr{C}(H_j' \mid E_s)$,

where

$$H_j = P_j a_{s+1}, \quad H'_j = P_j a_{s+2}, \quad \text{and} \quad 0 < \mathscr{C}(H'_j \mid E_s) < 1.\text{[49]}$$

This axiom implies

(9) $\eta_{jj} > 1$

for each attribute index j. Moreover, the similarity between P_j and P_h is maximal when $j = h$; hence

(10) $\eta_{jj} > \eta_{jh}$ for every $j \neq h$.

(10) is termed the principle of *self-similarity*: Any basic attribute is more similar to itself than to any other attribute.[50] Since

(11) $\mathscr{C}(P_j a_2 \mid P_h a_1)/\mathscr{M}(P_j a_2) = \mathscr{C}(P_h a_1 \mid P_j a_2)/\mathscr{M}(P_h a_1)$

(according to the basic axioms), (6) implies

(12) $\eta_{hj} = \eta_{jh}$,

and according to (5),

(13) $\mathscr{C}(P_j a_2 \mid P_h a_1) = \eta_{jh} \gamma_j$.

Since $\mathscr{C}(P_j a_2 \mid P_h a_1) < 1$, (11) implies that $\eta_{jh} < 1/\gamma_j$ and $\eta_{hj} < 1/\gamma_h$. η is always positive; hence (12) implies

(14) $0 < \eta_{jh} < 1/\max(\gamma_j, \gamma_h)$.

(14) expresses the possible range of variation of η_{jh} in terms of γ_j and γ_h.[51]

Carnap lays down e.g. the following rules for the choice of η-values: If no quantitative distance function for U is available, but we see no reason to regard P_j as more (or less) similar to P_m than to P_n, we should take $\eta_{jm} = \eta_{jn}$. If comparative judgments of relative similarity are possible, we should choose $\eta_{jm} > \eta_{jn}$ if and only if P_j is more similar to P_m than to P_n.[52] If a normalized distance function d for U is available, the η-values may be determined in terms of a function f such that

(15) $\eta_{jm} = f(d_{jm})$.

This function is termed the η-function for F. (15) is always positive, and according to the principle of self-similarity, it assumes its maximum when $d_{jm} = 0$; this maximum is always greater than 1 (cf. [9]). Carnap discusses the shape of the η-function in detail; here it suffices to note that it seems plausible to assume that for sufficiently long distances the similarity influence has no noticeable effect. This means that for long distances d, say $d > d^*$, the η-function has a constant value η^*.[53] The choice of d^* seems to be partly subjective; the problem of the choice of η^* will be discussed below in Section VIII. It is also important to observe that η_{jh} may exceed 1 not only at $d = 0$, but also for small positive values of d, that is, for pairs P_j, P_h with $j \neq h$. Observation of an attribute P_j may be positively relevant to a hypothesis about the occurrence of some other attribute sufficiently similar to P_j. This is genuine (positive) 'similarity influence'.

VII

Carnap discusses in detail certain special kinds of families, namely

(i)   families with $\gamma_j = \gamma_h$ (for every j, h), and
(ii)   families with $\eta_{jh} = \eta_{mn}$ (for all pairs $j \neq h$ and $m \neq n$).

Families of type (i) are termed *families with γ-equality*; those of type (ii) are termed *families with η-equality*.[54] Carnap's early work is restricted to families with γ-equality, for instance, in the system of the *Continuum* $\gamma_j = = 1/k$ for every j; in this system η-equality holds as well. The principle of self-similarity (10) can be regarded as an axiom of inductive logic, thus is it also satisfied by families with η-equality. In these families we thus have $k + 1$ η-parameters, η_{jj} for each attribute P_j and $\eta_{jh} (j \neq h)$. The parameter η_{jh} will be abbreviated η.

According to (11),

(16) $$\sum_{m_j} \mathscr{C}(P_m a_2 \mid P_j a_1) = \sum_m \eta_{mj} \gamma_m = 1,$$

where

(17) $$\sum_m \eta_{mj} \gamma_m = \eta_{jj} \gamma_j + \sum_{m \neq j} \eta_{mj} \gamma_m.$$

If F is a family with η-equality and $\eta = \eta_{mj}$, (16) and (17) imply

(18) $$\eta_{jj} \gamma_j + \eta \sum_{m \neq j} \gamma_m = \eta_{jj} \eta_j + \eta(1 - \gamma_j) = 1;$$

hence

(19) $$\eta = \frac{1 - \eta_{jj} \gamma_j}{1 - \gamma_j}.$$

The axiom of instantial relevance implies that $\eta_{jj} > 1$; consequently $\eta < 1$.

375

In the present case the axiom of instantial relevance implies the principle of self-similarity. (19) implies

$$(20) \qquad \eta_{jj} = \eta + \frac{1 - \eta}{\gamma_j}.$$

(19) and (20) show that $0 < \eta < 1$ and $1 < \eta_{jj} < 1/\gamma_j$, and η and η_{jj} are inversely related to each other. γ_j and η determine the value of η_{jj}.

For families with η-equality it is possible to define $\mathscr{C}$-functions which satisfy the following λ-*principle*:

> For any $s = \langle s_1, s_2, ..., s_k \rangle$, $C_j(s)$ (i.e., $\mathscr{C}(P_j a_{s+1} \mid E_s)$) depends only on s_j and s, but is independent of the other $k - 1$ members of s.[55]

This principle is inapplicable if the $\mathscr{C}$-values depend on similarity influence. However, if we accept η-equality, the λ-principle may be accepted too. $\mathscr{C}$-functions satisfying the λ-principle are termed $\lambda - \mathscr{C}$-*functions*. In Carnap's earlier systems a principle equivalent to the λ-principle was taken as an axiom.[56] However, it excludes similarity influence and is thus not generally acceptable; in Carnap's New System it is merely a principle satisfied by certain special $\mathscr{C}$-functions.

If C is a C-function for a family of k attributes with η-equality, and if the k γ-values and η are given, all values of C for $s = 0$ and $s = 1$ are determined in accordance with (4) and (6). However, if C satisfies the λ-principle, the values of C and $\mathscr{C}$ are determined for any s, and consequently all values of $\mathscr{C}$ are determined. If C is a representative function for a $\lambda - \mathscr{C}$-function,

$$(21) \qquad C_j(s) = \frac{s_j + \gamma_j \eta/(1 - \eta)}{s + \eta/(1 - \eta)}$$

for an arbitrary k-tuple s. The representative function of the *Continuum* is obtained as a special case of (21): if we define a new parameter λ:

$$(22) \qquad \lambda = \eta/(1 - \eta),$$

and assume that the family of attributes under discussion satisfies attribute symmetry (γ-equality), (21) simplifies to

$$(23) \qquad C_j(s) = \mathscr{C}(P_j a_{s+1} \mid E_s) = \frac{s_j + \lambda/k}{s + \lambda},$$

which is the representative function of the λ-system.[57] The λ-continuum is a special case of Carnap's Basic System (for one family of predicates) in which

(i)   The family of attributes under discussion has η-equality and $\mathscr{C}$ satisfies the λ-principle, and

(ii)   all primitive attributes of the family have the same relative width $1/k$, and consequently the family has γ-equality.

It is of interest to observe here the dependence of λ on η. The range of admissible values of η is $0 < \eta < 1$; according to (22), $\eta = 0$ implies $\lambda = 0$ or the 'straight rule' of induction; this value is excluded by the coherence requirements on $\mathscr{C}$. $\eta = 1$ implies $\lambda = \infty$ and $\eta_{jj} = 1$; this value is excluded by the axiom of instantial relevance and the principle of self-similarity. The difference between η and η_{jj}, and hence the value of η (the larger η is, the smaller is the difference between η and η_{jj}), indicates the *differential relevance* of P_j-observations to a hypothesis $P_j a_{s+1}$ in comparison to other observations. η-equality implies that *only* P_j-observations have positive differential relevance to $P_j a_{s+1}$. If $\eta = 1$, P_j-observations possess no differential relevance to $P_j a_{s+1}$, and consequently all observations are irrelevant. On the other hand, very small values of η overemphasize the differential relevance of P_j-observations.

VIII

The definition (22) puts the choice of λ into a new prespective. In the *Continuum of Inductive Methods* Carnap distinguished two types of c-functions and inductive methods. In methods of the first kind, the value of λ is constant, independent of the size of the family. In methods of the second kind, the value of λ for a given family is a function of its size, that is, a function of k. The simplest function of this type is obviously $\lambda(k) = k$; this yields the c-function c^*.[58]

In Carnap's Old System width is defined in terms of size; the size of a family is thus an important parameter in inductive logic, and it is perhaps not entirely implausible to regard λ as dependent on k. However, in the New System the widths of attributes and distances between attributes are not based on the size of a family, but on separate width- and distance-functions for the underlying attribute space. The size of a family has little theoretical significance; it indicates merely how the attribute space has been partitioned into basic attributes. In this case it is not reasonable to define λ as a function of k. In the Basic System inductive methods of the second kind (including c^*) do not seem plausible.

$\lambda - \mathscr{C}$-functions are applicable to families with η-equality, that is, to families in which the distances between different attributes are equal, and to 'long-distance families' in which the distances exceed d^* and $\eta_{jh} = \eta = \eta^*$. η is a decreasing function of d_{jh}; consequently η and λ should be the larger, the smaller is the distance between different attributes. On the basis of this relationship, it is possible to make in certain special cases objective comparisons between the λ-values appropriate to different families. If the average distance between different attributes in a family F'' is larger than that in F', we should use a larger η-value (and hence a larger λ-value) for F'' than for F'. An objective comparison between the average distances in the two families is possible e.g. if the attribute spaces of F' and F'' are subspaces of a common attribute space U.[59] However, such comparisons are possible only in exceptional cases, and they cannot determine the absolute (numerical) value of η and λ. According to Carnap's assumptions concerning the shape of the η-function, all long-distance families with η-equality should have the same η-value η^*, but the choice of the numerical value of η^* seems partly subjective. There seem to be no compelling *a priori* grounds for choosing any particular numerical value for η^*.

In the 'Basic System' Carnap argues, however, that the range of acceptable λ-values for long-distance families can be narrowed down to $\lambda = 1$ or $\lambda = 2$ on the basis of *a priori* considerations. ($\lambda = 1$ corresponds to $\eta = \frac{1}{2}$, $\lambda = 2$ to $\eta = \frac{2}{3}$.) Let E_1 and E_2 be two sample propositions and let S_1 and S_2 be the corresponding structure propositions. (The structure proposition corresponding to a given sample proposition specifies the k-tuple of the sample proposition.)[60] According to the λ-system, $\mathscr{M}(E_1) > \mathscr{M}(E_2)$ if the degree of order or uniformity of E_1 exceeds that of E_2, that is, if the distribution of individuals among different attributes is according to E_2 more even than according to E_1. Carnap suggests that this inequality can be extended to the structure propositions S_1 and S_2, that is,

(24)   (i) $\mathscr{M}(S_1) \geq \mathscr{M}(S_2)$ or
          (ii) $\mathscr{M}(S_1) > \mathscr{M}(S_2)$,

if S_1 exhibits a higher degree of uniformity than S_2. If the same λ-value is used for all long-distance families, including those with $k = 2$, (24.i) holds for all families only if $\lambda = 2$, and the stronger condition (24.ii) holds only if $\lambda < 2$. If (for reasons of simplicity) we consider only integers as possible values of λ, (24.ii) implies $\lambda = 1$. According to (24.i), $\lambda = 2$ is also an admissible value of λ.[61]

Carnap shows that λ-values $\lambda < 1$ lead to intuitively unacceptable $\mathscr{C}$-values, for instance, the values of $C_j(s_0^j)$ (i.e., $\mathscr{C}(P_j a_2 \mid P_j a_1)$) seem too high or 'overoptimistic'.[62] However, the values $C_j(s_0^j)$ may seem too high even for $\lambda = 1$ and $\lambda = 2$; for instance, if F is a family of three basic attributes with $\gamma_1 = \gamma_2 = \gamma_3$, $\lambda = 1$ implies $C_j(s_0^j) = 0.67$ and $\lambda = 2$ implies $C_j(s_0^j) = 0.56$. Even the latter value may seem too high. Carnap's argument for $\lambda = 1$ is not very convincing. The postulate (24) expresses a strong *a priori* belief in a high degree of uniformity in our universe of individuals. It may be justified in special cases (for instance, on the basis of suitable background information), but is hardly acceptable as a general principle of inductive logic. In Carnap's New System an argument of this type (i.e., an argument based on the degree of order of an attribute partition) is also weakened by the fact that the degree of order of structure propositions depends on the choice of the attribute partition, but λ should not depend on it.

In the 'Basic System' Carnap says that the definability of λ in terms of the similarity parameter η supports an 'objectivist' viewpoint in the choice of λ. The value of η does not depend only on idiosyncratic, personal factors, but on the structure of the attribute space, viz. on the distance between different attributes. The concept of distance is based on judgments of similarity; these judgments are, in a certain sense, 'subjective', but they are not arbitrary. Moreover, the similarity judgments of different persons show a high degree of interpersonal consistency.

As was mentioned above, the η-parameter indicates the differential relevance of P_j-observations to a hypothesis $P_j a_{s+1}$ in comparison with other observations. In normal circumstances our judgments of differential relevance are based on extensive information about past inductions and on knowledge about the world (including theoretical knowledge). From this standpoint, the question of the *a priori* differential relevance of observations seems impossible to answer or even meaningless. Carnap's proposal to use innate perceptual spaces as a basis of *a priori* judgments of differential relevance is an ingenious and highly plausible idea. In the absence of factual information, innate similarity judgments form a natural basis for judgments of relevance.

How are relevance judgments based on perceptual similarity? The concept of similarity is notoriously elusive; Nelson Goodman has remarked that "as it occurs in philosophy, similarity tends under analysis either to vanish entirely or to require for its explanation just what it purports to explain".[63] If we think of similarity from the behavioral viewpoint, the 'inductive behavior' associated with relevance judgments provides a convenient and natural method of measuring similarity.[64] If the structure of the perceptual space is determined in this way, it is almost analytic (true by definition) that the value of λ is determined by similarity judgments, and all 'long distance families' have the same λ-value. The value of λ cannot be determined by rational *a priori* considerations, in fact, there can be no question of 'choosing' (or 'justifying') the value of λ. η and λ reflect an innate, natural standard of 'similarity' which is given, not chosen. On this interpretation of 'similarity', there is an analytic connection between similarity and relevance, and Carnap's proposal to take similarity as the basis of relevance judgments (and his conjecture (C) as well) becomes a truism. On the other hand, if perceptual similarity is measured by psychophysical methods which require explicit verbal (or 'introspective') similarity judgments on the part of the subject, it is not obvious that these alone should determine relevance judgments. In this case relevance judgments may depend, not only on perceptual similarity, but on other factors as well. This viewpoint leaves room for 'personalism' (in Carnap's sense) in Carnap's New System, and makes the conjecture (C) a problematic hypothesis.

Similarity is a context-dependent and theory-dependent concept; similarity judgments depend on one's theoretical standpoint. The interest of Carnap's proposal depends on the existence of non-theoretical, 'natural' similarity relations. However, quite apart from this question, the general theory-dependence of similarity relations has interesting consequences for Carnap's inductive logic. In the 'Basic System' Carnap accepts the 'two-level-conception' of the language of science. According to Carnap, the Basic System is constructed only for an observational language. However, if an inductive logic is constructed for the theoretical language, the transition from the observational level to the theoretical language may involve a total restructuring of the similarity relations, and consequently abrupt changes in the $\mathcal{M}$- and $\mathcal{C}$-values. As an example of this possibility Carnap mentions the family of colors: in the observational language, similarity relations between colors are based on perceived similarity; on the theoretical level they are defined in terms of the wavelength of light. The similarity relations (i.e., the distances) between the same colors are entirely different in the two cases.[65] This result can be generalized: in the same way as Carnap's 'observational' and 'theoretical' attribute spaces may involve different similarity relations, different theories may involve different concepts of similarity. This implies that strict probabilistic comparisons between theories are in some cases in principle impossible. Thus Carnap's New System vindicates the skepticism expressed by some philosophers about the probabilistic comparisons between different theories.[66]

IX

In Carnap's Old System the measure function and confirmation function are defined for the sentences of a formalized language $\mathcal{L}$; in the New System they are defined for the propositions (sets of models) of a language. This formal difference reflects a deeper philosophical difference between the two systems. In the New System Carnap discusses mainly monadic predicate languages, but the confirmation functions defined for these languages depend on parameters that cannot be expressed in these languages, for instance, on the widths of attributes and distances between attributes. The width function and the distance function are not reflected in the structure of the language. In the Old System the width of a predicate can be shown in the language itself by transforming the predicate into normal form (into a disjunction of Q-predicates), but in the New System such procedure is not possible.

As was mentioned above, Carnap's aim was to define a 'strong extension' of probability theory in which the structure of the sample space determines a unique probability measure or a restricted class of 'acceptable measures'.[67] In the Old System the structure of the sample space is determined by the structure of the language $\mathcal{L}$; in the New System it is determined by the structure of the attribute space U, that is, by the topological and metric properties of U (in the case of one family of attributes; in languages with several families, the structure of the probability space is determined by the spaces $U^1, ..., U^n$). This structure is, in general, much richer than the structure of the language $\mathscr{L}$. According to the Old System, the conceptual factors important for inductive logic are properties of language; in the New System they are rather properties of the reality to which the language is applied. It is perhaps appropriate to say that in the New System the focus of Carnap's interest has shifted from methodological questions to the epistemological foundations of probabilistic reasoning. In this respect Carnap's use of the η-parameter is of great interest. His discussion of similarity influence and the structure of perceptual space brings his inductive logic into contact with his own early epistemological work in which the structure of the world is constructed on the basis of a single undefined relation of similarity between 'elementary experiences',[68] and also with recent discussions of the epistemological importance of innate standards of similarity.[69]

In his methodological writings Carnap often says that the purpose of inductive logic is to "help people make decisions in a rational way," and emphasizes the application of inductive logic to practical decision problems. These statements should perhaps not be understood too literally. Carnap's work has mainly foundational interest; it is an analysis of the foundations of probabilistic reasoning. Many of his axioms and conditions are not applicable to actual knowledge situations: they do not concern the *credence-functions* used in actual knowledge-situations, but only *credibility-functions* which are independent of factual information and from which the credence functions are derived.[70]

As was observed above, Carnap was even in his latest works optimistic about finding new axioms and conditions for rational probability measures and thus narrowing down the range of admissible $\mathscr{C}$-functions – even though he, in his characteristically cautious spirit, wanted to 'keep an open mind' in this respect. In the 'Basic System' he writes:

I have repeatedly experienced a development of the following kind, similar to that described in this section in connection with the problem of choosing a λ-value. At the beginning of the investigation there is a great number, sometimes even an embarrassing abundance of possibilities to choose from. But then, with the gain of deeper insight into the situation, the range of choice is gradually narrowed down.[71]

However, in many respects the development of Carnap's inductive logic exhibits also the opposite pattern: not only did he find new axioms and conditions that restrict the range of acceptable probability functions, but he also had to reject conditions previously accepted. For instance, the condition of attribute symmetry and the λ-principle were first tentatively accepted as axioms, but on the basis of new insights and developments it was realized that these conditions are satisfied only by certain special – though very important – probability measures. Thus, new insights into the foundations of inductive inference not only restricted the possibility choice, but also opened up completely new possibilities. This is especially clear in the case of Carnap's discovery of η and Hintikka's discovery of inductive methods for general hypotheses.[72] These developments do not detract from the value and importance of Carnap's research. Carnap, perhaps more than anyone else, has contributed to our understanding of the complex factors underlying rational probability assignments – and this, of course, is the principal task of the philosophy of induction.

University of Turku

NOTES

[1] These publications include Carnap [6], [8], [9], [10], and [11]. Certain developments of Carnap's theory of inductive probability are described in Carnap and Stegmüller [14], Appendix B, and in Carnap [7].

[2] Part I (the first 13 sections) of this work has appeared in 1971 in Vol. I of *Studies in Inductive Logic and Probability* [13B]; Part II will be published in the forthcoming Vol. II of *Studies in Inductive Logic and Probability* [13C].

[3] Carnap [4], pp. 164–175.

[4] [4], pp. 165–167.

[5] [4], pp. 168–175.

[6] This conception of the application of inductive logic has been presented in detail by Carnap in [6] and [12]. ([12] is a revised and expanded version of [6].)

[7] Some philosophers have assumed that betting on general hypotheses and theories does not make sense, and the betting interpretation (or belief-interpretation) and the confirmation-interpretation of inductive logic are therefore somehow imcompatible. See Hacking [17], pp. 215–216, and Lakatos [29], p. 361. This prejudice against betting on general hypotheses appears unfounded, however; see Hintikka [25], pp. 339–340 and Pietarinen [32], p. 26. See also note 16 below.

[8] Cf. Carnap [4], pp. 19–25.

[9] Cf. Nagel [31], and Popper [33], pp. 251–276 and [34], pp. 213–216.

[10] Carnap [4B], pp. xv–xix. See also Carnap [7], p. 967.

[11] The confusions involved in the 'Popper-Carnap-controversy' are discussed in detail in Michalos [30]; see especially Chapter III. Cf. also Bar-Hillel [1].

[12] Cf. Carnap [7], pp. 967–969.

[13] Several measures of support proposed in recent literature are discussed and compared in Kyburg [28]. See also Hintikka [22], pp. 328–329.

[14] Cf. Carnap [4], pp. 570–571.

[15] Cf. Hintikka [24].

[16] However, the inductivist or confirmation-theoretic applications of inductive logic

can also be studied from the decision-theoretic viewpoint; see Hilpinen [19]. The confirmation-theoretic and the decision-theoretic conception of inductive logic are not incompatible.

[17] Carnap [13A], section 18 ('Lambda-families').

[18] Carnap [3], p. 76. Cf also [12], p. 15.

[19] See Carnap [3], section 6, and [4], p. ix and pp. 562–563.

[20] Cf. Carnap [5], pp. 56–79.

[21] Carnap [13A], section 18 ('Lambda-families').

[22] Cf. Carnap [11], pp. 313–314, and [13A], section 18.

[23] [13A], section 18 ('Lambda-families').

[24] Cf. Suppes [37], chapter 3, pp. 100–101.

[25] Cf. section IX of this paper.

[26] Carnap [13A], section 18 ('Lambda-families').

[27] [13A], 'Lambda-families'. 'Objectivism' here does not mean 'empiricism', but 'rationalism', according to which it is possible to restrict the class of $\mathscr{C}$-functions by rational *a priori* considerations.

[28] Cf. Hintikka [20], [21], and [23].

[29] Pietarinen [32] has studied the problems of analogy and lawlikeness by the methods developed by Carnap and Hintikka.

[30] Carnap [13B], pp. 40–41.

[31] [13B], p. 43.

[32] [13B], pp. 43–44.

[33] [13B], pp. 117–118.

[34] [13B], p. 121.

[35] For Q-predicates, see Carnap [4], pp. 122–126.

[36] This reinterpretation of the Old System is described in Carnap [7], p. 973–974.

[37] [13B], pp. 121–122.

[38] [13B], p. 62.

[39] Representative *MI-*, *C$_J$*- and C-functions are discussed in detail in [13B], sections 11–12 (pp. 131–160).

[40] Carnap [13C], section 15.

[41] [13C], section 15.

[42] [13C], section 14B ('Distance and Width').

[43] [13C], section 14B.

[44] [13C], section 14B.

[45] [13C], section 16A ('The Analogy Influence: Gamma- and Eta-parameters').

[46] Cf. Carnap [7], p. 975, and Carnap and Stegmüller [14], p. 244.

[47] Carnap [13C], section 14B.

[48] [13C], section 16A. Cf. also note 72.

[49] [13B], p. 161 This principle is not independent of the other axioms of inductive logic; it can be proved from the basic axioms, the axiom of symmetry (for individuals), and the axiom of convergence ('The Reichenbach Axiom'); cf. Humburg [26].

[50] [13C], section 16C ('The Search for Principles of Analogy by Similarity').

[51] [13C], section 16A.

[52] [13C], section 16B ('Rules for the Eta-parameter').

[53] [13C], section 17A ('The Eta function: General Considerations').

[54] [13C], section 18 ('Some Special Kinds of Families').

[55] [13A], section 18 ('Lambda-families'). I. J. Good ([15], p. 26) has termed this principle 'W. E. Johnson's sufficiency postulate'; it has been accepted by Johnson in [27].

[56] The axiom (C9) on p. 14 of [5] is equivalent to the λ-principle.

[57] [13A], 'Lambda-families'.

[58] Carnap [5], sections 11 and 15.

[59] In [13A], section 18 ('Lambda-families') Carnap discusses an example of this type in which the families F' and F'' are subfamilies of the family of colors.

[60] For structures and structure-propositions, see Carnap [13B], p. 123.

[61] This argument is presented in Carnap [13A], section 18 ('Lambda-families').

[62] [13A], section 18 ('Lambda-families').

[63] Goodman [16], p. 29.

[64] Cf. Quine [35], p. 123.

[65] See Carnap [13B], pp. 51–52.

[66] Cf. e.g. Bar-Hillel [1].

[67] Cf. Section III of this paper.

[68] Cf. Carnap [2].

[69] Cf. Quine [35] and Stemmer [36].

[70] These concepts are explained in Carnap [12].

[71] Carnap [13A], 'Lambda-families'.

[72] Additional new possibilities of this kind are created by the extension of inductive methods to relational languages; cf. Hilpinen [18]. (*Added in proof*) Note that the η-parameter discussed here must be distinguished from that defined in [14], p. 251, and in Carnap's article in the present volume. The latter parameter concerns analogy between different families (inter-family analogy); the η-parameter described here is related to intra-family analogy.

BIBLIOGRAPHY

[1] Yehoshua Bar-Hillel, 'Popper's Theory of Corroboration', in *The Philosophy of Karl R. Popper* (ed. by P. A. Schilpp), Open Court Publ. Co., La Salle, Ill., 1974, pp. 332–348.

[2] Rudolf Carnap, *Der logische Aufbau der Welt*, Felix Meiner Verlag, Leipzig 1928. (English translation *The Logical Structure of the World* by Rolf A. George, University of California Press, Berkeley and Los Angeles, 1967)

[3] Rudolf Carnap, 'On Inductive Logic', *Philosophy of Science* **12** (1945), 72–97. Reprinted in *Probability, Confirmation, and Simplicity* (ed. by M. H. Foster and M. L. Martin), The Odyssey Press, New York, 1966, pp. 35–60.

[4A] Rudolf Carnap, *Logical Foundations of Probability*, The University of Chicago Press, Chicago, 1950.

[4B] Rudolf Carnap, *Logical Foundations of Probability*, 2nd Edition, The University of Chicago Press, Chicago, 1962.

[5] Rudolf Carnap, *The Continuum of Inductive Methods*, The University of Chicago Press, Chicago, 1952.

[6] Rudolf Carnap, 'The Aim of Inductive Logic', in *Logic, Methodology and Philosophy of Science* (ed. by E. Nagel, P. Suppes and A. Tarski), Stanford University Press, Stanford, Calif , 1962, pp. 303–318.

[7] Rudolf Carnap, 'Replies and Systematic Expositions V. Probability and Induction', in *The Philosophy of Rudolf Carnap* (ed. by P. A. Schilpp), Open Court Publ. Co., La Salle, Ill , 1963, pp. 966–998.

[8] Rudolf Carnap, 'Probability and Content Measure', in *Mind, Matter and Method*

(ed. by P. K. Feyerabend and G. Maxwell), University of Minnesota Press, Minneapolis, 1966, pp. 248–260

[9]   Rudolf Carnap, 'Inductive Logic and Inductive Intuition', in *The Problem of Inductive Logic* (ed. by I. Lakatos), North-Holland Publ. Co., Amsterdam, 1968, pp. 258–267.

[10]   Rudolf Carnap, 'On Rules of Acceptance', in *The Problem of Inductive Logic* (ed. by I. Lakatos), North-Holland Publ. Co., Amsterdam, 1968, pp. 146–150.

[11]   Rudolf Carnap, 'Reply to J. Hintikka', in *The Problem of Inductive Logic* (ed. by I. Lakatos), North-Holland Publ. Co., Amsterdam, 1968, pp 312–314.

[12]   Rudolf Carnap, 'Inductive Logic and Rational Decisions', in *Studies in Inductive Logic and Probability*, Vol. I (ed. by R. Carnap and R. Jeffrey), University of California Press, Berkeley and Los Angeles, 1971, pp. 5–31.

[13A] Rudolf Carnap, 'An Axiom System of Inductive Logic' (unpublished; a preliminary version of 13B-C)

[13B] Rudolf Carnap, 'The Basic System of Inductive Logic, Part I', in *Studies in Inductive Logic and Probability*, Vol. I (ed. by R. Carnap and R. Jeffrey), University of California Press, Berkeley and Los Angeles, 1971, pp. 33–165.

[13C] Rudolf Carnap, 'The Basic System of Inductive Logic Part II', in *Studies in Inductive Logic and Probability*, Vol II (ed by R Jeffrey), University of California Press, Berkeley and Los Angeles (forthcoming).

[14]   Rudolf Carnap and Wolfgang Stegmüller, *Induktive Logik und Wahrscheinlichkeit*, Springer Verlag, Wien, 1959.

[15]   Irving John Good, *The Estimation of Probabilities*, Research Monograph No 30, The MIT Press, Cambridge, Mass , 1965.

[16]   Nelson Goodman, 'Seven Strictures on Similarity', in *Experience and Theory* (ed. by L. Foster and J. W. Swanson), The University of Massachusetts Press, Amherst, 1970, pp. 19–29.

[17]   Ian Hacking, *The Logic of Statistical Inference*, Cambridge University Press, Cambridge, 1965.

[18]   Risto Hilpinen, 'Relational Hypotheses and Inductive Inference', Synthese 23 (1971), 266–286.

[19]   Risto Hilpinen, 'Decision-Theoretic Approaches to Rules of Acceptance', in *Contemporary Philosophy in Scandinavia* (ed. by R. E. Olson and A. Paul), The Johns Hopkins Press, Baltimore, 1972, pp. 147–168.

[20]   Jaakko Hintikka, 'A Two-Dimensional Continuum of Inductive Methods', in *Aspects of Inductive Logic* (ed. by J. Hintikka and P. Suppes), North-Holland Publ. Co., Amsterdam, 1966, pp. 113–132.

[21]   Jaakko Hintikka, 'Induction by Enumeration and Induction by Elimination', in *The Problem of Inductive Logic* (ed. by I. Lakatos), North-Holland Publ. Co , Amsterdam, 1968, pp. 191–216.

[22]   Jaakko Hintikka, 'The Varieties of Information and Scientific Explanation', in *Logic, Methodology and Philosophy of Science* III (ed. by B. van Rootselaar and J. F. Staal), North-Holland Publ. Co., Amsterdam, 1968, pp. 311–332.

[23]   Jaakko Hintikka, 'Inductive Independence and the Paradoxes of Confirmation', in *Essays in Honor of Carl G. Hempel* (ed. by N. Rescher et al.), D. Reidel Publ. Comp., Dordrecht, 1969, pp. 24–46.

[24]   Jaakko Hintikka, 'Statistics, Induction and Lawlikeness: Comments on Dr. Vetter's Paper', *Synthese* 20 (1969), 72–83.

[25] Jaakko Hintikka, 'Unknown Probabilities, Bayesianism, and de Finetti's Representation Theorem', in *Boston Studies in the Philosophy of Science*, Vol. VIII (ed. by R. C. Buck and R. S. Cohen), D. Reidel Publ. Comp., Dordrecht, 1971, pp. 325–341.

[26] Jürgen Humburg, 'The Principle of Instantial Relevance', in *Studies in Inductive Logic and Probability*, Vol. I (ed. by R. Carnap and R. Jeffrey), University of California Press, Berkeley and Los Angeles, 1971, pp. 225–233.

[27] W. E. Johnson, Appendix (ed by R. B. Braithwaite), to 'Probability: Deductive and Inductive Problems', *Mind* 41 (1932), 421–423.

[28] Henry E. Kyburg, Jr., 'Recent Work in Inductive Logic', *American Philosophical Quarterly* 1 (1964), 249–287.

[29] Imre Lakatos, 'Changes in the Problem of Inductive Logic', in *The Problem of Inductive Logic* (ed. by I. Lakatos), North-Holland Publ. Co., Amsterdam, 1968, pp. 315–417.

[30] Alex C. Michalos, *The Popper-Carnap Controversy*, Martinus Nijhoff. The Hague, 1971.

[31] Ernest Nagel, 'Carnap's Theory of Induction', in *The Philosophy of Rudolf Carnap* (ed. by P. A. Schilpp), Open Court Publ. Co, La Salle, Ill., 1963, pp. 785–825.

[32] Juhani Pietarinen, *Lawlikeness, Analogy, and Inductive Logic*, Acta Philosophica Fennica 26, North-Holland Publ. Co., Amsterdam, 1972.

[33] Karl R. Popper, *The Logic of Scientific Discovery*, Hutchinson, London, 1959.

[34] Karl R. Popper, 'The Demarcation Between Science and Metaphysics', in *The Philosophy of Rudolf Carnap* (ed. by P. A. Schilpp), Open Court Publ. Co., La Salle, Ill., 1963, pp. 213–226.

[35] W. V. Quine, 'Natural Kinds', in *Ontological Relativity and Other Essays* (by W. V. Quine), Columbia University Press, New York and London, 1969, pp. 114–138.

[36] Nathan Stemmer, 'Three Problems of Induction', *Synthese* 23 (1971), 287–308.

[37] Patrick Suppes, *Set-Theoretical Structures in Science*, Institute for Mathematical Studies in the Social Sciences, Stanford University, Stanford, 1967.

The Unimportance of Semantics[1]

Richard Creath

Arizona State University

Our deepest commitments about history are reflected in how we break it down into periods. (Cf. Galison 1988) By drawing a break at a certain point we emphasize the novelty and importance of a new development. It is also how we contain and dismiss certain work as no longer relevant. Thus, in the history of physics we break the story with Newton, both to emphasize his roles in bringing previous developments to a close and in initiating new lines of work, and also to suggest that the ongoing practice of physics thereafter can appropriately in large measure ignore what preceeds Newton. Periodizing history is essential to understanding it, including when we periodize the work of a given writer. In philosophy, anyone who did not see a gulf between Kant's early work and his critical philosophy or between Wittgenstein's *Tractatus* and his *Philosophical Investigations* would be missing something enormously important. But periodization can also be dangerous in blinding us to the continuities between periods and in erroneously suggesting that it is safe to ignore what has come before. Nowhere is this more true than in standard treatments of the work of Rudolf Carnap.

Carnap's work is typically divided into four periods: There is the Carnap of the *Aufbau* (1928) which lasts until the early 1930s; there is the Carnap of syntax which lasts till the ink is dry on *The Logical Syntax of Language* (1934), say 1935; there is the semantical period which lasts till the mid- or late-1940s; and finally there is a period devoted to confirmation and probability theory and other broadly pragmatic matters in the philosophy of science. Some would add a Kantian or pre-critical period on the front; others would draw the lines in slightly different ways. But usually Carnap's work is thus drawn—and quartered.

Even Carnap encouraged this periodization; in fact, he appealed to it to avoid annoying questions about his earlier work. For example, in the late 1940s Carnap visited Minnesota. Wilfrid Sellars was at the time teaching a seminar on the *Aufbau* and so peppered Carnap with questions about it. Carnap couldn't escape; they were riding together in a car from the airport. When Carnap could contain his exasperation no further, he cut off discussion of the *Aufbau* by saying "But that book was written by my grandfather!" (Sellars 1975, p. 277) On the present occasion I have no wish to argue against periodization per se; it is essential to the historian's task. Moreover, I have no

PSA 1990, Volume 2, pp. 405-416
Copyright © 1991 by the Philosophy of Science Association

wish to argue (my title notwithstanding) either that semantics is unimportant to us or that Carnap's work in his so-called semantical period is uninteresting or unimportant. Instead, what I shall urge is that there is no great cleavage either between Carnap's syntactical and semantical work or between his semantical and later work. Of course, there are differences, but they are not as important as the continuities. And we will mislead ourselves about all of it if we are guided by the standard periodization.

The first thing I want to do, however, is to review, perhaps a bit cynically, what I take to be the reasons or causes behind the standard periodization. Thereafter, I can review what is new and imporatant in the *Logical Syntax* that binds the rest of Carnap's work together and then use that to argue against the standard periodization more specifically.

It is easy to see why someone now would want there to be a wall after the *Logical Syntax*. After all, it says that all logic (and for that matter all philosophy) is syntax, and we know this to be false. In addition, it rejects a truth predicate and with it reference and designation, all of which we know to be central to our enterprise. Besides, it is so *hard*! It seems to have hundreds of pages of incredibly complicated English-free notation, all in defense of some thesis that we know to be wrong anyway.

By contrast, *Meaning and Necessity* (1947), which is emblematic of semantics, is so readable, so easy, and so accessible to anyone who has had standard contemporary graduate training in philosophy. We may quarrel with this or that claim in the book, but its view is still in the mainstream.

The argument for a cleavage between the semantics and the later work is similar in form. In the later period Carnap's work becomes wholly absorbed with philosophy of science and especially with probability and confirmation theory, a concern which seems absent in the semantical period. In this metaphysical age, worrying about the epistemology of science seems peripheral to the mainstream of philosophers. It is rather an odd taste to be indulged in members of the PSA, but not at all approved for New England Cartesians or for modal logicians on either coast. So the semantics seems relevant to contemporary concerns, but the later work does not. Moreover, we have it on excellent authority that Carnap's probability theory is wrong, and besides the probability books are so *hard*. Life is short and sweet, so we had better concentrate on the most promising stuff, namely the semantics.

All in all, then, the reasons for the standard periodization hinge on a determination to focus on that which seems accessible, relevant, and largely right to *us*. Insofar as this exemplifies a principle of charity it is not at all disreputable. I think, however, we should resist the urge to be thus charitable for two reasons: First, by looking at what we thought was wrong we might learn something. We might learn that it wasn't so wrong, or we might learn that scattered among the falsehoods were a lot of important truths. Second, we might come to understand quite differently and perhaps a bit better what already seemed familiar and right.

To make my case for a different periodization, or at least for less of a wall on either side of semantics, I need now to go back to the *Syntax* to see what was interesting and new about that. In fact I will spend a good deal of time on the development of the *Syntax*, but if we understand the importance of that clearly enough, the rest of my argument will fall neatly and fairly quickly into place.

To find out what was new in the *Syntax* we have to ask what Carnap was committed to just before he wrote that book. Well, his deepest commitment was to a thoroughgoing empiricism, which consisted of two parts. The first of these was a rejection of intuition. This supposed transempirical mode of knowing independent matters of fact was much loved by Platonic and Cartesian rationalists and apparently by Frege and Russell as well. The rejection of intuition is what the elimination of metaphysics is all about. The second part of Carnap's empiricism is the conviction that the meaning of a claim is somehow the mode of its verification. This link between meaning and knowing or justification derives more or less straightforwardly from Hume, but it was not dependent on any particular form of verification or testing. He could change his mind about how to test at the drop of a counterexample without giving up the crucial linkage to meaning. Whatever is not appropriately linked to experience is unintelligible. Metaphysics, thus, is not false, not unjustified, but utterly without cognitive meaning. Even so, the slogan that meaning is the mode of verification is at best only a strategy for a theory of meaningfulness and only hazily suggests a full fledged theory of meaning that would include synonymy, implication, and confirmation.

Now the severest problem of a traditional empiricism, such as Carnap's, is the question of what to do about mathematics and logic. (We might add epistemology to that list.) This is where logicism, which Carnap appropriated from Frege and Russell, comes in. If mathematics can be reduced to logic as logicism says, then the severest problem for empiricism is narrowed to what to do about logic. For this Carnap accepted Wittgenstein's doctrine in the *Tractatus* that logic is no news at all. Thus, empiricism is saved.

Carnap had a variety of other commitments before *Syntax* which should be briefly mentioned. He had always defended a kind of conventionalism, but on closer inspection this comes to nothing more than a Duhemian underdetermination thesis. He was also committed to something that might be called the possibility of alternative *Aufbaus*. Even before that book was published Carnap agreed that one could set up a constructional system in various ways. One could do it on an autopsychological basis, as he in fact did following the empiricist tradition and especially Russell. Or one could do it on a physical basis as Neurath had insisted. Now Carnap also seemed to think that some constructional systems were more "correct" than others, but what the possibility or the unequal correctness of the alternatives came to was not very well worked out.

A third early commitment beyond empiricism was to a philosophy of geometry according to which alternative mathematical geometries are really differing implicit definitions of the terms they contain. Thus, they do not disagree. This very early belief (It is expressed in Carnap's disertation.) undoubtedly derives from Hilbert, but it was also reinforced by Carnap's association with Schlick. If Quine is right that there is no fathoming the subdoctoral mind (Quine in Dreben 1990), then there is no hope of finding out what prompted Carnap to accept this in the first place. Finally, Carnap is committed to a full scale fallibilism. Not only are theories uncertain, but so too is the observational basis on which they rest. This idea is not yet there in the *Aufbau*, but it pre-dates the *Syntax*.

So much for background. When Carnap sat down to write what was to become *The Logical Syntax of Language* he most certainly did not have the whole actual outcome in mind. In fact, all he intended to say was that one could indeed describe the logical form of a language and do so within that language itself. That one can talk about logical form is directly contrary to Wittgenstein's claim that logical form could be shown but not said, and Carnap proposed to accomplish it via Godel's technique of arithmetization. That one could talk about logical form within that language was very important to Carnap, von Neumann, Neurath, and others apparently because it was confusedly connected with the question of the unity of the language of science. (Contrary to Carnap's first impression, separating the object and metalanguages would appear to be no more insidious than type theory.) In order for a language to be able to describe itself that language has to be fairly weak. Stronger languages, including those in which classical mathematics could be expressed, must be ruled out. So there was to be no discussion or defense of these stronger languages, no principle of tolerance, and no discussion of general syntax. If Carnap had left Syntax at this point it would have been rather tame stuff.

In the process of writing *Syntax*, however, Carnap came to see that the forms of a variety of stronger languages could be described, albeit in a metalanguage stronger than the object languages themselves. Even more importantly he saw that by transfinite means (essentially by allowing an omega rule) the incompleteness established by Gödel of any language sufficiently strong to express classical mathematics could in important respects be repaired. Given the utility of these stronger languages and the metamathematical means to deal with them, such stronger languages could no longer be dismissed. This discovery was as electrifying to Carnap as the discovery that there are non-Euclidian systems was to geometers in the 19th century. (And for similar reasons.)

In order to cope with the discovery, Carnap made use of two old commitments from the pre-*Syntax* days. First, he used the idea that philosophic issues or doctrines can be absorbed into matters of logical form. The claim was originally Wittgenstein's, but he was concerned with a single language. The point still applies when we have a variety of languages. Importantly, among the philosophic commitments embedded in a language are the epistemic ones. I shall return to this later in greater detail. Given that an epistemology is embedded in each of the languages, there can be even in principle no non-question-begging epistemic grounds for preferring one language or logic or philosophy over another. Second among the pre-Syntax commitments to be used, Hilbert's implicit definition approach to geometry was waiting in the wings. Now, Carnap could generalize it. Alternative sets of logical rules could be thought of as implicit definitions of the philosophic terms they contain. Alternative philosophic presuppositions are not in conflict; they are just different ways of speaking. Like any theory of implicit definition, Carnap's theory of meaning here (insofar as we can call it that) is both *functionalist* and *holist*. It is functionalist because it specifies a system of relations that the words must bear to one another without saying more concretely what will do the job. It is holist because it is the totality of rules that defines each term. If one were going to use the word 'meaning', the meaning of a term would be its function within the whole system of rules.

Because the alternative sets of logical rules are thought of as definitions we get a thoroughgoing conventionalism with respect to philosophy and certain basic parts of science. This is encapsulated in the Principle of Tolerance. And the conventionalism we get is vastly more powerful than the Duhemian underdetermination thesis that had gone before. Underdetermination, remember, imagines that questions of what counts as evidence and what logical relation theory must bear to evidence are *settled* in advance. It then notices that, given plausible answers to those questions, for any amount of evidence, more than one theory will bear the appropriate relation to it. Carnap's new conventionalism, by contrast, ultimately says that what counts as evidence and what the appropriate logical relations are (even what the logical consequence relation itself is) *are all up for grabs too*. Carnap's is a very radical view.

By making it a *linguistic* conventionalism Carnap avoids the result that anything goes. What is conventional is meaning. The word 'unicorn' considered merely as a noise has no meaning. But we can go on to make it meaningful by specifying linguistic conventions in any way we choose. In these conventions are embedded epistemic standards. If the conventions are chosen in one way it will be contingently correct to say 'There are unicorns' and chosen in another way (i.e., in a way that gives 'unicorn' a different meaning) it will be contingently correct to say 'There are no unicorns'. Once the meanings of the various words have been fixed, however, the logical and epistemological standards have likewise been fixed by being embedded in the conventions that gave the words meaning. What we get is not "ways of world making" but "ways of word making".

Carnap's conventionalism is tempered in another way, too, by his pragmatism. There may be no non-question-begging epistemic grounds for choosing among the alternative languages (logics, philosophies), but some languages are more convenient than others. Inconsistent languages are pragmatic disasters, and so are languages without inductive rules. It is not necessary to establish that a language is maximally or even minimally convenient before using it, but philosophic discussion (where it is not wholly misguided) must be pragmatic. Qua pure logicians our job is merely to trace out the consequences of this or that convention. This is an engineering conception of philosophy. Of course, we may make proposals and defend them as useful, but here we go beyond pure philosophy and enter the empirical. Carnap's pragmatism is encouraged but not contained in his association with the Bauhaus school of artists and architects. This association has recently been illuminatingly explored by Peter Galison. (1990) Carnap's pragmatism was likewise encouraged but not contained in Neurath's attempts at social engineering and by political and social changes within Europe more generally. The contrast with the Platonic ideal of philosophy, or for that matter with Kant's, couldn't be greater. The development of Carnap's radical conventionalism and pragmatism really is a watershed and really does begin a new period in his philosophy.

Before proceeding to his later work, however, I want to explore the *Syntax* still further. The first thing to do it to look at his pre-*Syntax* views to see how much they changed, and among these the place to begin is with logicism. Under the new conventionalism one no longer has to reduce mathematics to logic in order to resolve the epistemic problems of empiricism. Mathematical axioms can be treated as implicit definitions quite directly. Mathematics would then be true in virtue of meaning and hence by convention even without any reduction to logic. Such a reduction is still possible and desirable. Any reduction is an economy and thus pragmatically attractive. This is like reducing the number of axioms in propositional logic; it is nice but hardly earthshattering. In the (very weak) sense of thinking that the reduction is both possible and desriable Carnap is still a logicist, but this is logicism virtually in name only. In fact it had been absorbed into formalism. The reduction no longer carries with it any special epistemic benefits. Both logic and mathematics can be conventional or analytic with or without the reduction.

So what happens to the rest of Carnap's pre-*Syntax* views? As expected, he remains an empiricist, but perhaps surprisingly, empiricism itself becomes a convention, albeit one for which there are powerful pragmatic reasons. (Carnap 1936-37, p. 33) His anti-metaphysical stance, that is his rejection of intuition, is retained and systematized. The most basic ontological commitments are matters of convention. Carnap also gives the appearance of being a nominalist, especially in the last section of the book. But this section seems to have been written early, probably before his conventionalist breakthrough. In any case, whatever the appearances, Carnap is a neutralist, even a noncognitivist, about basic ontological claims. Outside the language there is nothing to say; inside a language those questions have been trivially settled. The nominalist appearance is given by the fact that the language that Carnap prefers and would propose is a nominalist one, but that is hardly ontological commitment. Even Carnap's verificationism (better called confirmationism) is retained and deepened. He now has more nearly a full theory of meaning, and by identifying the epistemic rules as implicit definitions of the terms they contain, he is able to give some substance to what had been only a slogan: that meaning is the mode of verification.

A variety of other pre-*Syntax* views should be mentioned as well. His Duhemian conventionalism (the underdetermination thesis) is retained, but it pales beside his radical new conventionalism. There is still the possibility of alternative *Aufbaus*, but now this can be given a systematic account, not a mere suggestion. Gone, however, is the idea that one of these constructional systems is more correct, though some may be more convenient. The early Hilbertian philosophy of pure geometry is obviously generalized and moved to center stage. Finally, the fallibilism is still there, but it, too, is now a convention. Like empiricism itself, though, it is a convention for which there are powerful pragmatic reasons.

The actual text of *The Logical Syntax* proceeds first by example. It takes a weak language and then a stronger language and shows how to describe their logical forms, presumably on the grounds that showing is easier than saying how to do this. The last part of the book is a discussion of the philosophic import of syntax, but between the examples and the finale is a discussion of general syntax which is the real heart of the book. Here Carnap's primitive terms are 'is a sentence' and 'is a direct consequence of'. The first of these exhausts what we call syntax. Plainly, Carnap means to go beyond that. Not only is the logical consequence relation itself semantical, as we use the term, but so are truth tables, interpretation, and analyticity, all of which Carnap discusses. The treatment especially of the last of these is surprisingly close to a full semantical account. No doubt Carnap is still wrong in thinking that all philosophy is syntax, even in his broader sense, but he is not nearly as wrong as we might have thought given our use of the word.

All of this raises the question of why Carnap rejected the concept of truth. There is a rumor afoot, propagated by Hartry Field, that it was because truth did not admit of physicalist defintion. (Field 1972) This cannot be right about Carnap. Field need not worry, though, because the historical question does not bear directly on Field's own program. Carnap does discuss truth in *Logical Syntax*, and what he says is very odd. Most of the discussion concerns the semantical paradoxes, as though he were worried that any concept of truth were unavoidably inconsistent. That would be respectable but still a mistake. But Carnap goes on to show just how to avoid the contradiction, namely by formulating the truth predicate in a metalanguage distinct from the language to which that predicate applies. So why doesn't Carnap accept this? Well, it might have been because he still wanted to do the logic within the language it described. But he doesn't say this. And it would be a rather weak reason. And more importantly, he has plainly renounced this want in the rest of the book. What he does say is that this approach would not yield a genuine syntax: "*For truth and falsehood are not proper syntactical properties*; whether a sentence is true or false cannot generally be seen by its design, that is to say, by the kinds and serial order of its symbols." (Carnap, 1937, p.216) What kind of reason is this?! Of course truth is not syntactical in this sense, but the question is why philosophy should be restricted to syntax in so narrow a way. Coffa interprets the argument as verificationism, but it is not even that. In technical terms this is just plain 'goofy'. It is as though 'and' is not a logical term on the ground that it is not a purely logical matter whether birds sing *and* Caesar marched. Or that 'two' is not properly definable in logic because it is not a purely logical matter whether there are two toads in Transylvania.

The argument against truth is so bad that it is plausible to assume that Carnap was antecedently prejudiced against the concept of truth. If so, he might have thought that his cavalier attitude about it in *Syntax* was harmless. Certainly Carnap was so predisposed, for under the pernicious influence of Neurath truth would have been called "metaphysical" and "absolutist". It was metaphysical because its acceptance somehow committed one to a domain of Russellian facts or Kantian things-in-themselves. It was absolutist because it was somehow confused with certainty, which in non-analytic cases Carnap's and Neurath's fallibilism forbids. This hypothesis about the underlying causes of the rejection of truth is confirmed when Carnap finally accepts the notion soon after *Syntax*. He does not say: now I see that truth is physicalistically definable. Instead, he says in effect: now I see that truth and confirmation must be distinguished and also that truth is not metaphysically loaded. (Carnap 1936)

What is sad about the whole episode is not only that truth is in fact entirely compatible with the conventionalism and pragmatism of *Logical Syntax*. Rather, it is that the background prejudices against truth actually *fly in the face* of the central lessons of that book, namely, its epistemic conventionalism and it ontological neutrality. Obviously, these are hard lessons to learn and ones which even Carnap had not fully absorbed.

It may be a blunder to identify truth with certainty, but it is no mistake at all to recognize that the enterprise of *Logical Syntax* is open to an epistemic interpretation. Carnap did not care for the word 'epistemology' because he thought it to have been preempted by psychologists and by philosophers practicing the work that Frege dismissed as psychologism. But that need not mislead us. What Carnap is investigating is the pure structure of epistemology. His most basic relation, that of direct consequence, is after all the relation that reasons bear to that which they immediately justify. If you want to find out what some foreign speaker means by some term, go find out what arguments the speaker accepts. (Carnap 1950b, p.37) Something is conventional just in case it could have been otherwise, and there is no epistemic reason for choosing among the alternatives. Carnap emphasizes the centrality of analyticity, but something is analytic just in case its justification can be traced back to conventions and nothing more. Thus, to be analytic is to have a special epistemic status. Analyticity is no mere substitute for truth within logic and mathematics, a substitute which is needed when one lacks a concept of truth but which is obsolete when one has a real truth concept. In effect, the analytic is the epistemically conventional. By seeing meaning as provided by implicit definitions and these as exhibiting the epistemic structure of the language, Carnap's verificationism is built in at the very foundation. The epistemic dimension of all this never changed. Later, when he addressed the issue of what we could believe about theoretical entities, it boiled down for him to the question of whether we should adopt by convention a rule of inference that would justify theoretical claims on observational evidence. He himself was willing to adopt such a convention, but instrumentalists were not. (Realists on the other hand often spoke as though the inference rule had been ordered by God, and there was nothing left for human convention to decide.) In any case, my point is not about scientific realism, but rather merely to illustrate that Carnap's notion of logic, from *Syntax* onward, is broad enough to include what we call epistemology. Certainly Carnap makes quite a point of calling his enterprise the logic of science.

If we count *The Logical Syntax of Language* as fundamentally epistemic, as we should, then it may seem that there are two overwhelming ommisions in the book. There is only the most rudimentary discussion of induction and confirmation, and there is almost nothing said of observation. Perhaps these omissions can be forgiven in a work devoted primarily to classical logic and mathematics. But still we should want more. As far as induction is concerned, it seems that Carnap did not know what to say at this point.

What he does say is sadly deductivist. What is required is that his most basic relation, that of direct consequence, be generalized in a natural way to include partial implication. This would then be an account of inductive inference, of confirmation, and of logical probability. Indeed, all of Carnap's discussions of probability and later philosophy of science can be thought of as attempts to carry out this very program. Whatever the success of Carnap's theory of logical probability it must be seen as continuous with the program of Syntax and in fact as trying to develop a proposal for a workable convention, perhaps one that would be an explication of our usual conventions.

As far as observation is concerned the story is a bit more complicated. Carnap did talk about this in "On Protocol Sentences" (1932), published while he was writing Syntax and in "Testability and Meaning" (1936-37) published just after. What he had to say was important, perhaps even on the right track. But not only do we recognize it as inadequate, Carnap did too. Unfortunately, in the nearly forty years remaining to him he did not see how to develop these ideas further in a satisfying way. In any case from the early 1930s Carnap is a conventionalist about observation as well. The protocol sentences and reports are not themselves conventions, but rather what we take as a protocol is conventional. Empirical psychology will help with the pragmatic question of how to revise or formulate the observation language, but there is no convention independent fact of the matter about whether protocols will concern sensory experiences, or physical objects, or both. Carnap hoped that the observable features of things could be marked by a special vocabulary and hence that observationally justified beliefs could be picked out syntactically. Then the fact that protocols need not be justified by inferring them from other beliefs might somehow be marked via the direct consequence relation. We know that this won't work and that observation is a vastly more complicated affair than just a special vocabulary. Carnap knew it too, but he did not know just how to improve the account. Moreover, he was more interested in saving both observation and theory while damning metaphysics than in exploring the limits or nature of observation. Though he hadn't yet figured out how he would carry it off, Carnap is quite explicit about being a fallibilist even about observation reports. Now I think that something along the lines that Carnap wanted (but never found) can be devised. This, however, is neither the time nor the place. What we can say now is that the conventionalist aspects of Carnap's approach to observation have for the most part been neither appreciated nor explored. We can also say, despite the omissions on observation and induction, that Carnap is giving us in Syntax the structure of an epistemology, or at least trying to do so.

Now it may seem odd in a paper nominally devoted to semantics that so far I have talked almost exclusively about the so-called syntax period. Of course, my thesis is that by misunderstanding Syntax we are led to exaggerate the differences between semantics and the work before or after and also to misunderstand Carnap's semantics itself. A correct understanding here will show Carnap's last four decades are, if not a seamless fabric, then at least a more or less continuous development. Carnap is not lurching from one misguided enthusiasm to another. Rather he has one broadly consistent leading idea, which demands our attention and which is too little understood.

With a clearer picture of Syntax in mind we can now address the supposed break between Syntax and semantics. As we saw there is a lot of what we call semantics in Syntax. It is in the consequence relation, in the truth tables, in the work on interpretation, and especially clearly in the discussion of analyticity. Of course, a concept of truth is officially rejected in Syntax, but when Carnap finally does accept truth, this is fully consistent with his conventionalist program. The confusions that had prevented the acceptance of truth are avoided not by abandoning that program but by carrying it through. (The conventionalism by itself will guarantee that the language is ontologically neutral. It will also guarantee that the protocols can have any desired degree of justification; if, as Carnap thinks, it is not handy to let them be certain, then don't. If truth can be defined in the way that *Syntax* outlines or that Tarski shows, then truth is a completely separate matter.) The move to semantics does not change the content of Carnap's conventionalism, but it does change the form. Now Carnap speaks of truth, reference, designation, and the like, and shows how such notions are interdefinable. The theory of truth and reference amounts to an implicit definition in the metalanguage not to a physicalist reduction. That Carnap now talks cheerfully "of" propositions, properties, meanings, etc. is taken by some to show that he has given up his old nominalism and has become a Platonist. This is twice wrong. He wasn't a nominalist before, and he isn't a Platonist now. Instead he remains a metaphysical non-cognitivist, and this is founded on his ongoing conventionalism. He doesn't have to repeat himself. After all, he thinks he is in the friendly and thoroughly pragmatic U.S. and that all this can be taken for granted.

To see that his conventionalist epistemological theme is sustained, let us very quickly review the major writings of the semantical period. In "Truth and Confirmation" (1936) Carnap admits, indeed insists, that the confusion of truth and certainty is just a mistake. There is certainly no repudiation of his conventionalism. The first book where he is working out his semantical views is *Foundations of Logic and Mathematics*. (1939) It title notwithstanding this is largely about the interpretation of science. He reiterates and defends his conventionalism for logic (no doubt in response to Quine's "Truth by Convention" (Quine 1936)) and reaffirms that his logic is a logic of science. The books *Introduction to Semantics* (1940) and *Formalization of Logic* (1943) can be considered as a pair. *Formalization* was written first and contains a result sufficiently dramatic that Carnap decided to preceed it with *Introduction to Semantics* both to prepare the reader for *Formalization* and as a textbook on semantics. What is so surprising? It is that logic too is open to non-standard interpretations. This arises out of his functionalist approach to definition; if logic proceeded by latching onto Platonic (convention independent) entities it is unlikely that this problem would ever have come up. Even *Introduction to Semantics* is at great pains to stress its continuity with what has gone before.

Meaning and Necessity is the book that we now tend to think of when we think of Carnap's semantics, no doubt because it is so delightfully readable. On this topic it is interesting to note that Church complained that the book was mere informal prolegomena and that publication should await the development of a strictly formal system. (In other words, don't publish until it has been made totally unreadable.) In any case the book arose out of the Quine-Carnap correspondence, that is, it was prompted by Quine's worries about meaning and analyticity. I have already argued that analyticity is at bottom an epistemic notion. At this point in the debate, however, Quine had not yet brought forward his alternative theory of knowledge, and Carnap was still reluctant to use the word 'epistemology'. So Carnap did not emphasize this aspect, even though it is there. Instead he concentrated on presenting clarification and technical improvement. His method of intension and extension, as Carnap was quick to agree, is really a method of intension. (The intensions do all the work.) But on closer inspection it turns out that expressions have the same intension if and only if they are alike in point of epistemic functioning. There are also non-essential changes in the presentation of analyticity; the definitions are now given in terms of state descriptions. These changes are the direct products of his emerging theory of inductive confirmation, work on which is occuring simultaneously. Finally, even the theory of modality that is given in the book is designed to make it empiricistically acceptable by making it a linguistic, i.e., conventional, affair.

The last major publication of the so-called semantical period is "Empiricism, Semantics, and Ontology" (1950a). This is basically a reiteration of his epistemic conventionalism and pragmatism and the ontological non-cognitivism that follows from it. The tone of the paper is very much: I cannot believe I need to repeat this stuff; I've been saying it all along. And indeed he had.

We are now in a position to evaluate the periodization of Carnap's work that puts a sharp break between *Syntax* and semantics. The case for the break is not very convincing. The radical conventionalism and pragmatism that was the main message of *Syntax* is still there. It may be altered in form, but not in substance. The ontological neutrality is still there, and the epistemic interpretation which makes sense of *Syntax* is still every bit as illuminating as before. In the end, Carnap's logic is still a logic of science. If our concerns are no longer epistemological, then so much the worse for us. As historians we should avoid a periodization which imposes our limitations on Carnap. I do not deny that there are changes over time in Carnaps's work. Certainly his program broadened. Certainly he emphasized, as every scholar does, what was new rather than old. Certainly the move from Prague to Chicago, and with it the change of context and critics, altered how he chose to present his ideas. But just as certainly there is no break in Carnap's fundamental conventionalism.

As for the supposed break between semantics and the work on probability and philosophy of science, very little needs to be said that has not already been covered. The probability work, though it is presented as describing a purely logical relation, is of obvious epistemic import. It is essentially an attempt to carry out the generalization of the direct consequence relation that we earlier described as the unfinished business of *Syntax*. Even the accounts of meaning now offered are overtly epistemological. Although I haven't the room to demonstrate this here, when we come to see the probability as an attempt to formulate conventions, the whole set of which give the meaning for the terms of the language, then many of the criticisms of *Probability* (1950b) can be blunted. If the epistemic reading of *Syntax* is acceptable then *Syntax* and the late work on probability and philosophy of science are of a piece. If the program of *Syntax* is continued in both semantics and probability, then it is unlikely that there is a break between the latter two. It is even hard to specify just when the break is supposed to have occurred. The probability work began well before *Meaning and Necessity* was published, and indeed it is reflected in the text of that work as well. In fact, Carnap lists *Logical Foundations of Probability* (1950b) as Volume Four of his Studies in Semantics. I think he knew what he was doing.

What then shall we make of all of this? Certainly, we are entitled to conclude that the standard periodization of Carnap's work with which we began is at best an exaggeration. It seems to me, however, that the moral goes far deeper than this. The legend that Carnap's thought made a radical shift in the mid-1930s is all that has licensed us in isolating and ignoring *The Logical Syntax of Language*. But what is really radical and well worth studying in Carnap's work is precisely the epistemic conventionalism and pragmatism that he announced there. It is also what is constant throughout the rest of his work. Once we see this, we have a whole new way of approaching his semantics and especially the topic of analyticity. Analyticity becomes an epistemic notion, and this fact can be used to respond to the criticisms of Quine among others. The work on probability will likewise have to be reevaluated, keeping in mind that conventionalism and a functionalist approach to meaning are being presupposed. If this is done, it seems likely that at least some of the objections to Carnap's probability theory can be answered.

How we periodize our history (and Carnap's) matters. But getting the historical record straight is only secondary. What is primarily at stake here is how we can learn from the past and how we can set that to work in our ongoing philosophy. Carnap's conventionalism and pragmatism is a rich and powerful and largely neglected tool to be used by us—now. It is the future that matters most about the past. And by reexploring our past we can hope to shape that which is to come.

Notes

[1] I would like to thank colleagues Jane Maienschein and Michael White and fellow symposiasts Burton Dreben and Michael Friedman for comments on an earlier version of this paper. I would also like to thank the College of Liberal Arts and Sciences of Arizona State University for a research travel grant.

References

Carnap, R. (1928), *Der Logische Aufbau der Welt*. Berlin-Schlachtensee: Weltkreis-Verlag.

_____. (1932), "Über Protokollsätze", *Erkenntnis*, III: 215-228. (Translated as Carnap 1987.)

_____. (1934), *Logische Syntax der Sprache*. Wien: Julius Springer. (Translated as Carnap 1937.)

_____. (1936), "Wahrheit und Bewährung", *Actes du Congrès international de philosophie scientifique*, Sorbonne, Paris, 1935. 4. *Induction et probabilité. Actualités scientifique et industrielles*, 391, Paris: Hermann & Cie. 18-23. (Translated as Carnap 1949.)

_____. (1936-37), "Testability and Meaning", *Philosophy of Science* 3: 419-471, 4: 1-40.

_____. (1937), *The Logical Syntax of Language*. London: Kegan Paul Trench, Trubner & Co. (Translation of Carnap 1934.)

_____. (1939), *Foundations of Logic and Mathematics. International Encyclopedia of Unified Science* I,3. Chicago: University of Chicago Press.

_____. (1940), *Introduction to Semantics*. Cambridge, MA: Harvard University Press.

_____. (1943), *Formalization of Logic*. Cambridge, MA: Harvard University Press.

_____. (1947), *Meaning and Necessity: A Study in Semantics and Modal Logic*. Chicago: University of Chicago Press.

416

_____. (1949), "Truth and Confirmation", in *Readings in Philosophical Analysis*, Herbert Feigl and Wilfrid Sellars (eds.), New York: Appleton-Century-Crofts, 119-127. (Translation of Carnap 1936.)

_____. (1950a), "Empiricism, Semantics, and Ontology", *Revue internationale de philosophie*. IV,11: 20-40.

_____. (1950b), *Logical Foundations of Probability*, Chicago: University of Chicago Press.

_____. (1987), "On Protocol Sentences", *Nous*, XXI: 457-470. (Translation of Carnap 1932.)

Dreben, Burton, (1990), "Quine", in *Perspectives on Quine*, Robert B. Barrett and Roger F. Gibson (eds.), Cambridge, MA: Basil Blackwell, 81-95.

Field, Hartry, (1972), "Tarski's Theory of Truth", *Journal of Philosophy*, LXIX: 347-375.

Galison, Peter, (1988), "History, Philosophy, and the Central Metaphor", *Science in Context*, II: 197-212.

_____,(1990), "*Aufbau*/Bauhaus: Logical Positivism and Architectural Modernism", *Critical Inquiry*, XVI: 709-752.

Quine, W.V., (1936), "Truth by Convention", *Philosophical Essays for A.N. Whitehead*, O.H. Lee, ed., New York: Longmans, 90-124.

Sellars, Wilfrid, (1975) "Autobiographical Reflections" in *Action, Knowledge and Reality: Critical Studies in Honor of Wilfrid Sellars*, Hector-Neri Castaneda (ed.), Indianapolis, IN: The Bobbs-Merrill Co., 277-293.

Acknowledgments

Tarski, Alfred. "The Semantic Conception of Truth and the Foundations of Semantics." *Philosophy and Phenomenological Research* 4 (1944): 341–75. Reprinted with the permission of Brown University.

Carnap, Rudolf. "Modalities and Quantification." *Journal of Symbolic Logic* 11 (1946): 33–64. Reprinted with the permission of the Association for Symbolic Logic. All rights reserved. This reproduction is by special permission for this publication only.

Reichenbach, Hans. "On Probability and Induction." *Philosophy of Science* 5 (1938): 21–45. Reprinted with the permission of Williams & Wilkins Co.

Carnap, Rudolf. "The Two Concepts of Probability." *Philosophy and Phenomenological Research* 5 (1945): 513–32. Reprinted with the permission of Brown University.

Reichenbach, Hans. "Philosophical Foundations of Probability," In J. Neyman, ed., *Proceedings of the Berkeley Symposium on Mathematical Statistics and Probability* (University of California, Berkeley Symposium on Mathematics, Statistics and Probability, Proceedings 1949): 1–20. Reprinted with the permission of the University of California Press.

Popper, Karl R. "The Propensity Interpretation of the Calculus of Probability, and the Quantum Theory." In S. Korner, ed., *Observation and Interpretation* (New York: Academic, 1957): 65–70. © by Karl R. Popper. Reprinted with the permission of A.R. Mew and Melitta Mew, Executors of the Estate of Prof. Sir Karl Popper, and the Colston Research Society.

Reichenbach, Hans. "On the Justification of Induction." *Journal of Philosophy* 37 (1940): 97–103. Reprinted with the permission of the Journal of Philosophy, Inc., Columbia University, and the author.

Carnap, Rudolf. "On Inductive Logic." *Philosophy of Science* 12 (1945): 72–97. Reprinted with the permission of Williams & Wilkins Co.

Hempel. Carl G. "Studies in the Logic of Confirmation." *Mind* 54 (1945): 1–26, 97–121. Reprinted with the permission of Oxford University Press.

Goodman, Nelson. "A Query on Confirmation." *Journal of Philosophy* 43 (1946): 383–85. Reprinted with the permission of the Journal of Philosophy, Inc., Columbia University, and the author.

Hempel, Carl G. "A Note on the Paradoxes of Confirmation." *Mind* 55 (1946): 79–82. Reprinted with the permission of Oxford University Press.

Goodman, Nelson. "On Infirmities of Confirmation Theory." *Philosophy and Phenomenological Research* 8 (1947): 149–51. Reprinted with the permission of Brown University.

Carnap, Rudolf. "Reply to Nelson Goodman." *Philosophy and Phenomenological Research* 8 (1948): 461–62. Reprinted with the permission of Brown University.

Kemeny, John G. "The Use of Simplicity in Induction." *Philosophical Review* 62 (1953): 391–408. Reprinted with the permission of AMS Press, Inc.

Carnap, Rudolf. "The Aim of Inductive Logic." In E. Nagel, P. Suppes and A. Tarski, eds., *Logic, Methodology and Philosophy of Science* (Stanford, CA: Stanford University Press, 1960): 303–18. Reprinted with the permission of Stanford University Press.

Hempel, Carl G. and Paul Oppenheim. "Studies in the Logic of Explanation." *Philosophy of Science* 15 (1948): 135–75. Reprinted with the permission of Williams & Wilkins Co.

Nagel, Ernest. "The Meaning of Reduction in the Natural Sciences." In Robert C. Stauffer, ed., *Science and Civilization* (Madison: University of Wisconsin Press, 1949): 97–135. Reprinted with the permission of University of Wisconsin Press.

Braithwaite, R.B. "The Nature of Theoretical Concepts and the Role of Models in an Advanced Science." *Revue Internationale de Philosophie* 8 (1954): 34–40. Reprinted with the permission of Editions Universa.

Hilpinen, Risto. "Carnap's New System of Inductive Logic." In Jaakko Hintikka, ed., *Rudolf Carnap, Logical Empiricist* (Dordrecht: Reidel, 1975): 333–59. Reprinted with the permission of D. Reidel Publishing Company.

Creath. Richard. "The Unimportance of Semantics." In Arthur Fine, et al., eds., *PSA 1990: Proceedings of the 1990 Biennial Meeting of the Philosophy of Science Association*, Vol. 2 (East Lansing: Philosophy of Science Association, 1991): 405–16. Reprinted with the permission of the Philosophy of Science Association.

Milton Keynes UK
Ingram Content Group UK Ltd.
UKHW031136141024
449569UK00006B/147